In Memoriam

Imre G. Izsak

(1929–1965)

Imre Iszak, who died of a heart attack while attending the Symposium in Paris, was a celestial mechanician of the first order, an outstanding theoretician and a brilliant oberservational analyst. During the few short years of the space age and its new disciplines, he made himself a world authority on satellite geodesy. His achievement was already great; his potential was enormous.

The Smithsonian Astrophysical Observatory was honored in his work.

Imre Gyula Izsak (in Hungarian Imre is the same name as Amerigo) was born in the small town of Zalaegerszeg, some 200 miles from Budapest. At the University of Budapest, he studied astronomy under the late Karoly Lassovsky (who afterwards also joined the Smithsonian) and concentrated on study of variable stars and globular clusters.

Like so many other of his countrymen, he fled during the 1956 revolt, and resumed astronomic studies at the Zürich Observatory. In december, 1958, he moved to the United States, to the Observatory of the University of Cincinnati, and in the next year he joined the staff of the Smithsonian Astrophysical Observatory. On February 24, 1964—three days after his birthday—he became a United States citizen.

In Cambridge he lived with his wife and young son, Andrew, at 37 Concord Avenue.

His interest had shifted to celestial mechanics and mathematics both theoretical and applied, and at the Smithsonian he concentrated on study of the Earth's shape and gravity field. Working with tens of thousands of precise observations of artificial satellites obtained by the observatory's network of 12 Baker-Nunn satellite tracking cameras, he was the first to determine from satellite observations the ellipticity of the Earth's equator, he improved positional accuracy of the tracking station coordinates and continually improved estimates of tesseral harmonics. He had brought with him to the symposium the most accurate mapping of the geoid surface ever made.

He arrived in Paris the evening of April, 19. During the afternoon of April 21 he became ill. He died, quietly, in his bed, that night.

He was 36.

F. L. Whipple

Director of the Smithsonian
Astrophysical Observatory
Cambridge, Massachusetts, U.S.A.

COSPAR-IAU-IUTAM

TRAJECTORIES OF ARTIFICIAL CELESTIAL BODIES

AS DETERMINED FROM OBSERVATIONS

TRAJECTOIRES DES CORPS CELESTES ARTIFICIELS

DÉTERMINÉES D'APRÈS LES OBSERVATIONS

PROCEEDINGS OF A SYMPOSIUM
HELD IN PARIS, APRIL 20-23, 1965

EDITED BY
JEAN KOVALEVSKY

WITH 55 FIGURES

SPRINGER-VERLAG
BERLIN/HEIDELBERG/NEW YORK
1966

Proceedings of a symposium held in Paris, April 20-23, 1965, organized jointly by the committee on Space Research (COSPAR), the International Astronomical Union (IAU) and the International Union for Theoretical and Applied Mechanics (IUTAM).

Edited by

JEAN KOVALEVSKY

with the assistance of

BRUNO MORANDO

ISBN 978-3-642-49328-7 ISBN 978-3-642-49326-3 (eBook)
DOI 10.1007/978-3-642-49326-3

Softcover reprint of the hardcover 1st edition 1966
Library of Congress Catalog Card Number: 66-26599

Titel Nr. 1345

Avant-propos

C'est une chance rare et dont j'apprécie tout le prix que de se trouver en même temps appelé à présider COSPAR, Comité scientifique de ICSU, et IUTAM, Union scientifique adhérant à ICSU, et d'avoir ainsi l'honneur de représenter en même temps ces deux organisations, COSPAR et IUTAM, patronnant avec l'Union Astronomique Internationale un Symposium sur la Détermination des Trajectoires des Corps célestes artificiels à partir de leur observation.

Le Pr. J. Kovalevsky, dont le nom est aussi connu à COSPAR qu'à IAU, et qui n'est pas non plus un étranger pour IUTAM puisqu'il est un éminent spécialiste de la Mécanique céleste, expose très complètement, dans la préface qu'il donne au présent volume, l'origine, les buts et les faits marquants du déroulement de ce Symposium, tenu à Paris du 20 au 23 Avril, et malheureusement endeuillé par le décès tout à fait soudain du regretté Dr. Imre G. Izsak auquel ce livre se trouve dédié selon un désir unanime.

Je ne doute pas que le présent ouvrage apportera une importante contribution aux progrès du sujet si fondamental qu'il traite, et qu'il sera ainsi apprécié à sa juste valeur par tous ses lecteurs.

Que le Pr. Kovalevsky soit particulièrement remercié de la part qu'il a prise personnellement à la préparation et au succès justifié de ce Symposium, et que soient remerciés également ici tous les auteurs dont ce volume publie les travaux.

Maurice Roy

Preface

In May 1962 the International Union of Theoretical and Applied Mechanics (IUTAM) organized a symposium in Paris on the theory of the movement of the Earth's artificial satellites designated to confront the results of the first spatial experiences with the mechanical point of view. The papers submitted during this meeting have been published in 1963 in a collection entitled "Dynamics of Satellites" by Springer-Verlag.

During the last three years the importance of studies pertaining to the dynamics of artificial satellites has continued to increase, and many results due to observations have led to a deeper knowledge of the field of forces in which these objects move, particularly the field of gravitation of the Earth, the forces due to pressure of radiation, friction of the atmosphere, etc... A new symposium seemed therefore suitable to determine these recent advances. However, this time it appeared appropriate to consider it more from the point of view of the interpretation obtained from the results of observations (determination of orbits and forces existing) than from the point of view of the theory of motion. For this reason the complete title of the second symposium is as follows: "Trajectories of Artificial Celestial Bodies, as Determined from Observations".

The interest of this second meeting has been acknowledged by many international scientific bodies, and if the initiative is due to the IUTAM through the medium of its president Professor MAURICE ROY, the International Committee of Space Research (COSPAR) as well as the International Astronomical Union (IAU) proclaimed their great interest regarding this project and took an active part in the organization and financing of this symposium.

These three organizations were unanimous in advising to restrict the number of participants, according to a formula currently applied at the IUTAM and which had been that of the 1962 symposium. This restriction was a great handicap to the work of the scientific committee entrusted with the dispatch of invitations and choice of communications.

This committee under the presidency of Professor MAURICE ROY, was composed as follows:

Mrs. A. G. MASSEVITCH and Professor E. BUCHAR for the COSPAR,
Professors S. HERRICK and G. C. MCVITTIE for the IUTAM,
Professor D. BROUWER and the undersigned for the IAU.

Finally, and also taking into account several missing participants who failed to appear at the last minute, 26 invited participants were present. They were joined by six observers who were in Paris during the symposium.

The sudden death of our colleague I. G. IZSAK astronomer of the Smithsonian Observatory cast a shadow on the symposium. Acting on the decision of the symposium this volume is dedicated to the memory of this scientist who was loved by all those who knew him. His numerous original works of great importance, particularly in the fields of celestial mechanics, astronomy, applied mathematics or in the study of the Earth's field of gravitation with the aid of the satellites accurate observations, are highly esteemed by all, astronomers, geodesists, and specialists in the field of space research. Professor F. WHIPPLE, Director of the Smithsonian Observatory, sent us a short biography of I. G. IZSAK, which is to be found at the beginning of the book.

The meetings were held on 20, 21, 22 and 23 April in the rooms put at the disposal of the symposium at the Centre Universitaire International. The eight sessions were presided over respectively as below:

Tuesday 20 April:	opening session:	Prof. D. BROUWER
	afternoon:	Mrs. A. G. MASSEVITCH
Wednesday 21 April:	morning:	Dr. D. G. PORTER
	afternoon:	Prof. M. ROY
Thursday 22 April:	morning:	Prof. G. N. DUBOSHIN
	afternoon:	Mr. W. M. KAULA
Friday 23 April:	first session:	Prof. S. HERRICK
	second session:	Prof. G. A. CHEBOTAREV

This volume contains all communications presented during the sessions of work. 9 from the USA, 7 from the URSS, 4 from the UK, and respectively, one from Canada, France, Hungary, Japan, New-Zealand and Czechoslovakia. With two exceptions they are given in the sequence heard: it was endeavoured to group them by subject. Summaries of the three communications which have already been made into a publication are given with all necessary references. Two articles submitted in Russian have been translated into English by Mr. B. MORANDO and myself. Each original article is preceded by a summary in English and in French. An analytical index is given to facilitate the use of this volume, indicating the place where the principal problems are dealt with, and which have been evoked during these days in the communications as well as during the discussions. I wish to thank Mr. B. MORANDO here who has given me important help by doing a large part of the work necessary in noting, selecting and editing the principal phases of the numerous and fruitful discussions which followed the communications and in which all participants took an active part. We hope thus that the work contains the essential of the ideas exchanged on this occasion.

The reader will look up the main part of the volume for details pertaining to the works given; let us mention only the main directions of research which are apparent either on hearing or reading them.

One of the problems presently preoccupying the COSPAR and more particularly its Working Group No. 1 (W.G.1) is the utilization for scientific use of the numerous visual observations of the satellites of rather poor accuracy. The sub-group of dynamics of the W.G.1 was particularly assigned to promote research in this field. The five first articles here show the first results of this research, proving that under certain conditions highly interesting results can be obtained without necessitating important calculating facilities: this research is therefore possible for countries not in a position to devote important equipment for space research.

Other communications show to what extent a particularly thorough analysis of the observation material will allow for the considerable improvement in accuracy of the results, in comparison with procedures brutally utilizing the method of the mean squares. It seems indeed that future improvement of parameters representing forces influencing the satellites depends not less on the clarification of the methods of analysis of the measurement (whether of short or long arcs) and their numerical treatment, than on the accuracy of the observations. This is one of the principal conclusions of these meetings.

The effects of radiation pressure attracted the attention of several participants, either in the case of a direct effect or one due to radiation reflected by the Earth according to two possibilities. Of particular interest for theoretical and applied mechanics was the evidence of the effect that solar heat can have on the rotation of plastic satellites.

Geodesy is one of the sciences which has benefitted the most by the launching of satellites and their observation. It is curious to behold the importance a good determination of the orbits can have in the success of geometric methods. Of course dynamic methods have been dealt with at length: the knowledge of the Earth's gravitational field has considerably increased since three years. I. G. IZSAK made out a very detailed map of the altitude of the geoid with respect to the ellipsoid, resulting from 26,000 observations. The detailed analysis relating to radial velocity by the DOPPLER-FIZEAU radio electric effect now make it possible to evaluate certain harmonics of the order 12 or 15.

Finally among the artificial celestial bodies, the space probes are getting more and more important, and it was quite natural that they would not be ignored by a symposium devoted to the determination of the orbits of artificial celestial bodies. The sensational results in trajectography made by the radio-tracking of Mariner II are indicated in this volume. However, as projects in this field are even more numerous than their realisations, it was necessary to discuss the problems remaining: lunar probes, cometary probes and the satellites around the Moon have been studied in this sense during the symposium.

We are very glad that Springer-Verlag accepted to ensure the edition of this symposium through the medium of this volume which thus constitutes the logical continuation to the work "Dynamics of Satellites" already mentioned, the contents of which by the way are still up to date.

I wish to thank here the three unions or international scientific committees, COSPAR, IAU and IUTAM who made possible a good organization of this symposium in financing the travelling expenses of certain delegates who otherwise would not have been able to participate in these meetings.

I also thank the Centre Universitaire International du Ministère de l'Education Nationale and in particular its director Mrs. ZALCBERG who welcomed the symposium, and the Direction des Relations avec la Communauté et l'Etranger du Ministère de l'Education Nationale whose subvention made it possible for the delegates and their families to gather at a friendly dinner. Finally, the Bureau des Longitudes has also contributed to the organization of these meetings, through payment of part of the expenses and the loan of material as well as the help of its personnel.

Paris, 1965 **Jean Kovalevsky**

Préface

En Mai 1962, l'Union Internationale de Mécanique théorique et appliquée (IUTAM) organisait, à Paris, un symposium sur la théorie du mouvement des satellites artificiels de la Terre destiné à confronter les résultats des premières expériences spatiales du point de vue de la mécanique. Les travaux exposés au cours de cette réunion ont été publiés en 1963, dans un recueil intitulé «Dynamics of Satellites» par les éditions Springer-Verlag.

Au cours des trois dernières années, l'importance des études de dynamique des satellites artificiels n'a fait que croître et de nombreux résultats d'observations ont permis d'approfondir la connaissance du champ de forces dans lequel se meuvent ces objets, notamment en ce qui concerne le champ de gravitation de la Terre, les forces dues à la pression de radiation, au frottement atmosphérique, etc. . . Un nouveau symposium, faisant une mise au point de ces récents progrès, semblait donc souhaitable, mais, cette fois, il paraissait utile de l'envisager plus du point de vue de l'interprétation des résultats d'observation (détermination des orbites et des forces en présence) que du point de vue de la théorie des mouvements. C'est pourquoi, le titre complet de ce second symposium est «Trajectoires des Corps Célestes Artificiels, Déterminées d'après les Observations».

L'intérêt de cette seconde rencontre a été reconnu par plusieurs organismes scientifiques internationaux et, si l'initiative en est due à l'IUTAM par la voix de son président, Monsieur le Professeur Maurice Roy, aussi bien que le Comité International de Recherches Spatiales (COSPAR) que l'Union Astronomique Internationale (IAU) se sont déclarés très intéressés par ce projet et ont activement participé à l'organisation et au financement de ce symposium.

Ces trois organismes ont été unanimes à conseiller de restreindre le nombre de participants, selon une formule couramment appliquée à l'IUTAM et qui avait été celle du symposium de 1962. Cette restriction a rendu très difficile le travail du comité scientifique chargé de lancer les invitations et de choisir les communications.

Placé sous la présidence du Professeur MAURICE ROY, ce comité comportait:

Mme A. G. MASSEVITCH et le Professeur E. BUCHAR pour le COSPAR,
les Professeurs S. HERRICK et G. C. McVITTIE pour l'IUTAM,
le Professeur D. BROUWER et le signataire de ces lignes pour l'IAU.

En définitive, et compte-tenu de plusieurs défections de dernière minute qui ont été regrettées, il y a eu 26 participants invités présents auxquels se sont joints six observateurs présents à Paris pendant que se tenait le symposium.

Le symposium a été endeuillé par le décès subit de notre collègue I. G. IZSAK, astronome à l'Observatoire Smithsonian. Sur décision du symposium, le présent volume est dédié à la mémoire de ce savant aimé de tous ceux qui le connaissaient. Tous, astronomes, géodésiens et spécialistes de la recherche spatiale, ont la plus grande estime pour les nombreux travaux originaux fort importants dont il fut l'auteur, en particulier dans les domaines de la Mécanique Céleste, de l'Astronomie, des Mathématiques appliquées ou de l'étude du champ de gravitation de la Terre à l'aide des observations précises de satellites. Le Professeur F. WHIPPLE, directeur de l'Observatoire Smithsonian, nous a envoyé une courte biographie de I. G. IZSAK, que l'on trouvera en début de ce livre.

Les réunions se sont tenues les 20, 21, 22 et 23 Avril dans les locaux mis à la disposition du symposium du Centre Universitaire International. Les huit séances ont respectivement été présidées par:

Mardi 20 Avril:	séance d'ouverture:	Prof. D. BROUWER
	après-midi:	Mme A. G. MASSEVITCH
Mercredi 21 Avril:	matin:	Dr. D. G. PORTER
	après-midi:	Prof. M. ROY
Jeudi 22 Avril:	matin:	Prof. G. N. DOUBOCHINE
	après-midi:	Mr. W. M. KAULA
Vendredi 23 Avril:	première séance:	Prof. S. HERRICK
	deuxième séance:	Prof. G. A. CHEBOTAREV

Ce volume contient l'ensemble des communications présentées au cours des séances de travail. On en compte 9 d'origine USA, 7 d'origine URSS, 4 d'origine UK et, respectivement, une canadienne, française, hongroise, japonaise, néo-zélandaise et tchécoslovaque. A deux exceptions près, elles sont données dans l'ordre dans lequel elles ont été entendues: on s'est efforcé de les grouper par sujet. Des résumés des trois communications qui ont déjà fait l'objet d'une publication sont donnés avec toutes références utiles. Deux articles qui ont été remis en russe ont été traduits, du russe en anglais, par M. B. MORANDO et moi-même. Un résumé en anglais et en français précède chaque article original. Un index analytique est destiné à faciliter l'usage de ce volume en indiquant l'endroit où se trouvent traités les principaux problèmes évoqués au cours de ces journées tant dans les communications qu'au cours des discussions. Je tiens à remercier ici M. B. MORANDO qui m'a fourni une aide importante en faisant une grande partie du travail nécessaire pour relever, choisir et rédiger les principales phases des nombreuses et fructueuses discussions qui ont suivi les communications et auxquelles tous les participants ont pris une part active. Nous espérons ainsi que l'essentiel des idées émises à cette occasion se trouve dans cet ouvrage.

Le lecteur se reportera au corps du volume pour le détail des travaux exposés; signalons seulement les principales directions de recherche qui se dégagent de leur audition ou de leur lecture.

Un des problèmes qui préoccupent actuellement le plus le COSPAR et plus particulièrement son sous-groupe de travail n° 1 (W.G.1) est l'utilisation à des fins scientifiques des nombreuses observations visuelles peu précises de satellites. Le sous-groupe de dynamique du W.G.1 était particulièrement chargé de promouvoir des recherches dans ce domaine, recherches dont les cinq articles de tête donnent ici les premiers résultats, prouvant que, dans certaines conditions, des résultats d'un grand intérêt peuvent être obtenus sans nécessiter d'équipement important en calculateur: ces recherches sont donc à la portée de pays ne pouvant pas consacrer des équipements importants à la recherche spatiale.

D'autres communications montrent dans quelle mesure une analyse particulièrement approfondie du matériel d'observation permet d'améliorer considérablement la précision des résultats, en comparaison avec les procédés utilisant brutalement la méthode des moindres carrés. Il semble bien que l'amélioration future des paramètres représentant les forces agissant sur les satellites artificiels dépend au moins autant de l'affinement des méthodes d'analyse des mesures (qu'il s'agisse d'arcs courts ou longs) et de leur traitement numérique que de l'augmentation de la précision des observations. C'est là une des conclusions principales que l'on peut tirer des ces réunions.

Les effets de la pression de radiation ont retenu l'attention de plusieurs participants, qu'il s'agisse de l'effet direct, ou celui dû à la radiation réfléchie par la Terre selon deux modes possibles. D'un intérêt particulier pour la Mécanique théorique et appliquée a été la preuve de l'effet que l'échauffement solaire des satellites plastiques peut avoir sur leur rotation.

La géodésie est une des sciences qui a le plus profité du lancement des satellites et de leur observation. Il est curieux de constater l'importance qu'une bonne détermination des orbites peut avoir dans le succès des méthodes géométriques. Bien entendu, les méthodes dynamiques ont été longuement traitées: la connaissance du champ de gravitation terrestre a fait depuis trois ans, des progrès considérables. I. G. IZSAK a construit une carte très détaillée de l'altitude du géoïde par rapport à l'ellipsoïde de référence déduit de 26000 observations. L'analyse détaillée des observations de vitesse radiale par effet DOPPLER-FIZEAU radio électrique permet désormais d'évaluer certaines harmoniques d'ordre 12 ou 15.

Enfin, parmi les corps célestes artificiels les sondes spatiales jouent un rôle de plus en plus grand, et il était naturel qu'un symposium consacré à la détermination des orbites des corps célestes artificiels ne pouvait les ignorer. Les résultats du sensationnel succès de trajectographie qu'a été la poursuite radio de Mariner II figurent en bonne place dans cet ouvrage. Mais, les projets en ce domaine étant pour le moment plus nombreux encore que les réalisations, il était nécessaire de discuter les problèmes qui restent posés: les sondes lunaires, les sondes cométaires et les satellites autour de la Lune ont été étudiés dans cet esprit au cours de ce symposium.

Nous nous réjouissons beaucoup du fait que la maison Springer-Verlag ait accepté d'assurer l'édition de ce symposium sous forme de ce volume qui constitue ainsi la suite logique de l'ouvrage « Dynamics of Satellites » déjà cité et dont le contenu reste d'ailleurs encore tout à fait d'actualité.

Je remercie ici les trois unions ou comités scientifiques internationaux, COSPAR, IAU et IUTAM qui ont permis une bonne organisation de ce symposium en finançant les frais de déplacement de certains délégués qui n'auraient pas pu autrement participer à ces réunions. Je remercie aussi le Centre Universitaire International du Ministère de l'Education Nationale et particulièrement sa directrice Madame ZALCBERG qui a accueilli le symposium, ainsi que la Direction des Relations avec la Communauté et l'Etranger du Ministère de l'Education Nationale grâce à la subvention de laquelle, les délégués et leurs familles ont pu se réunir en un dîner amical. Enfin, le Bureau des Longitudes a également contribué à l'organisation de ces réunions, tant par la prise en charge d'une partie des dépenses, et le prêt de matériel que par le concours de son personnel.

Paris, 1965 **Jean Kovalevsky**

List of Participants

R. J. Anderle	U.S. Naval Weapons Laboratory, Dahlgren, Virginia 22448, U.S.A.
J. D. Anderson	J.P.L., 4800 Oak Grove Drive, Pasadena, California 91103, U.S.A.
R. F. Arenstorf	NASA, Marshall Space Flight Center, Huntsville, Alabama 35812, U.S.A.
R. M. L. Baker, Jr.	Computer Sciences Corporation, 650 N. Sepulveda Boulevard, El Segundo, California 90245, U.S.A.
F. Barlier	Observatoire de Meudon, 92-Meudon, France
Yu. V. Batrakov	Institute for Theoretical Astronomy, Leningrad, B-164, USSR
J. C. Blaive	C.N.E.S., B.P. No. 4, 91-Brétigny s/Orge (Essonne), France
D. Brouwer	Yale University Observatory, Box 2023, Yale Station, New Haven, Connecticut 06520, U.S.A.
G. A. Chebotarev	Institute for Theoretical Astronomy, Leningrad, B-164, USSR
B. H. Chovitz	U.S. Coast and Geodetic Survey, Washington Science Center, Rockville, Maryland 20852, U.S.A.
P. Contensou	O.N.E.R.A. (Office National d'Études et de Recherches Aérospatiales), 92-Chatillon, France
A. H. Cook	Standards Division, National Physical Laboratory, Teddington, Middlesex, England
G. N. Duboshin[1]	University of Moscow, Korpus ,,M", kw. 173, Moscow, B-234, USSR
S. Herrick	4731 Boelter Hall, University of California, Los Angeles 24, California 90024, U.S.A.
M. Ill	Observatoire de Baja, Tóth Kálmán u. 19, Baja, Hungary
W. M. Kaula	Institute of Geophysics and Planetary Physics, University of California, Los Angeles, California 90024, U.S.A.
D. G. King-Hele	Royal Aircraft Establishment, Farnborough, Hants, England
J. Kovalevsky	Bureau des Longitudes, 3, rue Mazarine, 75-Paris 6e, France
Y. Kozai	Tokyo Astronomical Observatory, Mitaka, Tokyo, Japan
B. Lago	C.N.E.S., B.P. No. 4, 91-Brétigny s/Orge (Essonne), France
R. S. Long	Mathematics Department, University of Canterbury, Private Bag, Christchurch, New Zealand
J. Mar	Defence Research Telecommunications Establishment, Ottawa 4, Ontario, Canada
A. G. Massevitch	Astronomical Council of the Academy of Sciences of the USSR, Vavilova 20, Moscow, B-312, USSR
W. G. Melbourne	J.P.L., 4800 Oak Grove Drive, Pasadena, California, U.S.A.
R. H. Merson	Royal Aircraft Establishment, Farnborough, Hants, England
W. H. Michael, Jr.	NASA, Langley Research Center, Hampton, Virginia, U.S.A.
B. Morando	Bureau des Longitudes, 3, rue Mazarine, 75-Paris 6e, France
J. G. Porter	Hempstead Lane, Hailsham, Sussex, England
L. Sehnal	Československá akademie věd, Astronomicky Ustav, Ondřejov, ČSSR
H. G. Walter	ESDAC, Havelstr. 16, 61 Darmstadt, Germany

[1] French transcription: G. N. Doubochine

Contents

On the Use of the Results Obtained from Synchronous Observations of the Artificial Satellites of the Earth from the INTEROBS Programme for Scientific Purposes

By

I. D. Zhongolovich

Institute for Theoretical Astronomy, Leningrad, U.S.S.R.

Abstract. A method for obtaining the inclination of the orbit of an artificial satellite, as well as the longitude of its ascending node and the nodal period, from synchronous observations of at least two stations is proposed in this paper.

Résumé. L'auteur propose, dans cette communication, une méthode permettant de déterminer l'inclinaison de l'orbite d'un satellite artificiel, la longitude de son nœud ascendant et la période nodale à partir d'observations synchrons effectuées en deux stations au moins.

I. Introduction

The INTEROBS programme has supplied many synchronous visual observations of artificial satellites of the Earth. These observations may as well be used in the same manner as nonsynchronous ones, together with the observations made from isolated stations in a general program of determination of the elements of a satellite from all available observations. Yet, doing this, we loose the added favorable quality inherent in the synchronous observations.

That is why it is logical, together with the general use of all observations from isolated stations, including the non-synchronous ones, to devise methods of special reduction for the synchronous observations only in order to obtain from them the results that are the most suited to these observations, with the hope of getting these results with a higher accuracy.

Below are given the methods for obtaining from these synchronous observations three elements: the inclination of the orbit i, the longitude of the ascending node Ω, and the nodal period T_{Ω}. These particular elements may be obtained with the greatest accuracy from the *special* synchronous observations of the INTEROBS programme for which a whole series of topocentric positions of the satellite is always given for successive near instants for each observation.

II. Geocentric Rectangular Coordinates of the Satellite

It is essential, for further computations, to obtain for each instant of the synchronous observation, the geocentric rectangular coordinates of the satellite x_s, y_s, z_s in a system of coordinates fixed with respect to the Earth (the origin is the center of mass, the z axis is the axis of rotation, the x axis is parallel to the meridian of Greenwich, the y axis is directed towards the west).

Rigorous methods based on the least squares method lead to the following system of formulae for obtaining the coordinates of the satellite:

$$\left.\begin{aligned} m_1 &= \cos\delta_1 \cos(S-\alpha_1), & m_2 &= \cos\delta_2 \cos(S-\alpha_2), \\ n_1 &= \cos\delta_1 \sin(S-\alpha_1), & n_2 &= \cos\delta_2 \sin(S-\alpha_2). \\ p_1 &= \sin\delta_1, & p_2 &= \sin\delta_2; \end{aligned}\right\} \qquad (1)$$

$$\gamma = m_1 m_2 + n_1 n_2 + p_1 p_2; \qquad (2)$$

$$\left.\begin{aligned} \Sigma_1 &= (x_2 - x_1) m_1 + (y_2 - y_1) n_1 + (z_2 - z_1) p_1, \\ \Sigma_2 &= (x_2 - x_1) m_2 + (y_2 - y_1) n_2 + (z_2 - z_1) p_2; \end{aligned}\right\} \qquad (3)$$

$$\varrho_1 = \frac{\Sigma_1 - \gamma\Sigma_2}{1-\gamma^2}, \qquad \varrho_2 = \frac{\gamma\Sigma_1 - \Sigma_2}{1-\gamma^2}; \qquad (4)$$

$$\left.\begin{aligned} x_s &= \tfrac{1}{2}(x_1 + x_2 + m_1\varrho_1 + m_2\varrho_2), \\ y_s &= \tfrac{1}{2}(y_1 + y_2 + n_1\varrho_1 + n_2\varrho_2), \\ z_s &= \tfrac{1}{2}(z_1 + z_2 + p_1\varrho_1 + p_2\varrho_2). \end{aligned}\right\} \qquad (5)$$

Here, x_1, y_1, z_1 and x, y_2, z_2 are the coordinates of the two points of observation which are supposed to be known, S is the Greenwich sideral time of the synchronous observations; α_1, δ_1 and α_2, δ_2 correspond to the topocentric observations of the direction of the satellite; m_1, n_1, p_1 and m_2, n_2, p_2 are the direction cosines of the topocentric direction of the satellite; ϱ_1, ϱ_2 the topocentric distances of the satellite at the time of the observations; γ, Σ_1, Σ_2 are auxiliary quantities.

III. Determination of the Period of the Satellite

Having the geocentric rectangular coordinates of the satellite x_s, y_s, z_s for each of the instants of the synchronous observations, we determine the corresponding values of the geocentric spherical coordirates of the satellite:

$$\left.\begin{aligned} x_s &= r\cos\varphi\cos\lambda, \\ y_s &= r\cos\varphi\cos\lambda, \\ z_s &= r\sin\varphi, \end{aligned}\right\} \qquad (6)$$

where r is the geocentric radius vector, φ the latitude of the subsatellite point (or the geocentric declinations), λ the longitude (from Greenwich) of the subsatellite point.

Thus for the series of times T_j (in universal time) of the set of synchronous observations we have the following series of quantities (the longitude λ will not be used any more):

$$\left.\begin{array}{l} T_1, T_2, T_3, \ldots, T_n; \\ \varphi_1, \varphi_2, \varphi_3, \ldots, \varphi_n; \\ r_1, r_2, r_3, \ldots, r_n. \end{array}\right\} \quad (7)$$

The next thing to do is to obtain with the greatest accuracy from all these quantities, the corresponding time T_0 at which the satellite crosses a certain mean celestial parallel at the latitude φ_0.

For this we rely on the following correction $T_0 - T_j$ for each time T_j:

$$T_0 - T_j = \left(\frac{r^2}{k\sqrt{p}}\right)_{\text{sp}} (\mu_0 - \mu_j), \quad (8)$$

where,

$$\sin\mu_0 = \frac{\sin\varphi_0}{\sin i}, \qquad \sin\mu_j = \frac{\sin\varphi_j}{\sin i}. \quad (9)$$

Here,

$$k = \left(\sqrt{398\,603} = 631.35\right) \text{km}^{3/2}\,\text{sec}^{-1},$$

i is the inclination of the orbit, $p = a(1 - e^2)$ is the parameter of the orbit. If we wish to have $T_0 - T_j$ in seconds, we have to express r and p in kilometers, μ_0 and μ_j in radians.

Formula (8) is obtained from the integration of the corresponding expression for the velocity of variation of the latitude of the subsatellite point in the unperturbed mouvement of the satellite.

Example. On the 27th of August 1963. Observations at the stations of Arkhangelsk and Vologda; Object: 1960 ε_3.

T_j	φ_j	$T_0 - T_j$	T_0
$19^h02^m33\overset{s}{.}9$	$64\overset{\circ}{.}5158$	$125\overset{s}{.}33$	$19^h04^m39\overset{s}{.}23$
03 03.9	64.0897	94.96	38.86
03 04.9	64.0835	94.61	39.51
03 05.8	64.0607	93.20	39.00
04 02.7	62.9429	37.02	39.72
04 03.6	62.9244	36.23	39.83
04 04.8	62.9058	35.45	40.25
04 31.4	62.3108	11.58	42.98
05 03.1	61.4092	− 20.72	42.38
05 04.0	61.3725	− 21.96	42.04
05 04.9	61.3362	− 23.17	41.73
06 03.0	59.4681	− 80.80	42.20
06 04.0	59.3527	− 83.95	40.05
06 05.0	59.3024	− 85.38	39.62
06 39.3	58.1709	−116.09	43.21
			mean $T_0 = 19^h04^m40\overset{s}{.}71$
			± 0.39

In this example, the latitude was taken equal to 62°. Then, one determines in the same manner the instant of transit of the satellite at the same parallel φ_0 at the next revolution and one can then obtain, by combining the results, the corresponding period with sufficient accuracy.

This period should be called quasi-draconitic (or quasi-nodal) since it differs from the draconitic or nodal period which is the interval of time between two successive crossings of the celestial equator. These periods differ only by a very small quantity due to the *variation* of small perturbations undergone by the satellite during its motion from the equator to the parallel φ_0 from one revolution to another. The principal perturbations of each instant T_0 due to the main harmonic, J_2, of the gravitational field of the Earth are given by the following formula:

$$\delta T_0 = \frac{3}{2} J_2 \frac{a_0^2 r^2 \cos^2 i}{k p^{5/2}} \left[\tilde{u} + \frac{e}{2} \cos\omega \sin u - \frac{3}{2} e \sin\omega \cos u - \right.$$
$$\left. - \frac{1}{2} \sin 2u - \frac{e}{6} \cos\omega \sin 3u + \frac{e}{6} \sin\omega \cos 3u - \frac{4}{3} e \sin\omega \right], \tag{10}$$

where a_0 is the radius of the Earth, $u = v + \omega$ is such that v is the true anomaly and ω the argument of perigee, and e is the eccentricity. For the example given above, the difference between the two quantities δT_0 in ten days amounts approximately to 0.00006 minute and can be totally neglected.

IV. Determination of the Inclination and of the Longitude of the Node

In order to obtain the inclination i of the orbit and the longitude (right ascension) Ω of the node, it is necessary to transform the coordinates x_s, y_s, z_s of the satellite in a system of coordinates $(\bar{X}, \bar{Y}, \bar{Z})$ that does not rotate with the Earth. If $\bar{Z}$ is the axis of rotation of the Earth, $\bar{X}$ is directed towards the vernal equinox, and the $\bar{Y}$ axis is positive towards the *east*, the transformation of coordinates is given by formulae:

$$\left. \begin{aligned} \bar{X} &= x_s \cos S + y_s \sin S, \\ \bar{Y} &= x_s \sin S - y_s \cos S, \\ \bar{Z} &= z_s; \end{aligned} \right\} \tag{11}$$

where S is the Greenwich sideral time. Then, we determine the quantities:

$$\left. \begin{aligned} M &= \cos\delta \cos\alpha = \frac{\bar{X}}{R}, \\ N &= \cos\delta \sin\alpha = \frac{\bar{Y}}{R}, \\ P &= \sin\delta \qquad\quad\; = \frac{\bar{Z}}{R}, \end{aligned} \right\} \tag{12}$$

where $R = \sqrt{\bar{X}^2 + \bar{Y}^2 + \bar{Z}^2}$, α and δ are the geocentric equatorial coordinates of the satellite, and its right ascension and declination. Those quantities M, N and P are computed in this manner for each instant of synchronous observations.

In the spherical triangle determined by the position of the satellite at this given instant, the ascending node of the orbit and the intersection of the celestial meridian of the satellite with the equator, we can write the relation:

$$\tan i \sin(\alpha - \Omega) = \tan\delta. \tag{13}$$

If we write this equation in the following form:

$$\cos\delta \sin\alpha \tan i \cos\Omega - \cos\delta \cos\alpha \tan i \sin\Omega = \sin\delta, \tag{14}$$

introducing

$$A = \tan i \cos\Omega, \quad B = -\tan i \sin\Omega, \tag{15}$$

and using expressions (12), we obtain for each instant T_j of the set of synchronous observations, the following relation:

$$N_j A + M_j B = P_j. \tag{16}$$

The solution of the n similar equations ($j = 1, 2 \ldots n$) by the method of least squares, gives us the most probable values of A and B from which we can determine i and Ω from the formulae:

$$\tan i = \sqrt{A^2 + B^2}; \quad \tan\Omega = -\frac{B}{A}. \tag{17}$$

For the example already treated in the preceeding paragraph, one can get:

$$\begin{aligned} i &= 64^\circ\!.98 \\ \Omega &= 187^\circ\!.19. \end{aligned} \tag{18}$$

The quantities i and Ω are actually instantaneous values of these elements for the mean instant of the given set of observations.

Bibliography

Zhongolovich, I. D.: The computation of mean coordinates of artificial satellites of the Earth using synchronous observations from two known points on the surface of the Earth. Bull. of the Stations for Optical Observations, No. 46, 1966.

Preliminary Analysis of INTEROBS Programme Observations

By

M. Ill and **I. Almár**

Baja Observatory and Konkoly Observatory, Hungary

Abstract. The authors give a short survey of the material obtained within the INTEROBS—programme and they discuss in some detail the analysis made in Hungary. The determination of changes in the period of a satellite, using approximately known orbital elements, is in progress. The preliminary analysis of the data proved that reliable dP/dn values can be obtained from simultaneous visual observations even during one-week intervals of cooperative work.

Résumé. Les auteurs décrivent sommairement les observations obtenues dans le cadre du programme INTEROBS et discutent en détail l'analyse effectuée en Hongrie. Des déterminations de variation de période d'un satellite utilisant des éléments orbitaux approchés sont en cours. L'analyse préliminaire des données montre que des valeurs dignes de confiance de dP/dn peuvent être obtenues à partir d'observations visuelles simultanées même pendant un intervalle de temps d'une semaine correspondant à la durée d'une campagne de coopération.

The INTEROBS programme (the name comes from the words INTERnational OBServation) is a cooperative programme of observation, the aim of which is to determine the orbital elements of the observed satellites by simultaneous visual observations, or rather the variations in the air density from the variations of the orbital elements.

The first measurements used to test the method within the INTEROBS programme were made by the stations No. 1113 (Baja, Hungary) and No 1120 (Bautzen, Germany) in 1961. Further measurements were made to gather experience with the participation of stations No 1111 (Budapest, Hungary) and No 1185 (Rodewisch, Germany) in 1962. The authors gave an account of these experiences on the conference of observers (1962, Leningrad). It was suggested by the conference to the observers of satellites to participate in the INTEROBS programme and the conference intrusted the coordination of the programme to the station No 1113 (Baja). Since that time, 23 stations from the following countries: Bulgaria, Germany, Hungary, Poland, Rumania, Soviet Union took part in the programme. Later, stations from Finland and Italy joined the programme.

The observations of the INTEROBS programme are taking place within the so-called cooperation weeks. The participating stations endeavour—during the cooperations weeks—to make as many measurements as possible on the basis of the predictions of the computing centre COSMOS (Moscow). The results of observations are sent to the cooperation centre (Station No 1113, Baja), where they are summarized, and those suitable to further elaboration are selected (observations suitable for further elaboration are those which are simultaneously observed by another station). The material suitable for elaboration is published, and placed at the disposal of the participating stations. The material obtained within the INTEROBS programme between August 1963 and October 1964, is administered as a common propriety, and is used to test the method chosen by every station. So, an opportunity is given to compare the results obtained by different methods. The aim of this work is to choose the adequate methods. The comparison of the results obtained will be made in October 1965 at a conference in Budapest. It will be discussed then, on the basis of the experience obtained, whether it is worth to go ahead with the cooperation, and in what form.

So far, cooperation weeks of the INTEROBS programme were held in the following months: August 1963, February, March, April, May, June, July, August, September, October 1964. These measurements were published in [*1*] and [*2*].

7024 position measurements were sent in the coordinating centre. These data were obtained from the observations of 328 transits of 10 different satellites. From this material, 143 simultaneous transits could be chosen with 4103 positions, i.e. 143 such transits, where the same satellite was observed simultaneously by at least two observing stations, independently of the length of the simultaneously observed arc of trajectory.

30% of the observing material apply to 1960-ε 3.
13% of the observing material apply to 1963-17A.
45% of the observing material apply to 1963-47A.
So, 88% apply to three satellites.

The experiences on the first year showed that the number of simultaneously observed transits depends considerably on the meteorological factors, and, in this respect, the computing center (the task of which is to supply uniformly all the participating stations with predictions of proper quality and quantity during the cooperation-weeks) is also an important factor.

The mass treatment of the data was started in Hungary only in 1964. The first step was to determine the coordinates of the satellite at a given moment (within the observed time interval) measured simultaneously from at least two stations. This we made graphicaly: we draw the values α, δ or A, h observed from two or more stations in function of time (on a large scale) and from each curve, smoothing the observed values, we can read the coordinates referring to a selected moment. The coordinates obtained by this graphical interpolation are the so-called "simultaneous points".

Because of the large number of simultaneous points we had to choose a programme which is suitable for computer reduction. Such a method was suggested by I. Almár and E. Illés [3]. The aim of the authors was the elaboration of a procedure which provides directly from the simultaneous points at each transit the orbital elements, together with the errors. For the computing of the space positions, a very simple range of formulae of "cosmical triangulation" is used in the system of coordinates fixed relatively to the stars, starting from the simultaneous positions. After this, from the obtained XYZ coordinates, we compute with the method of least squares the components of the normal vector of the plan which contains the center of the Earth and which lies at best between these points. We obtain also the errors of the determined components. Then, projecting the space positions on this determined plan, we reduce the question to a two dimensional problem. If the observations have a sufficient accuracy and if the observed arc of the orbit is long enough, then it is possible to determine the parameters of the ellipse which passes at best through the points, and one of the focuses of which coincides with the zero-point (the center of the Earth). The method gives us also the errors of the orbital elements.

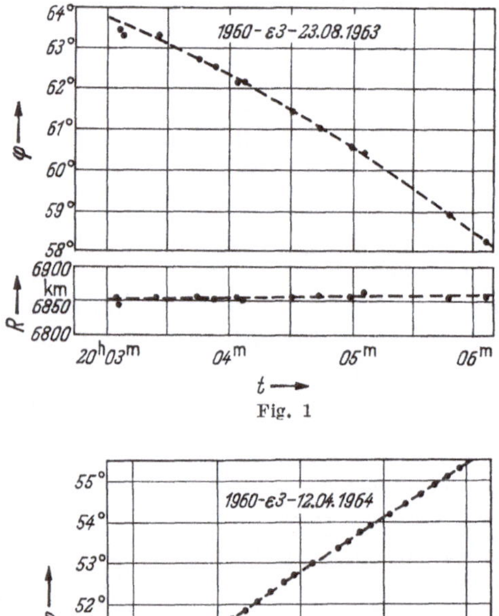

Fig. 1

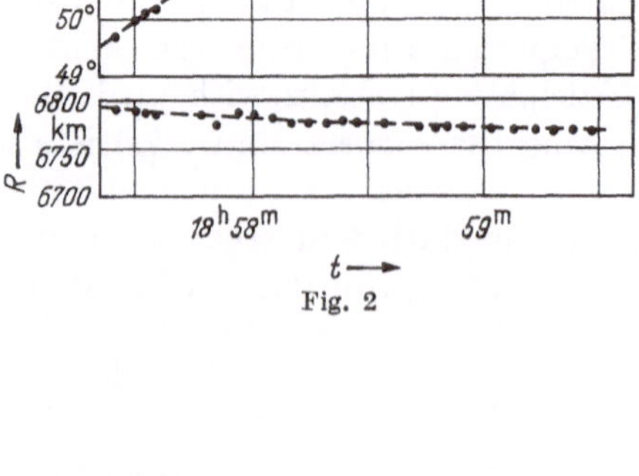

Fig. 2

In the first phase of the treatment, the data referring to 1960-ε 3 and 1963-43 A (the data published in [1]) were elaborated jointly by the Baja and Budapest stations, with an Elliott-803 electronic computer. As a first step, we determined the space coordinates (XYZ) and radius-vectors (R) of the satellites and the geographical latitudes of the subsatellite points (φ). The radius-vectors being drawn in function of time, we can estimate their quality from the spreading of the data. On the basis of the curves obtained, we established that many of the measurements show such a great scattering, that orbital elements cannot be possibly computed from them. The transits for which the single measurements were scattered in time, have been considered the worst, because the fixing of the simultaneous points by graphical interpolation raised difficulties in such cases. Good results were given generally by transits whose positions were separated only by a few seconds of time. The $R(t)$ curves from such better transits show scattering of the order of a few kilometres (Figs. 1—5).

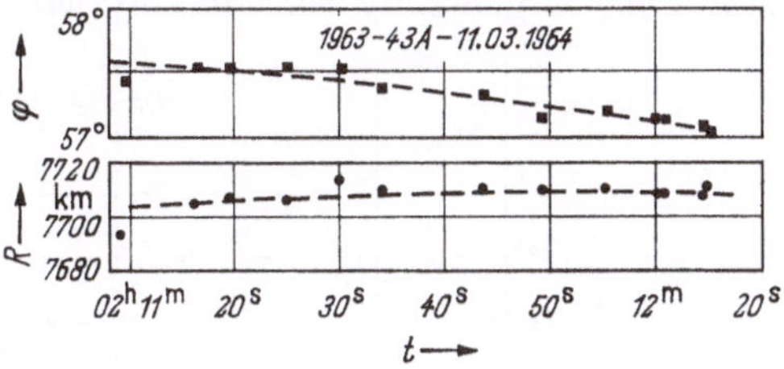

Fig. 3

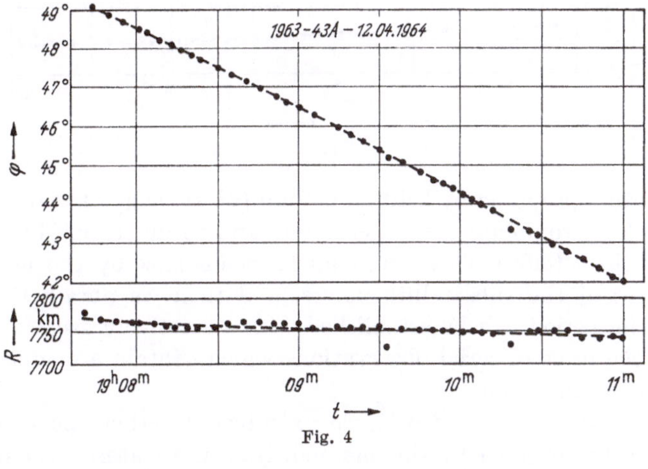

Fig. 4

Looking at the curves, we can establish that the first and last points of the curve may be off compared with the general course of the curve. This can be explained by the undefined character of the graphical interpolation on these points.

Taking in view the scattering of the radius-vectors, it becomes clear, even on the curves showing the best results, that in case of methods, where the orbital elements are computed from radius-vectors, the actually computed radius-vectors can often give rise to unreliable orbital elements because of the scattering (mainly on short arcs of orbit). It is practical in these cases to draw the curve radius-vectors in function of time, and further, to perform the calculations omitting the deficient points.

The $\varphi(t)$ curves have the same aspect as the $R(t)$ curves, but the scattering is generally considerably less. This encouraged us to apply

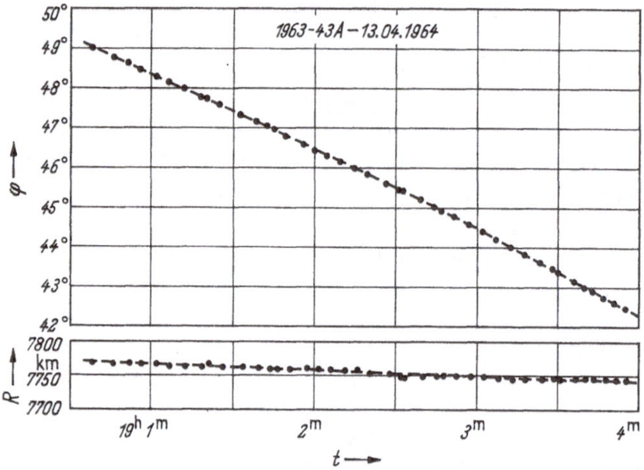

Fig. 5

the procedure proposed by ZHONGOLOVICH [4]. He suggested to establish the use of approximate orbital elements and to determine by this the period of the satellite. This can be made for instance by the use of the coordinates of the subsatellite points, and then we obtain the nodal period. The period can be established with a relatively great accuracy, even on the basis of serial observations which contain a comparatively short arc of the orbit.

At this stage of our work, we examined whether the changes of period can be revealed by this method, from the material got from the INTEROBS programme. The treatment has been done in the following way: we chose a reference latitude on the basis of the obtained coordinates of the subsatellite points, and then we calculated, using the approximate orbital elements, for each subsatellite point the time of crossing the reference latitude. The average of this times calculated in this way was named the observed time: O. We determined the change in period per revolution as follows: if the period of the satellite is constant, the time of the n-th crossing of the reference latitude, measured from a time t_0 is:

$$C = t_0 + n P, \tag{1}$$

where P is the nodal period of the satellite at t_0. But the observed time O differs from this calculated time, because the satellite has a decrease of period and therefore (if we consider that the decrease of the period is constant) the time of the n-th crossing of the reference latitude is:

$$O = t_0 + n P - \frac{n(n+1)}{2} \varDelta \tag{2}$$

where $\varDelta$ is the change of period per revolution ($= dP/dn$).

From (1) and (2), we get,

$$\varDelta = \frac{-2(O-C)}{n(n+1)} \tag{3}$$

So we obtain for each observed crossing of the reference latitude a value of $\varDelta$. If $\varDelta$ is not constant, but scatters around a constant value, then it is suitable to determine the period belonging to the last observed point and to use the same procedure starting from the last point. In this fashion, so we obtain a new value $\varDelta'$. By weighting the obtained values $\varDelta$ and $\varDelta'$ we obtain:

$$\bar{\varDelta} = \frac{n(n+1)\varDelta + n'(n'+1)\varDelta'}{n(n+1) + n'(n'+1)}.$$

The weighting is justified by the fact that the effect of the error in 0 on $\varDelta$ is inversely proportional to $n(n+1)$ and $n'(n'+1)$ respectively.

The $-2(O-C)$ values being drawn in function of $n(n+1)$, we obtain a straight line if $\varDelta$ is constant. If $\varDelta$ is not constant, i. e. it has a systematic deviation starting from a time t_k, we divide the interval into two parts: from $t_0 - t_k$ anf drom $t_k - t_n$. The treatment further on is the same as above. In the last case, we obtain graphically a broken straight line.

In the manner mentioned, I. Almár treated the material concerning to the object 1960-ε 3. He made use of 12 transits in 9 days (20—30 August, 1963). The approximate orbital elements used by him were those published ITA (Institute of Theoretical Astronomy, Leningrad). According to the values obtained the average value of $\varDelta$ in the time between 21—26 August, 1963 was (Fig. 6):

$$dP/dn = -0.0110 \text{ sec/Rev}$$

and between 26—30 August, 1963:

$$dP/dn = -0.0075 \text{ sec/Rev.}$$

(The scattering of is $dP/dn = 0.002$ sec/Rev.)

On the curve, we can see the change of dP/dn as a sudden break. The reality of the break is conspicuous also on the $O-C$ curve (Fig. 7). The obtained result being compared with the different indices of solar activity, it is conspicuous that the decrease of dP/dn happened at the same time when, as a consequence of the disappearence of a large sunspot group, the relative sunspot number and the sunspot areas decrea-

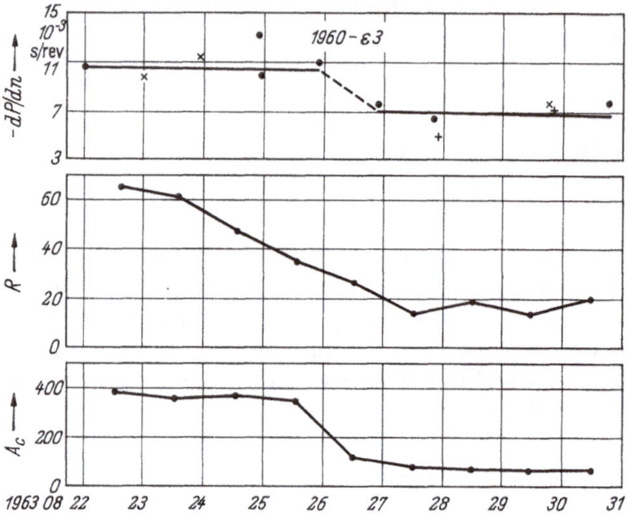

Fig. 6

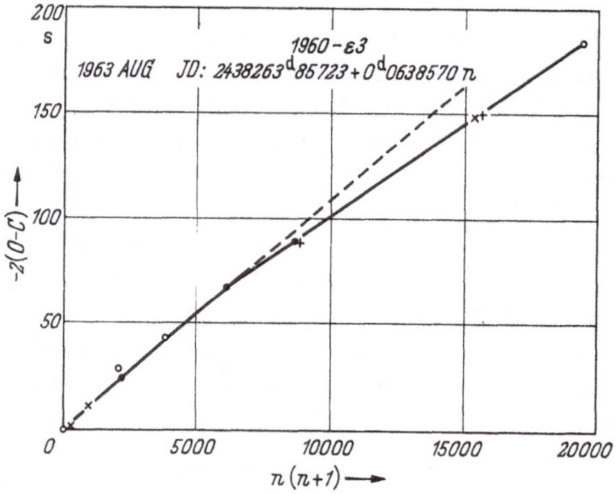

Fig. 7

sed suddenly (Catanian measurements). There is no correlation with radio-flux.

If we compute the life-cime t_L of the satellite on the basis of the present data and if we compute the theoretical values of Δ with the formula given by King-Hele, then the results obtained are: -0.0072 sec/Rev, and -0.0084 sec/Rev depending upon the constant used.

The material of the month of April 1964 is considerably smaller. It was not possible to determine an acceleration, but we got a reliable value of Δ:

$$dP/dn = -0.0128 \text{ sec/Rev.}$$

The theoretical value are considerably smaller: -0.0088 sec/Rev and -0.0103 sec/Rev, respectively (Fig. 8).

The material concerning 1963-43A (10—14 March, 1964, 10—15 April, 1964) is not of a good quality. The average standard deviation of the obtained latitude crossing times is about: 0.8 sec. We treated the material in the same manner but it was not possible to determine a reliable value of Δ (at each transit). Therefore we determined only the average nodal period concerning the two cooperation-weeks. These are:

$$P_n = 102.372 \text{ min.} \quad (10\text{—}14 \text{ March}),$$

$$P_n = 102.364 \text{ min.} \quad (10\text{—}15 \text{ April}).$$

From the change of period in this time interval we obtain an average value of 0.010 sec/Rev. The theoretical one computed with the formula given by KING-HELE is: 0.007 sec/Rev.

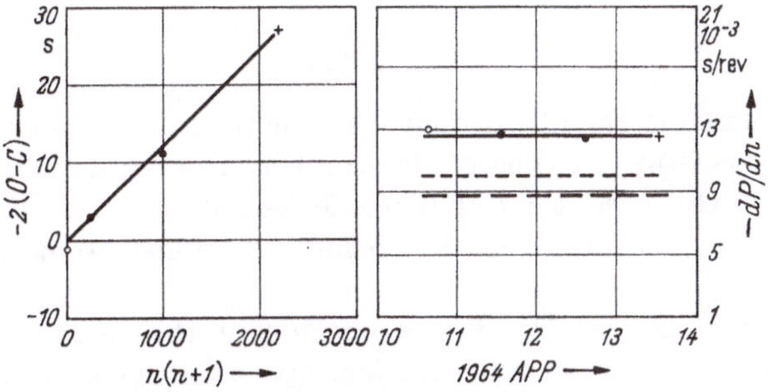

1960-ε3

1964 APR JD: 2438496^{d}82339 + 0^{d}0634370 n

Fig. 8

On the basis of the obtained results one can make the following conclusions: By suitable methods, the values of Δ can be determined with sufficient accuracy from simultaneous visual observations. The time of a possible sudden acceleration can be determined in a fortunate case with an uncertanity of some hours, or at the worst of 1—2 days. We can consider as important the fact that we obtain at each transit the values of Δ separately. There is no need of averaging the observing material collected from a longer time interval. This averaging would cause the sudden changes of air-density to disappear.

The computing of the other elements from the available material can be done only with difficulty, because the lengths of the observed arcs of orbit are too short. Because of this we proposed to observe, if possible, two successive transits of the same satellite during the cooperation-weeks of this year. The observed arcs of orbit are in such cases not quite identical and so they can partly prolongate each other. Another method is to observe satellites with large inclination and, by help of the newly joined stations (Italy, Finland, Sweden), we shall then prolongate considerably the observed length of the orbital arc.

References

[1] ILL, M.: Ergebnisse d. im Rahmen d. INTEROBS-Programms abgehaltenen Kooperationswochen (Folge I), Baja, 1964.
[2] ILL, M.: Ergebnisse d. im Rahmen d. INTEROBS-Programms abgehaltenen Kooperationswochen (Folge II), Baja, 1965.
[3] ILLÉS E., and J. ALMÁR: Zur Berechnung momentaner Bahnelemente künstlicher Erdsatelliten aus Basisbeobachtungen, Beobachtungen künstlicher Erdsatelliten, Nr. 3, 1964, Berlin: NKGG der DDR, 1965.
[4] ZHONGOLOVICH, I. D.: A Remark on the Treatment of Simultaneous Observations. Conference at Riga, 1—5 Febr. 1965; see also page 1 of the present book.

Discussion[1]

Prof. HERRICK remarked that this method is somewhat related to the GAUSS-GIBBS method of determination of orbits from positions only, but, in this case, the difference lies essentially in the fact pointed out by Mr. ILL that the period is being determined from observations of several passes.

Prof. HERRICK and Dr. BAKER considered that a sizeable amount of information is lost by using fixes or space positions rather than the observations themselves, the residuals for each observation reduced as such would certainly improve the precision of the determination of the orbit. Contamination of good data by forcing its use as a fix is possible in the proposed methods. The elimination of non simultaneous observations would also be avoided and more information would be used and precision would be improved. M. ILL and Mrs. MASSEVITCH pointed out that the presented method has the advantage of being much simpler and of requiring a much smaller amount of computing time, permitting anyhow to determine the period of a satellite within a single night of observations. Mr. MERSON considered that 16 observations in 3 or 4 passages are sufficient to provide a good orbit.

[1] This discussion also refers to the following two papers.

Determination of the Quasi-Nodal Period of the Satellite 1960 ε 3 from Simultaneous Visual Tracking Data

By

V. M. Amelin

Institute for Theoretical Astronomy, Leningrad, USSR

Abstract. Results of application of the method of I. D. Zhongolovich to a series of simultaneous visual observations of the satellite 1960 ε 3, observed from four stations. The quasi-nodal period is found to be decreasing during the five days of observations.

Résumé. Résultats de l'application de la méthode de I. D. Zhongolovich à une série d'observations visuelles simultanées du satellite 1960 ε 3 faites en quatre stations. La période quasi-nodale a décru pendant les cinq jours d'observations.

At the Institute for Theoretical Astronomy of the Academy of Sciences of the U.S.S.R. a test reduction of visual observations of satellite 1960 ε 3 (6005-3) obtained from the INTEROBS program, has been carried out.

The aim was to obtain a quasi-nodal period of the satellite. The method devised by Professor I. D. Zhongolovich has been used [*3*]. The material was taken from the first volume of results of observations of the INTEROBS program [*1*].

In this volume are collected observations made from August 1963 till April 1964. As it can be seen from these tables, the satellite is observed at each transit from at least two stations during 3 to 5 minutes. The number of observations for each transit was extremely different (from 4 or 5 to 15 or 20).

The calculation of the quasi-nodal period T_Ω was carried out in the following manner. Although the satellite was observed from both stations during the same interval of time, the instants of observation were different for each station. So, at the beginning, it was necessary to compute for both stations the topocentric coordinates of the satellite for the same given universal time t.

For the first station we get immediately the time of the observation t_j from Table 1 [*1*] and the corresponding topocentric equatorial coordinates of the satellite α_1 and δ_1. For the second station, using the data given in Table 1 [1], we construct two smoothed curves that give the observed topocentric coordinates of the satellite α and δ as functions of time. Using these curves, we determine the coordinates α_2 and δ_2 of the satellite at the second station at the time t_j mentionned above.

Thus, for the time t_j, the topocentric equatorial coordinates of the satellite α_1, δ_1 and α_2, δ_2 are now obtained for both stations. Then, one can calculate the geocentric rectangular coordinates of the satellite x_c, y_c, z_c with the formulae given in [2]. Knowing x_c, y_c, z_c we find the geocentric radius vector R of the satellite and the latitude φ of the subsatellite point:

$$R = \sqrt{x_c^2 + y_c^2 + z_c^2}, \qquad \tan\varphi = \frac{z_c}{\sqrt{x_c^2 + y_c^2}}. \tag{1}$$

The calculation of the quantities x_c, y_c, z_c, R and φ were carried out on the electronic computer BESM-2.

The quantities R and φ are computed for all the times t_j at which the satellite has been observed from the first station. Then, such procedure being repeated for a few transits, we choose a certain value φ_0 of the latitude in such a way that φ_0 belongs to the interval of observation for each transit.

For the following calculations it is necessary to know approximate values of the elements i, a, e of the satellite. We may notice that a value of the inclination i can be obtained with a precision of 0°,05 from the reduction of the same nearly synchronous visual observations. Knowing i (from the reduction of observations or from the ephemeris of the satellite) we compute the quantities μ_j and μ_0:

$$\sin\mu_j = \frac{\sin\varphi_j}{\sin i}, \qquad \sin\mu_0 = \frac{\sin\varphi_0}{\sin i}. \tag{2}$$

We shall call t_0 the universal time when the latitude of the subsatellite point was φ_0.

Then, applying a formula proposed by Professor ZHONGOLOVICH, we can calculate the interval of time during which the value of the latitude of the subsatellite point changes from φ_j to φ_0:

$$t_j - t_0 = \pm\left(\frac{R^2}{k\sqrt{p}}\right)_{\mathrm{Sp}} (\mu_j - \mu_0), \tag{3}$$

where $p = a(1 - e^2)$. The sign „+" in this formula is taken for the ascending branch of the trajectory of the satellite, and the sign „—" for the descending branch.

For each time t_j, we find the difference $t_j - t_0$ and consequently we get the value t_0.

Let n be the number of observations of the satellite for a given transit, the mean value T_0 of the time at which the satellite crosses the parallel φ_0 can be detected from the formula:

$$T_0 = \frac{[t_0]}{n}. \tag{4}$$

Let v be the difference $T_0 - t_0$. Then the mean quadratic error of the quantity T_0 will be:

$$M_{T_0} = \sqrt{\frac{[v v]}{(n-1)n}}. \tag{5}$$

If T_0' and T_0'' are the mean times at which the satellite crosses the parallel φ_0, we can compute the quasi-nodal period of the satellite T_Ω:

$$T_\Omega = \frac{T_0'' - T_0'}{k} \tag{6}$$

in which k is the number of revolutions of the satellite between the chosen transits.

The standard deviation of the quasi-nodal period T_Ω will be:

$$M_{T_\Omega} = \frac{\sqrt{M^2_{T_0''} + M^2_{T_0'}}}{k}. \tag{7}$$

As an example we have reduced the observations accomplished through the INTEROBS at Arkhangelsk, Vologda, Riga and Tartu from the 23rd to the 27th of August 1963.

Satellite 1960 ε_3 was taken, five transits being studied and φ_0 being given the value $\varphi_0 = 62°$N.

In Table 1 are given as examples, the following quantities for two transits (August 25 and 26): the times of observations t_j, the latitude of the subsatellite point φ_j, the time t_0 at which the satellite crosses the parallel φ_0 and the difference v.

It can be seen from Table 1 that the maximum divergence between the times t_0 for the 25th reaches 11.87^s but for the 26th it is 2.01^s only.

Such a divergence on the times t_0 depends on the precision of the observations at the different stations and on the distribution of the observations in time.

One may notice in particular that the smoothed curves which give α and δ as functions of time for the 26th can be established with better certainly than the curves for the 25th.

Generally better is the distribution of observations in time, better is the certainty with which such curves can be drawn.

In Table 2 are given the mean values of T_0 at which the satellite crosses parallel $\varphi_0 = 62°$N, the number k of revolutions between separate transits and, at last, the values of the quasi-nodal periods T_Ω.

We see, from Table 2, that the quasi-nodal period of the satellite decreases in jerks from the 23rd to the 27th. Owing to this in the most favorable case the period is determined with a precision of 0.0005^m or 0.03^s.

Table 1

N°	Stations	t_j	φ_j	t_0	v
		August, 25			
1	Arkhangelsk	$19^h33^m05^s.7$	$63°8588$	$19^h34^m28^s.56$	$0^s.51$
2	Vologda	33 11.8	63.8314	30.95	−1.88
3		34 04.4	62.6368	28.73	0.34
4		34 05.2	62.6138	28.83	0.24
5		35 03.4	61.1530	34.03	−4.96
6		35 04.3	61.1107	33.76	−4.69
7		35 05.2	61.0641	33.12	−4.05
8		36 03.4	58.9301	27.62	1.45
9		36 04.3	58.7535	23.44	5.63
10		36 05.3	58.6718	22.16	6.91
11		19 36 28.6	58.0300	19 34 28.59	0.48
		August, 26			
1	Arkhangelsk	$20^h04^m13^s.7$	$63°8490$	$20^h05^m34^s.64$	$0^s.29$
2	Tartu	04 45.6	63.2139	34.62	0.31
3		05 16.0	62.4771	34.00	0.93
4		05 40.4	61.8354	34.48	0.45
5		06 00.4	61.2679	34.99	−0.06
6		06 26.8	60.4363	34.86	0.07
7		07 39.0	57.9075	36.01	−1.08
8		20 07 55.7	57.2571	20 05 35.87	−0.94

Table 2

Stations	T_0	k	T_Ω
Arkhangelsk—Vologda	August, 23 $20^h04^m09^s.63 \pm 1^s.126$	16	$91^m.9501 \pm 0^m.0012$
Riga—Arkhangelsk	August, 24 $20^h35^m21^s.69 \pm 0^s.542$	15	$91^m.9415 \pm 0^m.0014$
Arkhangelsk—Vologda	August, 25 $19^h34^m29^s.07 \pm 1^s.163$	16	$91^m.9436 \pm 0^m.0012$
Arkhangelsk—Tartu	August, 26 $20^h05^m34^s.93 \pm 0^s.263$	15	$91^m.9398 \pm 0^m.0005$
Arkhangelsk—Vologda	August, 27 $19^h04^m40^s.71 \pm 0^s.395$		

Bibliography

[1] Ergebnisse der im Rahmen des INTEROBS-Programms abgehaltenen Kooperationswochen (1. Folge), 1964.

[2] Zhongolovich, I. D.: Artifical Satellites and Geodesy, Astr. J. U.S.S.R. XLI, No. 1 (1964).

[3] Zhongolovich, I. D.: see p. 1 of this book.

Evaluation of the Satellite Period on the Base of Simultaneous Visual Tracking from Two Given Stations

By

T. V. Kassimenko

The Astronomical Council of the USSR Academy of Sciences, Moscow, USSR

Abstract. First results of the reduction by the method proposed by Prof. Zhongolovich [1] of simultaneous visual tracking data for satellite 1960 ε-3 (the cabin of Soviet Space Ship I) are given in this paper.

Résumé. Cette communication présente les premiers résultats de la réduction d'observations visuelles simultanées du satellite 1960 ε 3 (cabine du premier vaisseau spatial soviétique) par la méthode du Prof. Zhongolovich.

Tracking data for satellite 1960-ε 3 obtained in August 1963 were reduced. Beginning from 1963, tracking stations of East European countries and the Soviet Union have been carrying out simultaneous observations of low satellites aimed to study short-periodic variations of the atmospheric density (INTEROBS Program). Results of these observations carried out from August 1963 to April 1964 have been published in a special catalogue [2]. During 10 days (August 20—30, 1963) stations in Arkhangelsk, Vologda and Tartu regularly observed the Cabin of the Soviet Space Ship I (1960-ε 3) according to the INTEROBS Program. Data on nine transits observed simultaneously by two stations were reduced. The reduction included the following stages:

1. Graphic Interpolation of Tracking Data to the Synchronous Moment

Values of topocentric satellite coordinates obtained at one station were drawn on a graph as a function of time, a smoothed curve was given (separately for each coordinate) from which then the values for the tracking time at the other station were read.

2. Evaluation of Levelled Rectangular Geocentric Coordinates

The evaluation was carried out with the aid of the formulae

$$\left.\begin{aligned} x_s &= \tfrac{1}{2}(x_1 + x_2 + m_1\varrho_1 + m_2\varrho_2), \\ y_s &= \tfrac{1}{2}(y_1 + y_2 + n_1\varrho_1 + n_2\varrho_2), \\ z_s &= \tfrac{1}{2}(z_1 + z_2 + p_1\varrho_2 + p_2\varrho_2), \end{aligned}\right\} \qquad (1)$$

where X_1, Y_1, Z_1 are the rectangular geocentric coordinates of the first station X_2, Y_2, Z_2—the coordinates of the second station, ϱ_1 and ϱ_2—the topocentric distance to the satellite; m, n, p—the angular coefficients of the topocentric directions to the satellite. Simultaneously the latitude of the subsatellite point φ_s and the geocentric radius-vector of the satellite r_s necessary for further reduction were computed.

3. Reduction of Tracking Data to a Reference Latitude and Evaluation of the Nodal Period P_Ω

An analytic reduction of tracking data to one reference parallel $\varphi_0 = 62°$ was carried out for all passages by the formula

$$(t_N - t_0)_s = \left(\frac{r_s}{\sigma}\right)^2 (M_N - M_0)', \tag{2}$$

where

$$\sin\mu_N = \frac{\sin\varphi_N}{\sin i},$$

$$\sin\mu_0 = \frac{\sin\varphi_0}{\sin i},$$

$$\sigma = 1473.24 \sqrt[4]{P},$$

i inclination, P parameter of the orbit.

Then the nodal period of the satellite P_Ω was determined from the relation $P_\Omega = \Delta T/n$, where ΔT—is the difference of the times of passages of the sub-satellite point through the chosen parallel for two following dates, n—the number of satellite revolutions corresponding to this difference. Results of the determination of the nodal period are given in the Table 1, where the number of satellite revolutions and the corresponding P_Ω for both dates as well as the daily variation of ΔP_Ω are given.

Table 1

Date	n	P_Ω	$\Delta P_\Omega/\Delta t$
August 19			$0^m.00$
	31	$91^m.9607$	
21			−29
	17	9521	
22			−16
	15	9490	
23			−10
	31	9461	
25			−12
	16	9425	
26			− 7
	15	9411	
27			−12
	32	9374	
29			

Table 1 shows that the nodal period can be determined with a high accuracy by the method proposed by Prof. ZHONGOLOVICH.

It is interesting to note that earlier data on several passages were reduced by us using another method developed for reduction of simultaneous visual tracking data [3]. Results of the determination of the major semi-axis for the same dates are compared below.

In Column I the determination of the major semi-axis has been performed by the formula $a = \frac{r_s(1 + e\cos v)}{1 - e^2}$, where the eccentricity e and the true anomaly v were taken from the orbital elements for the same period [4], and the geocentric radius-vector r_s of the satellite was determined from simultaneous visual tracking data. In Column II the major semi-axis has been determined by the formula $a = C\,P_\Omega^{2/3}$, where P_Ω is obtained by the method proposed by Prof. Zhongolovich. The value of a as given in the average orbital elements sequences [4] for August 23 and 26, is given in Column III. It can be seen from Table 2 that the first method gives rather rough results.

Table 2

Date	Major semi-axis		
	I	II	III
23	6745.0	6748.0	6746.2
25	6806.2	6747.9	
26	6748.4	6747.9	6746.2
27	6754.6	6747.6	

The results from Table 1 are also given graphically (Figure). The variation of the period is rather smooth except the "jump" on August 19—21 (this is perfectly seen on the graph). It is interesting to note

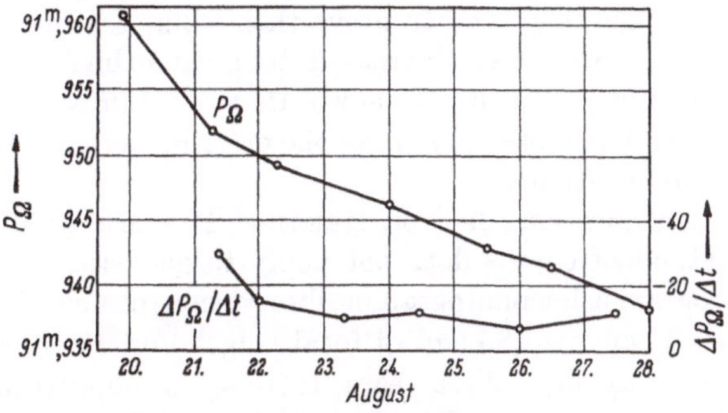

Fig. 1

that a solar flare and a magnetic storm following it, were observed on August 20, 1963 [5]. The "jump" in the period may possibly be a result of these phenomena.

References

[1] Zhongolovich, I. D.: p. 1 of the present book.

[2] Ergebnisse der im Rahmen des INTEROBS-Programms abgehaltenen Kooperationswochen (1. Folge), Baja, 1964.

[3] Ill, M.: Bahnbestimmung von künstlichen Erdsatelliten auf Grund visueller Beobachtungen, Baja, 1962.

[4] TCHEBOTAREV, G. A., and E. N. MAKAROVA: Average Orbital Elements of the Cabin of Soviet Space Ship I (1960 ε_3) for August—December 1963, Bull. of Stations for Optical Tracking, No. 40, 1964.
[5] Solar Geophysical Data, Series F, Part B, Central Ratio Propagation Laboratory, National Bureau of Standards, Boulder, 1963.

Discussion

The type of observations described in the last two papers may lead to the determination of fairly quick changes in the period of the satellite in spite of the fact that the visual observations are not very accurate ($0^\circ 1$ in position, and $0^s 1$ in time). This fact appears from a general discussion. Dr. COOK expressed the opinion that even if systematic errors are introduced in the value of the period, the *changes* in period should be estimated correctly. A value of the period may be determined on one day or even on one night at the best and it is agreed that short period perturbations due to the potential having small amplitudes, will not conceal the sudden changes in the period caused by modifications of the atmosphere by solar flares.

Mrs. MASSEVICH indicated that such methods are essentially devised to make use of very rough observations, as it was suggested by COSPAR in order to test the usefulness of visual observations. Since such observations can be made in a larger scale than precise photographic observations, and also on many more satellites, it is important to see what accuracy can be obtained and to prove their value for the study of the density of the atmosphere. Compared to good orbits obtained from precise observations, it can be shown that the jumps in period that are given in this paper are larger than the error in period as determined from visual observations.

Answering a question of Prof. HERRICK, Mrs. MASSEVICH indicated that this method of course does not apply to geodetic work, but that a geodetic program using photographic observations of Echo I and Echo II taken by small cameras (25 cm. of focal length and 10 cm of aperture) is being under way since three years. It involves cooperation of several countries (USSR, Bulgaria, Rumania, Poland, DDR, Hungary, Italia and Sweden). This purely geometrical method gives an accuracy of 4″ and 3 milliseconds on time — (precision of 50 metres in position).

Some Orbit Determinations Using Visual and other Observations

By

R. H. Merson

Royal Aircraft Establishment, Farnborough, Hants, England

Abstract. An analysis has been made of some 2000 visual observations of four satellites. An indication of the accuracy of determining orbital parameters is given, together with Tables of the accuracy of individual observing stations. A brief note on Minitrack observations of Ariel 2 is included.

Résumé. On a fait l'analyse d'environ 2000 observations visuelles de 4 satellites. On donne une indication de la précision obtenue par la détermination des paramètres orbitaux ainsi que des tables de la précision individuelle des stations. On a inclus une courte note sur les observations Minitrack d'Ariel 2.

1. Introduction

At the request of the Orbit Analysis Working Group of The British National Committee on Space Research and COSPAR Working Group One, work has continued at the R.A.E. on the assessment of the accuracy and usefulness of groundbased observations and, in particular, visual observations of earth satellites.

During the past year observations of five satellites have been studied, Transit 1B (1960 γ 2), Cosmos 2 (1962 ι 1), Samos 2 (1961 α 1), Starrad (1962 β K) and Ariel 2 (1964-15A). These satellites were chosen for a variety of reasons, including the availability of observations, the type of orbit and specific requirements for orbital data.

2. Ariel 2

A definitive orbit of Ariel 2 has been determined from Minitrack data supplied by N.A.S.A. This work was performed in connection with the on-board experiments, one of which (the Meteorological Office Ozone Experiment) required accurate satellite positions at certain times.

Orbital parameters of Ariel 2 were obtained at intervals of 25 nodes (about $1\frac{3}{4}$ days) between the launch date, March 27, 1964 and October 3, 1964. The results were circulated to the interested parties as they came to hand and a full report will be issued in the near future. The Minitrack network of stations, as has been reported before [1], provide a good orbit coverage and, having an observational accuracy of about one minute of arc, give a sufficiently accurate orbit for the results to be included in determinations of the zonal harmonics of the earth's gravitational field. In the case of Ariel 2 (eccentricity 0.073, inclination $51\overset{\circ}{.}64$, perigee height 290 km.) the orbit was sufficiently accurately determined for the satellite's position to be given to better than $\frac{1}{2}$ km. at all times.

3. Transit 1B

126 visual observations, mainly from U.K. observers, were used to determine orbital parameters of Transit 1B during the period June 2 to 10, 1962. A report on these has already been issued [2].

Transit 1B was chosen for study because of its inclination of $51\overset{\circ}{.}25$ and its moderately high perigee height (350 km). The former enabled U.K. observers to 'straddle' the apex of the orbit — in fact, the orbit coverage was from 9 minutes before to 5 minutes after the apex point. As a result of this, the inclination of the orbit was determined to within a standard deviation of $\pm 0\overset{\circ}{.}002$, which is very good considering the poor coverage of the orbit as a whole. Standard deviations of other parameters were: semi-major axis, ± 4 m; eccentricity, $\pm 9.10^{-5}$; R.A. of node, $\pm 0\overset{\circ}{.}01$; argument of perigee, $\pm 0\overset{\circ}{.}7$; the latter being relatively large because of the small orbital eccentricity of 0.02.

The accuracy of individual observing stations was assessed by taking the r.m.s. values of declination residuals and in all cases these were found to be slightly better than the observers' own a priori estimates, a somewhat unexpected result!

4. Samos 2

The satellite Samos 2 was chosen for three reasons. It has a retrograde orbit (inclination $97\overset{\circ}{.}4$) with a low eccentricity ~ 0.005 (perigee height about 475 km.), and was well-observed (in the Northern hemisphere) during the months of May, June and July of 1961 and 1962. A total of over 1,000 observations have been analysed, 550 from the U.K., 300 from France, 90 from Finland and 80 from Holland. The orbit coverage, however, was extremely poor, from argument of latitude 45° to 70° at most, i.e. no more than 7% of the orbit. As a consequence the orbital parameters are not as well-determined as had been hoped, the standard deviations of the elements being of the orders $\sigma(a) = 2$ m., $\sigma(i) = \sigma(\Omega) = 0\overset{\circ}{.}03$, $\sigma(e) = 10^{-4}$, $\sigma(t_0) = 0.2$ sec. For such a low eccentricity orbit, of course, the perigee position is not well-determined. A preliminary set of orbit determination runs was performed and the resulting perigee angles smoothed, taking into account the variation due to the major perturbing factors. Further runs were then performed with the perigee angles fixed at the smoothed values. An analysis of the consistency of the final orbital elements has not yet been completed, but an assessment of the individual observing stations has been made in the same manner as before, by taking the r.m.s. values of declination residuals. The results are shown in Table 1, which is almost self-explanatory. For each

Table 1. *Observations of Samos 2 (1961 alpha 1)*

Station	COS-PAR No.	Observations				$\sigma_A(°)$, σ_T(sec)		σ_{dec}	
		Total	Acc.	Rej. min.	Rej. maj.			before	after
U.K.									
Clifton	2219	5	5	—	—	0.1,	0.5	14′	11′
Coventry	2240	6	6	—	—	0.2,	0.5	18′	11′
Crowborough	2373	45	44	1	—	0.2,	0.3	14′	13′
E. Finchley	2374	7	6	1	—	0.1,	0.2	8′	12′
Farnham	2265	21	17	2	2	0.1,	0.2	8′	10′
Horsebridge	2280	14	12	1	1	0.1,	0.2	8′	9′
Kingston	2379	11	8	3	—	0.25,	0.2	16′	16′
Leeds	2291	11	9	—	2	0.2,	0.2	13′	14′
Newton Stewart	2312	21	18	1	2	0.5,	0.5	33′	32′
Selsey	2334	5	5	—	—	0.25,	0.3	16′	15′
Slough	2337	46	41	2	3	0.1,	0.4	12′	14′
Thames Ditton	2344	33	30	1	2	0.1,	0.3	10′	8′
Walthamstow	2354	43	34	8	1	0.1,	0.5	14′	11′
Willesden	2356	73	65	5	3	0.1,	0.2	8′	13′
Windsor	2357	44	35	5	4	0.1,	0.5	14′	12′
Windsor 2	2358	16	13	3	—	0.1,	0.5	14′	17′
Windsor 3	2359	6	6	—	—	0.2,	0.6	19′	13′
Winkfield	2360	19	18	1	—	0.1,	0.5	14′	8′
29 others		119	70	26	23				
France									
Meudon	2715	228	205	10	13	0.033,	0.1	3′	5′
Strasbourg	2726	31	20	10	1	0.033,	0.1	3′	5′
Besançon	2727	13	8	1	4	0.033,	0.1	3′	5′
3 others		23	15	3	5				
Holland									
Zwijndrecht	1808	37	28	1	8	0.1,	0.1	7′	7′
Oost-Soubourg	1819	35	30	4	1	0.25,	0.1	15′	13′
2 others		6	3	2	1				
Finland									
Jokioinen	6702	93	88	5	—	0.1,	0.1	7′	5′
Malta	2541	13	11	1	1	0.01,	0.05	$1\frac{1}{2}$′	2′
Totals		1024	850	97	77				

Notes: σ_A, σ_T are the listed angular and time accuracies;
σ_{dec}(before) $= \{\sigma_A^2 + (0.43\sigma_T)^2\}^{1/2}$ expressed in minutes of arc;
σ_{dec}(after) $=$ r.m.s. of declination residuals.

station the total number of observations is broken down into the number accepted (residuals less than 4.5σ), the minor rejections and the major rejections (residuals greater than about 5°). Two U.K. stations, Patcham and Bolton, had almost all their observations rejected and their station coordinates are suspect. The residuals of Strasbourg appear to be in two categories, one group very good and the other centred on 1° error (and this feature is not restricted to the Samos 2 observations). Two possibilities spring to mind. There may, in fact, be two observing points or, alternatively, refraction corrections are sometimes applied and sometimes not. These possibilities will be investigated.

In Table 1, the columns σ_A, σ_T hold the a priori estimates of angle and time errors, mostly taken from the COSPAR list of tracking stations. Orbit determination is based on angular errors only, observations being weighted according to a compounded estimate of declination error

$$\sigma_{\text{dec}} = \{\sigma_A^2 + \tfrac{1}{2}(\dot{\theta}\,\sigma_T)^2\}^{1/2},$$

where $\dot{\theta}$ is an approximation to the mean angular tracking rate at 45° elevation, given by

$$\dot{\theta} = n/\sqrt{2}\,(1 + \tfrac{1}{2}e^2 - a^{-1} R),$$

where n, e, a are mean motion, eccentricity and semi-major axis of the orbit, and R is the mean radius of the earth. This value of σ_{dec} is given under the column headed "before". The final column gives the r.m.s. value of the declination residuals of accepted observations.

5. Starrad

The satellite Starrad ($1962\beta K$) was selected for study because of its relatively high eccentricity (0.28). Some 500 visual observations from 30 stations made between November 1962 and December 1963 have so far been analysed. As far as the accuracy of orbital parameters is concerned the results have been rather disappointing, average figures being $\sigma(a) = 20$ m., $\sigma(e) = 5.10^{-4}$, $\sigma(i) = 0\overset{\circ}{.}03$, $\sigma(\Omega) = 0\overset{\circ}{.}04$, $\sigma(\omega) = 0\overset{\circ}{.}3$, $\sigma(t_0) = 3$ sec.

Since the orbit coverage is quite good, the reason for these rather high figures must be geometrical in nature and related to the large distance at which some of the observations were made (Starrad having a perigee height of 200 km, but going out to 5,000 km at apogee).

Observations from individual stations have been analysed as before, the results being shown in Table 2. In this case the average angular tracking rate $\dot{\theta}$ was so small that σ_{dec}('before') was taken equal to σ_A, the estimated angular error.

Table 2. *Observations of Starrad (1962 beta kappa)*

Station	COSPAR No.	Observations				σ_{dec}	
		Total	Acc.	Rej. min	Rej. maj.	before	after
Adelaide (Aust.)	0600	22	21	1	—	12′	14′
Jokioinen (Fin.)	1963	60	60	—	—	6′	8′
Besançon (Fr.)	2727	13	13	—	—	2′	6′
Malo (Fr.)	2555	19	18	—	1	12′	14′
Meudon (Fr.)	2715	18	17	1	—	2′	7′
Strasbourg (Fr.)	2726	12	12	—	—	2′	3′
Oost-Souburg (Holl.)	4119	13	12	—	1	15′	9′
Malmo (Swed.)	—	18	17	—	1	12′	9′
Bexhill 1 (U.K.)	2212	8	8	—	—	6′	11′
Cove (U.K.)	2244	7	7	—	—	6′	5′
Cowbeech (U.K.)	2392	32	29	—	3	4½′	5′
Crowborough (U.K.)	2373	21	20	1	—	6′	10′
Farnham (U.K.)	2265	6	4	2	—	6′	8′
Thames Ditton (U.K.)	2344	7	7	—	—	6′	2′
Willesden (U.K.)	2356	18	15	2	1	6′	3′
Windsor 2 (U.K.)	2358	15	15	—	—	6′	6′
Rochester (U.S.A.)	8564	7	5	2	—	6′	7′
Sacramento (U.S.A.)	8517	178	174	2	2	6′	7′
San Antonio (U.S.A.)	8634	14	13	1	—	6′	6′
Van Nuys B (U.S.A.)	8637	6	6	—	—	12′	20′
10 others		21	15	2	4		
Totals		515	478	14	13		

6. Cosmos 2

In order to assess the accuracy of observations from Russian stations it was decided to study the satellite Cosmos 2. This satellite was observed by the Baker-Nunn camera network (Field-reduced observations) during the first month of its life (April—May, 1962), by Russian observers during the second month (May, 1962) and by U.K. and West European observers during the latter part of May and early June, 1962. The orbit coverage is shown in Fig. 1. Here, the range of argument of latitude is shown for the observations used in each of the 35 determinations of orbit parameters, made at 25 node intervals. It will be seen that all the observations, except for one set, were made in only one of the two possible visual 'bands', giving a coverage from 5% to 18% of the orbit. The exception was one pass over the BN station Arequipa on May 30th. Three observations from this pass were used in the orbit determination at 4 nodes, and these 4 sets, as might be expected, were better than the others, having errors: $\sigma(a) = 3$ m., $\sigma(e) = 3.10^{-5}$, $\sigma(i) = 0.^\circ003$, $\sigma(\Omega) = 0.^\circ002$, $\sigma(\omega) = 0.^\circ01$, $\sigma(t_0) = 0.1$ sec.

In the case of the visual observations, the accuracy of determination of the orbit parameters is highly correlated with the orbit coverage. The best run with Russian observations (node 483, coverage 10%) gave $\sigma(a) = 8$ m., $\sigma(e) = 3.10^{-4}$, $\sigma(i) = 0\overset{\circ}{.}02$, $\sigma(\Omega) = 0\overset{\circ}{.}03$, $\sigma(\omega) = 0\overset{\circ}{.}5$, $\sigma(t_0) = 0.2$ sec. The overall best run with visual observations (node 808,

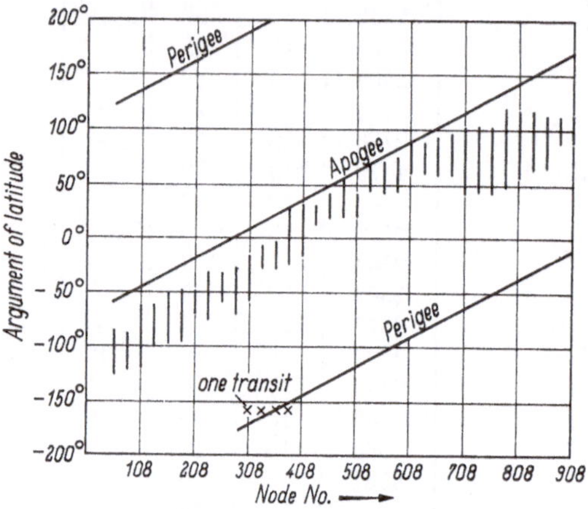

Fig. 1. Cosmos 2 – observation coverage.

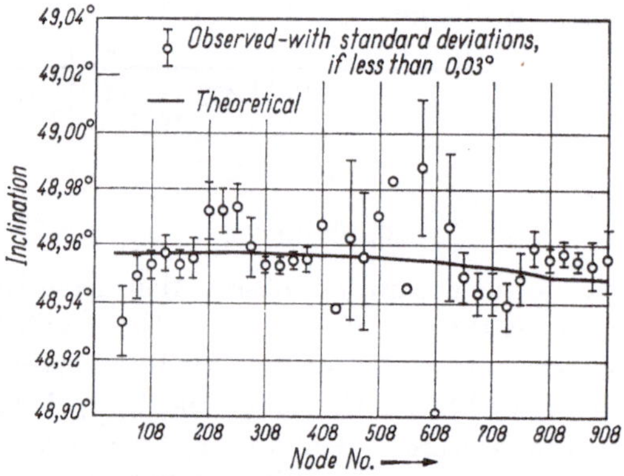

Fig. 2. Inclination of Cosmos 2.

coverage 16%) gave $\sigma(a) = 2$ m., $\sigma(e) = 3.10^{-4}$, $\sigma(i) = 0\overset{\circ}{.}004$, $\sigma(\Omega) = 0\overset{\circ}{.}02$, $\sigma(\omega) = 0\overset{\circ}{.}1$, $\sigma(t_0) = 0.6$ sec. Figs. 2, 3 and 4 give further details of the orbit determination. In Fig. 2 the observed values of inclination together with their standard deviations, are shown. Also shown is a smoothed curve whose variation was determined from the formulae for perturbations due to the Earth's field (J_2, J_3, J_4, J_2^2 terms), due to the Sun and Moon and due to the rotation of the atmosphere. Definite biasses can be seen in the observational points, a factor which has not been noticed in previous orbit determinations. In Fig. 3 the eccentricity observations are shown, together with a theoretical curve obtained from the perturbation formulae. Again, rather strong biasses in the observation points are apparent. In Fig. 4 are shown the residuals

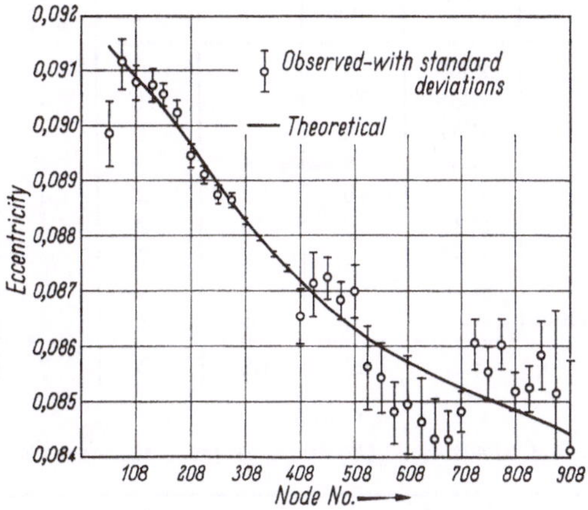

Fig 3. Eccentricity of Cosmos 2.

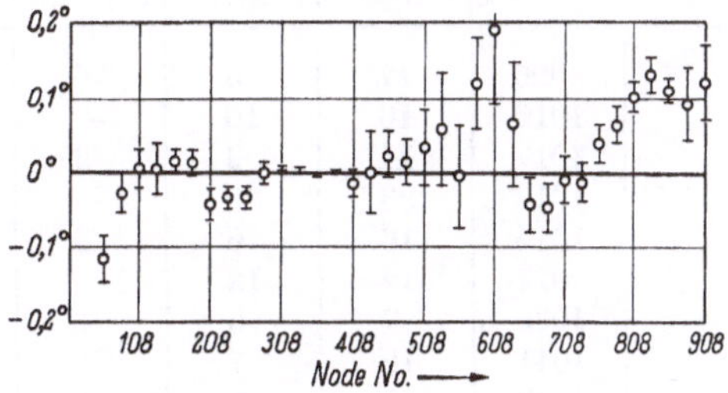

Fig. 4. Residuals in Ω of Cosmos 2.

of the observed values of Ω from the theoretical curve, and these have a similar pattern to the ones for i and e. Of course, successive orbit determinations are rather highly correlated in the sense that a large

Table 3. *Cosmos 2 — Field-Reduced Baker-Nunn Camera Observations*

Station	COSPAR No.	Observations			σ_{dec}	
		Total	Acc.	Rej.	before	after
Organ Pass	9001	22	22	—	3′	1′.5
Olifantsfontein	9002	45	44	1	3′	1′.5
Woomera	9003	37	36	1	3′	1′.0
Arequipa	9007	10	10	—	3′	0′.6
Jupiter	9010	3	3	—		
Maui	9012	3	3	—		
Totals		120	118	2		

Table 4. *Cosmos 2 — Visual Observations from U.K.-France-Japan-U.S.A.*

Station	COSPAR No.	Observations				σ_{dec}	
		Total	Acc.	Rej. min.	Rej. maj.	before	after
Farnham	2265	4	4	—	—	6′	5′
Walthamstow	2354	13	12	1	—	12′	12′
Windsor I	2357	8	7	1	—	9′	5′
Windsor II	2358	6	4	—	2	15′	12′
5 U.K. Stations		10	7	2	1		
Meudon (Fr.)	2715	18	15	2	1	2′.4	2′.6
Strasbourg (Fr.)	2726	9	2	7	—		
7 Jap. Stations		11	8	—	3		
2 U.S.A. Stations		3	3	—	—		
Totals		82	62	13	7		

Table 5. *Cosmos 2 — Observations from Russian Catalogues*

Station	COSPAR No.	Observations				
		Total	Acc.	Rej. min.	Rej. maj.	σ_{dec}
Ashkhabad I	1006	17	5	6	6	10′
Blagoveshchensk	1010	10	10	—	—	$4\frac{1}{2}'$
Dnepropetrovsk	1017	14	9	2	3	$5\frac{1}{2}'$
Erevan	1018	33	30	—	3*	$4\frac{1}{2}'$
Kishinev	1024	16	6	5	5	14′
Krasnodar	1027	14	13	—	1	13′
Odessa	1036	7	5	1	1	8′
Rostov	1041	15	7	1	7	9′
Samarkand	1043	11	5	1	5	16′
Saratov	1044	11	11	—	—	15′
Ushgorod	1055	11	7	4	—	9′
Chernovitsi	1062	8	6	2	—	14′
Orenburgh	1063	7	6	1	—	8′
Juzhno-Sakhalinsk	1065	10	8	2	—	7′
Changchun (China)	280	8	6	—	2	5′
Harbin (China)	284	7	7	—	—	9′
37 others (with less than 5 accepted)	—	149	61	28	60	—
Totals		348	202	53	93	

* All from one transit.

σ_{dec} is the r.m.s. of the Declination residuals of accepted observations in minutes of arc.

bulk of observations used in one are used in the next, but this does not explain why six successive residuals should have the same sign. This remains a mystery.

The observations themselves have been studied as before, details for individual stations being given in Tables 3, 4 and 5. There is not much comment required on these. The BAKER-NUNN observations appear to be twice as good as the a priori estimates and, as in the case of Samos 2, Strasbourg observations need further study to see why so many are rejected. In the case of the Russian observations, a priori estimates ($\sigma_A = 0.1$, $\sigma_T = 0.1$ sec) lead to σ_{dec} (before) $= 7'$. Table 5 shows that some Russian observations are better, and a few worse, than the nominal standard.

7. Summary and Conclusions

Some 2,000 visual observations of four satellites, with differing orbital characteristics, have been studied. As a result, certain conclusions can be drawn about the accuracy of orbital determination when Northern hemisphere visual observations are used and also about the accuracy of the observations themselves.

In reasonably favourable cases, i.e. 15 to 20% orbit coverage with apogee height less than 1,000 km., it should be possible to obtain accuracies of the order $\sigma(a) = 4$ m., $\sigma(e) = 10^{-4}$, $\sigma(i) = \sigma(\Omega) = 0.^\circ01$, $\sigma(t_0) = 0.2$ sec. Under special circumstances (apogee height less than 500 km., inclination about 50°) the standard deviation in inclination might be reducible to $0.^\circ002$, if care is taken in the smoothing of individual results.

The accuracy of individual observers varies from about $0.^\circ05$ (combined error), when using theodolites, to as much as $0.^\circ5$, though the latter in fact are not very useful. The bulk of observers, using binoculars, telescopes or naked eye—with stop-watches—have accuracies between $0.^\circ1$ and $0.^\circ2$ (combined angle/time error). In the case of U.K. and West European observers the claimed accuracies are nearly always justified. In the case of Russian observers, all apparently using the same equipment, there is some variation from station to station and further observations will have to be studied before the accuracies listed in Table 5 can be confirmed.

References

[1] MERSON, R. H., and R. T. TAYLER: The Orbital Elements of Ariel 1 for April—August, 1962, and a Comparison of R.A.E. and N.A.S.A. Ephemerides, R.A.E. Tech. Note Space 39, July 1963.

[2] MERSON, R. H., and A. T. SINCLAIR: Orbital Parameters of Transit 1 B (1960 δ 2), June 2—10, 1962, R.E.A. Technical Report No. 640036, October 1964.

Discussion

M. KING-HELE remarked that only a small number of satellites are observed by BAKER-NUNN, other accurate cameras or DOPPLER systems. For most satellites, only visual observations are available. These are not accurate enough for geodetic studies but, on the other hand, there is a great need for studies of atmospheric density at heights of 200 km, and visual observations are extremely useful for such studies. At present, most such satellites are poorly observed and the geophysical information that they could provide is lost. However, they are particularly well suited for such observations because the predictions are likely to be somewhat in error for photographic instruments, and the observations are usually made at distances less than 500 km, leading to positional errors of less than 0.5 km. Observers should concentrate their attention on the most suitable satellites; Dr. MERSON shows which type of orbit is likely to be useful. It is important to know that such observations, giving an inclination of Transit IB accurate to $0\overset{\circ}{.}002$ are precise enough for evaluating the rotational speed of the atmosphere from the change in inclination during a satellite's life-time, in addition to studies of air density and scale height.

A Method of Determination of the Major Axis of Artificial Earth Satellite Orbits Based on a Limited Number of Tracking Data

By

A. M. Lozinsky

The Astronomical Council of the USSR Academy of Sciences, Moscow, USSR

Abstract. The time at which a satellite crosses the plane of the parallel, on which a tracking station is situated, coincides with the time of its crossing the topocentric celestial equator. This can be used for the evaluation of the nodal period of satellites for which

$$\left(1 + \frac{H}{R}\right) \sin i > \sin \varphi .$$

In this case tracking data from only one station are needed. This method is considered.

A limited number of tracking data allows to evaluate the value of the semimajor axis.

Résumé. L'instant où un satellite traverse le parallèle sur lequel est située une station d'observation, coïncide avec son instant de passage sur l'équateur céleste topocentrique. Ce fait peut être utilisé pour évaluer la période nodale de satellites pour lesquels on a

$$\left(1 + \frac{H}{R}\right) \sin i > \sin \varphi .$$

Il suffit alors d'observations d'une seule station. L'auteur étudie cette méthode.

Une quantité limitée de données d'observation suffit pour évaluer le demi-grand axe.

As a rule, tracking of artificial satellites, both visual and photographic, is carried out independently of the problems that can be solved afterwards based on these tracking data. However, an expedient organization of satellite tracking for each specified problem simplifies the reduction, considerably decreases the number of necessary data and, above all, increases the precision of the final result. For example, during the first series of photographic simultaneous observations undertaken by the USSR stations in 1963, observations were carried out at each even minute; but many of those observations were wasted because of the rather disadvantageous tracking angles occurred at several participating stations. In 1964—1965, the observations were conducted according to the instructions of the coordinator, only at those even minutes, which could provide after reduction the necessary precision of triangulation.

Considerable amount of work is being done to establish correlations between the motion of a satellite along the orbit and the solar activity. As a rule, the orbital elements for these investigations are evaluated from all available tracking data, and then the correlation between variations of some orbital elements are correlated with the solar activity. However, the influence of the solar activity can apparently be determined from considerably less expediently collected tracking data. The semi-major axis of the satellite orbit a, connected with the nodal period P_Ω, subject to large variations depending on the character of solar activity. It is therefore very important to investigate how exactly the period is altered.

The method given below allows to estimate on the base of a limited number of satellite tracking data from only one station, a value, very close to P_Ω, namely, the period between two consecutive crossing of the satellite with the plane of the parallel where the tracking station is situated (only ascending or only descending passages are meant).

The Figure shows that an observer situated in A observes the crossing of this plane by a satellite as the crossing of a line of the topocentric celestial sphere by the satellite. It is possible to mark the time of crossing using a key, while observing the satellite with a small telescope (the optical axis of which is in the plane of the equator) at the moment of the crossing of a thread (in the plane of the equator).

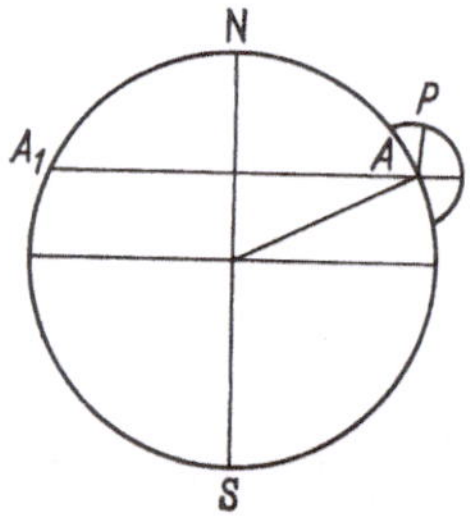

Fig. 1

The time of crossing of the celestial equator line is more accurately determined by photographic tracking. It is enough to obtain 8—10 short exposures in the time interval when the satellite approaches and then moves away from the celestial equator. The time of passage through the equator corresponds to the moment when the satellite crosses the plane of the Earth parallel, on which the tracking station is situated.

The reduction of the photographs in this case ought to be accomplished in the system of visible positions of reference stars. For each marking of the satellite passage it is sufficient to measure only the declination. Having compared the times coresponding to the exposure of each marking with the measured values of the declinations, an empiric function $t = f(\delta)$, which can be well approximated by a polynomial of the second degree $t = A_0 + A_1 \delta + A_2 \delta_2$ has to be evaluated.

When the coefficient A_0 is estimated by a least square method, the moment t of the satellite passage through the celestial equator is practically determined. Having obtained in that way the times of crossing of the Earth parallel plane for close dates and designating the time interval between them by Δt, the average nodal period for this time interval P_Ω can be determinated from the relation $P_\Omega = \Delta t/n$, where n — is the number of satellite revolutions between both observations. The period may be estimated to a precision of 0.05 sec. Photographic tracking may give even a greater accuracy.

Étude Fine d'un Nombre Restreint d'Observations

Par

F. Barlier

Observatoire de Paris-Meudon, France

Abstract. In the determinations of the trajectory of an artificial satellite, the mean quadratic residual is usually higher than what can be expected from the accuracy of the observations. The main reasons of this situation are given. In spite of all the interest that such a determination presents, it does not reveal all the information included in the observations. Among all possible methods to analyse the best this information, the author chooses to determine orbital parameters using very short arcs—and necessarily, therefore, a small number of observations. A satisfactory solution can be obtained, provided that some given restrictive conditions are satisfied. These conditions are studied in this article.

Résumé. Dans la détermination de la trajectoire d'un satellite artificiel, l'écart quadratique moyen des observations est en général supérieur à la valeur prévue. Les principales causes en sont énoncées. Pour intéressante que soit une telle solution, elle ne permet pas d'extraire toute l'information contenue dans les observations. Parmi toutes les méthodes possibles pour permettre d'analyser au mieux cette information, nous nous sommes attachés à déterminer les paramètres orbitaux sur des arcs très courts en n'utilisant nécessairement qu'un nombre restreint d'observations. Une solution très satisfaisante peut être obtenue mais osus des conditions restrictives précises qui sont étudiées.

I. Position du Problème

Lorsqu'on a déterminé la trajectoire d'un corps céleste, il est nécessaire d'en contrôler la validité en comparant les positions observées aux positions calculées. Si la comparaison est satisfaisante, c'est—à—dire si les résidus considérés comme variables aléatoires se répartissent autour d'une valeur moyenne conformément à la loi normale, nous sommes assurés d'avoir déterminé une solution qui représente bien les observations au sens de la méthode des moindres carrés. S'il en est ainsi, nous pouvons calculer pour chaque paramètre un intervalle de confiance dans lequel nous sommes assurés que se trouve la vraie valeur du paramètre avec la probabilité fixée à l'avance (Linnick [*1*]).

Mais dans le cas particulier des satellites artificiels l'écart quadratique moyen estimé à partir des résidus des observations est en général supérieur à la valeur attendue. Ceci est connu de tous ceux qui ont à la fois réduit les observations et déterminé les paramètres orbitaux. Ainsi le Smithsonian Observatory écrit dans ses préambules de publication de positions précises. «Bien qu'il soit démontré que les positions peuvent être déterminées avec une précision supérieure à 2″, une erreur quadratique moyenne de 4″ est attribuée à chaque position».

Il n'y a pas lieu de douter de la valeur attendue de l'écart quadratique moyen. Lorsqu'une plaque photographique a été réduite, il est en effet possible de tester la réduction; la méthode la moins sujette à critique consiste à redéterminer la position d'étoiles connues.

L'origine de cette différence est ailleurs. Nous allons énumérer ce que nous pensons être les principales raisons:

a) Le guidage. Lorsque le satellite est faible, il n'est pas possible en général de le photographier avec une caméra fixe; il faut suivre le satellite dans son mouvement. Ce guidage qui n'est jamais parfait, est une source d'erreurs systématiques surtout pour des positions rapprochées dans le temps de quelques secondes; il faut aussi ajouter les erreurs de scintillation ou de réfraction accidentelle.

b) Le rattachement géométrique des stations les unes aux autres. Les systèmes géodésiques ne sont pas encore rattachés de façon parfaite les uns aux autres et la position précise du centre de gravité de la Terre est une inconnue du problème.

c) Le raccordement des horloges des différentes stations. Par suite des fluctuations dans la propagation des ondes, une erreur de une ou plusieurs millisecondes est tout à fait possible en pratique, surtout dans des stations de campagne encore imparfaitement équipées.

d) Le potentiel terrestre. Il n'est pas encore parfaitement connu (anomalies locales en particulier).

e) Les fluctuations du mouvement par suite du frottement atmosphérique.

f) Les erreurs systématiques propres à chaque appareil. Dans la pratique, les éléments orbitaux déterminés à partir de telles observations ont néanmoins un grand intérêt. Leur utilisation l'a prouvé. Au surplus, il est possible d'augmenter le nombre d'inconnues et de demander aux observations de déterminer en même temps non seulement les éléments orbitaux et leurs variations mais encore les positions des stations et les autres paramètres qui paraissent nécessaires.

Cependant, étant donné l'enchevêtrement des effets et les corrélations existant entre les inconnues, nous pensons qu'il est possible de chercher à résoudre le problème sous un angle un peu différent. Puisque le rattachement des stations les unes aux autres, le raccordement des horloges des stations, les anomalies du potentiel terrestre etc. . . . sont une somme de grandeurs petites difficiles à déterminer, nous pouvons chercher provisoirement à les éliminer en calculant les paramètres orbitaux à partir d'une seule station sur un arc très court d'environ 10 degrés géocentriques. Il restera bien entendu dans une seconde étape à comparer les éléments entre eux pour étudier précisément l'effet de telles ou telles causes.

Nous avons déjà eu l'occasion de dire dans d'autres publications [2] qu'il était possible de déterminer les éléments orbitaux sur de petits arcs avec une excellente précision. Cependant le résultat est fonction essentiellement du matériel d'observation dont on peut disposer. Nous voudrions dans ce qui suit souligner, avec des exemples numériques les points délicats de ce problème, et cela, d'autant plus que dans la réalité l'on dispose en général d'un nombre restreint d'observations sur ces arcs petits.

II. Conditions Permettant L'Utilisation d'un Nombre Restreint d'Observations pour une Étude Fine de la Trajectoire

a) L'Influence d'Erreurs Systématiques

Ces erreurs existent de façon non négligeable pour des positions rapprochées dans le temps. Elles sont dues comme nous l'avons dit, au guidage, à la scintillation et à la réfraction accidentelle. Lorsqu'on dispose d'observations très rapprochées dans le temps, on observe que les résidus par rapport à une orbite moyenne ont une allure systématique, tantôt tous positifs, puis tous négatifs. A titre d'exemple nous donnons dans le tableau 1, la valeur de ces résidus pour un passage observé à Organ Pass (1961 δ_1) [3].

Il en résulte que les erreurs d'observations ne peuvent pas être considérées comme des variables aléatoires normales. Il existe une corrélation entre ces erreurs qui va, en général, en décroissant au cours du temps. Néanmoins ARBEY [4] a fait l'étude de l'utilisation de telles observations. On peut leur appliquer le critère des moindres carrés à condition de les grouper par paquet, en faisant par exemple un lissage polynomial. Seuls, les points moyens ainsi obtenus (on dit aussi

Tableau 1

Temps	Unité 2″ $(0-c)_\alpha \cos\delta$	Unité 2″ $(0-C)_\delta$
06ʰ40ᵐ49ˢ	−0.5	−1.1
53	+0.7	−0.9
57	+0.9	+1.6
41 01	+0.2	−0.8
05	+0.7	−0.1
43 23	−1.1	−1.2
45 07	+0.4	+1.0
11	+0.2	+1.2
50 08	+0.2	+0.7
10	+0.3	+0.9

lieux normaux) peuvent être considérés comme vraiment indépendants. C'est l'expérience qui doit fixer l'intervalle de temps sur lequel doit se faire le lissage. L'ordre de grandeur que nous donnerons est d'environ 15 secondes. D'ailleurs à Meudon, où nous avons une caméra à grand champ, nous faisons toujours les observations par séries espacées entre elles de 15 secondes environ. Ces séries correspondent bien à ce découpage. Nous avons aussi pu constater qu'on ne changeait pratiquement rien à la valeur des éléments en utilisant soit les points moyens soit toutes les observations. Par contre, il est indispensable d'en tenir compte dans le calcul d'erreur car on surestimerait toujours la précision des résultats.

La première condition est donc d'utiliser des techniques qui évitent au maximum le guidage. De ce point de vue, l'utilisation d'un grand télescope, comme le fait Hewitt en Angleterre, qui permet de photographier des satellites faibles sans guidage est surement très intéressante; ou alors, il parait nécessaire comme le fait Muller en France d'utiliser une lunette guide avec une grande distance focale et une grande ouverture permettant ainsi à un observateur de suivre le satellite.

b) Le Calcul des Erreurs à Craindre sur les Paramètres

De ce qui précède, il résulte que, en réalité, le nombre de degré de liberté du système sera en général assez petit. Il est égal à: $(N_r - n)$ où N_r est le nombre de points moyens, n le nombre d'inconnus, et non pas à: $(N - n)$ où N est le nombre d'observations.

De façon classique on identifie l'erreur quadratique σ avec la quantité suivante définie par:

$$s^2 = \frac{1}{N-n} \sum_{i=1}^{N} (0 - C)^2_{\alpha_i} \cos^2 \delta_i + (0 - C)^2_{\delta_j}$$

Ce point de vue est bon si $N - n$ est disons supérieur à 20.

Dans le cas où nous sommes placés $N - n$ sera compris très souvent entre 5 et 10; il est alors beaucoup plus préférable d'utiliser le fait rigoureux que la quantité:

$$t\,i = \frac{\tilde{E}\,i - E\,i}{\sqrt{c^{-1}_{i\,i}}\, s}$$

est une variable de Student a $N - n$ degrés de liberté. $\tilde{E}_i$ est la vraie valeur du paramètre E_i et C^{-1} est la matrice inverse des équations normales. Pour chaque paramètre, on pourra déterminer un intervalle de confiance (Linnick [*1*]), tel que

$$\tilde{E}_i - \gamma \sqrt{C^{-1}_{ii}} \leqq E_i \leqq \tilde{E}_i + \gamma \sqrt{C^{-1}_{ii}}$$

soit vérifiée avec une probabilité de $x\%$ (γ est une constante, fonction de x),

A titre d'exemple, nous donnons (voir Tableau 2) une série de déterminations des paramètres orbitaux moyens faits sur deux révolutions successives mais à partir d'une seule station. La cohérence des valeurs des paramètres orbitaux comparées aux erreurs à craindre prévues, permet de juger la validité du critère de précision et prouve qu'il est valable. Comme nous avons adopté un taux de probabilité de 95 % cette erreur est environ 4 fois l'erreur à craindre classique. A notre avis seul un critère de ce genre peut donner un ordre de grandeur de la précision qui ne soit pas trop optimiste, pour ce type de calcul.

Tableau 2

Date	Temps	a	e	i	σ	Nombre d'équations	Numéro de station
		7986	0.113	0.677			
05.06	04^h43^m	747 ± 10	506 ± 15	903 ± 05	5″15	71	9001
	06^h43^m	761 ± 07	498 ± 14	902 ± 07	4″99	82	9001
	18^h28^m	727 ± 29	456 ± 40	894 ± 09	2″18	12	9008
06.06	06^h08^m	721 ± 15	425 ± 09	911 ± 05	0″80	14	9012
	10^h32^m	769 ± 18	380 ± 45	953 ± 41	5″64	16	9012
	20^h19^m	727 ± 52	326 ± 64	888 ± 25	4″37	10	9008
07.06	20^h00^m	660 ± 31	316 ± 33	901 ± 17	3″73	12	9008
09.06	05^h23^m	560 ± 30	216 ± 29	890 ± 08	2″34	138	9001

c) Il Faut Renoncer à Déterminer tous les Paramètres Orbitaux

Nous donnons d'abord à titre d'indications deux calculs d'orbite d'Echo I, un premier en utilisant deux passages observés à Meudon, un deuxième en utilisant un seul de ces deux passages (Tableau 3 — calculs

Tableau 3

Calcul	a (Km)	$e \cdot p$ m	e (rad)	$e \cdot p$ (10^{-5} rad)	i (rad)	$e \cdot p$ (10^{-5} rad)
1	7883,060	6	0,040728	1,5	0,824808	2,8
2	7878,102	8682	0,041022	41,0	0,824843	28,0
3	7883,060	—	0,040687	4,9	0,824812	10,3
4	7883,100	—	0,040781	—	0,824828	—

Calcul	Ω (rad)	$e \cdot p$ (10^{-5} rad)	ω (rad)	$e \cdot p$ (10^{-4} rad)	M (rad)	$e \cdot p$ (10^{-4} rad)
1	3,40888	1,6	5,5536	3,1	1,9798	3,4
2	3,40891	3,3	5,5410	230,0	1,9922	231,0
3	3,40889	1,4	5,5532	3,2	1,9802	3,5
4	3,40892	—	5,5542	—	1,9790	—

1 et 2). La conclusion est immédiate, il faut renoncer ici à déterminer le demi-grand axe, c'est-à-dire la période du satellite avec un seul passage. Par contre en figeant a, on obtient des éléments tout à fait satisfaisants (Tableau 3 — calcul 3), mais une remarque très importante est à faire sur le calcul d'erreur. Si nous fixons la valeur de a, l'erreur sur les éléments orbitaux déterminés provient de deux sources

1°) des erreurs d'observation (déterminé par le calcul d'erreur),
2°) des erreurs commises sur a.

Il est donc nécessaire de faire varier a dans son intervalle de confiance entre les deux limites extrêmes et d'en étudier les conséquences sur les valeurs des autres paramètres orbitaux. Nous donnons dans le même tableau (calcul 4) la valeur des éléments orbitaux déterminés avec un demi-grand axe différent. On voit que ce sont essentiellement l'excentricité, l'argument du périgée et l'anamolie moyenne qui sont affectés par une erreur sur a.

d) Le Rejet des Observations Aberrantes

Lorsqu'il y a un nombre restreint d'observations, le rejet des observations aberrantes devient très difficile. De façon classique on rejette les observations ayant un résidu égal à 3 fois l'écart quadratique moyen car la probabilité pour qu'un résidu s'écarte de plus de 3 fois de cette valeur est pratiquement nulle (égale à 0,003). Ce point de vue est valable si le nombre de degrés de liberté est égal à 20 pour fixer les idées. Dans le cas contraire, il faut savoir que les risques courus sur la validité du résultat sont toujours grands lorsqu'on rejette des observations sans connaître les raisons matérielles autorisant le rejet. D'autre part, les mauvaises observations peuvent ne pas apparaître. Ainsi, à titre d'exemple nous donnons un cas d'espèce. Il s'agit d'une détermination d'une orbite reposant sur peu d'observations.

On se limite d'abord à 15 observations, (Tableau 4, calculs 1 et 2). Dans le tableau, on donne seulement la valeur des premiers résidus. Le calcul 1 ne parait pas satisfaisant car le premier résidu est important. Nous l'éliminons et cette fois dans le calcul 2, les résidus paraissent acceptables sauf peut être celui en déclinaison. Pourtant si au lieu de 15 observations, nous utilisons les 40 dont nous pouvons disposer, les résidus 3 et 4 qui paraissaient valables deviennent très grands et les valeurs des éléments orbitaux sont changés de façon importante (Tableau 4, calcul 3). Dans le cas présent si l'on n'avait pas disposé de plus de 15 observations, la détermination des éléments orbitaux aurait été mauvaise car le critère de rejet des observations aberrantes n'aurait pas pu conduire au rejet des observations 3 et 4.

Tableau 4

No. de l'Obs.	Unité 2''	calcul 1		calcul 2		calcul 3	
	Temps	$(0-C)_\alpha$	$(0-C)_\delta$	$(0-C)_\alpha$	$(0-C)_\delta$	$(0-C)_\alpha$	$(0-C)_\delta$
1	$04^h39^m34^s$	+5.0	+7.2	—	—	—	—
2	38	+3.1	+4.5	+1.5	+4.2	10.2	12.4
3	42	−0.6	+1.2	−1.9	+1.1	+6.8	+8.8
4	46	−0.5	−0.1	−1.5	−0.1	+6.7	+7.1
⋮	⋮	⋮	⋮	⋮	.	⋮	⋮
15							
	Moyenne des $(0-C)$	2.8		1.5		2.9	

Calcul	a	e	i
1	7986.726 ± 43	113.722 ± 66	0.677856 ± 21
2	725 ± 23	650 ± 52	880 ± 15
3	745 ± 11	511 ± 18	904 ± 06
	Ω	ω	M
1	2.378240 ± 26	4.486 ± 45	3.241 ± 58
2	8530 ± 19	491 ± 32	234 ± 41
3	9054 ± 23	496 ± 02	227 ± 02

III. Précision des Éléments Orbitaux ainsi Obtenus; Conclusion

Nous avons eu l'occasion de montrer ailleurs [*2*] que moyennant toutes les précautions indiquées ci-dessus, il est possible de déterminer avec grande précision les éléments orbitaux.

Nous donnerons seulement un exemple de la mise en évidence des termes à courte période sur un grand arc de trajectoire observé à Organ Pass [*3*]. Nous avons pu découper cet arc en quatre tronçons se couvrant deux à deux, en sorte que les déterminations des parties de trajectoire les plus extrêmes sont indépendantes. Les résultats sont donnés dans le Tableau 5.

Lorsqu'on corrige les valeurs osculatrices des termes à courte période, on retrouve bien une valeur constante conformément à la théorie, pour l'excentricité et l'inclinaison, aux erreurs probables près. Cela donne l'ordre de grandeur de la précision qu'on peut espérer.

Tableau 5

M (deg).	e (rad).	e'' (rad).	i (rad).	i'' (rad).
196	0,11282	0,11342	0,677762	0,677912
204	0,11278	0,11340	0,677771	0,677913
210	0,11276	0,11340	0,677784	0,677912
224	0,11274	0,11344	0,677847	0,677916

Pour détecter tout effet systématique de quelque sorte que ce soit, il est seulement important de faire un grand nombre de déterminations de trajectoire pour compenser en quelque sorte le fait que chacune soit déterminée avec peu d'observations. La précision peut atteindre la seconde de degré pour des paramètres comme l'inclinaison ou l'ascension droite du noeud. Pour l'excentricité, on peut espérer jusqu'à 10^{-5} radian. L'argument du périgée ou l'anomalie moyenne seront en général moins bien déterminés (10^{-4} radian par exemple).

Bibliographie

[1] LINNICK, Y. V.: Méthode des moindres carrés, chap. VI, No. 8, Dunod 1963 (ouvrage traduit du russe).
[2] BARLIER, F.: Recherches sur la détermination et l'utilisation des éléments osculateurs dans la mécanique des satellites proches, 1965, Bull. Astr., t. 25, fasc. 4 (sous presse).
[3] Catalog of Precisely Reduced Observations, Smith. Astron. Obs., Special Report No. 104, pp. 37–39, 40.
[4] ARBEY, L.: Bull. Astronom. 14, 75–78 (1949).

Discussion

If Dr. BARLIER's method were to be applied to DOPPLER technique data, Mr. ANDERLE thinks that the same selection principle would apply as to optical data. One might solve for mean anomaly to obtain information on drag or radiation effects, or one might solve for right ascension to determine even order zonal gravitational coefficients.

Long-Range Analysis of Satellite Observations

By

Yoshihide Kozai

Tokyo Astronomical Observatory, Mitaka, Tokyo, Japan

Abstract. Orbital elements determined from precisely reduced BAKER-NUNN observations of three satellites, 1959 α 1, 1959 η and 1960 ι 2 from their launching times through June 1963 are analyzed. Their residuals based on uniform systems show that the orbital elements of the three satellites are disturbed by resonance effects of higher order sectorial and tesseral harmonics of the earth's gravitational potential and by other sources still not known to us. An evidence of variation of J_2 is presented.

Résumé. On analyse les éléments orbitaux des trois satellites: 1959 α1; 1959 η et 1960 ι2. Ces éléments ont été obtenus après réduction d'observations BAKER-NUNN effectuées depuis le lancement de ces satellites jusqu'en juin 1963. Les résidus ramenés à des systèmes uniformes montrent que les éléments orbitaux des trois satellites sont perturbés par des effets de résonance dus à des coefficients tesseraux et sectoriaux d'ordre élevé du potentiel terrestre. D'autres perturbations sont dues à des causes encore inconnues. On expose aussi des preuves de la variation de J_2.

1. Introduction

In one of my previous papers (KOZAI, 1964), I redetermined numerical coefficients for zonal spherical harmonics in the gravitational field of the earth from the secular motions and the amplitudes of long- periodic perturbations with argument ω in satellite motions. One set of data consisting of the secular motions for the longitude of the ascending node and for the argument of perigee and the amplitudes in four orbital elements were derived from observations covering about one revolution period of the argument of perigee. Nine satellites were used in the analysis, and for some of the satellites more than five sets of data were adopted to eliminate accidental errors in the data. The zonal harmonic coefficients thus derived in the previous paper are given in Table 1 where the following values were adopted as the geocentric gravitational constant, GM, and the equatorial radius of the Earth, a_e;

$$GM = 3.986032 \times 10^{20}\ \text{cm}^3/\text{sec}^2,$$

$$a_e = 6{,}387{,}165\ \text{meters}.$$

Table 1. *Adopted Coefficients for Zonal Spherical Harmonics of the Gravitational Potential of the Earth* (KOZAI, 1964) (in Units of 10^{-6})

$J_2 = 1082.645$, $J_3 = -2.546$, $J_4 = -1.649$, $J_5 = -0.210$,
$J_6 = 0.646$, $J_7 = -0.333$, $J_8 = -0.270$, $J_9 = -0.053$,
$J_{10} = -0.054$, $J_{11} = 0.302$, $J_{12} = -0.357$, $J_{13} = -0.114$,
$J_{14} = 0.179$.

In the previous paper it was found that the data from different sets are not always consistent with each other even for one satellite, and the inconsistencies seem to be partly due to some systematic variations in orbital elements.

To find out systematic variations due to unknown anomalies in the earth potential and/or to other sources we must use satellites whose orbital elements are available for long periods of time. In this paper three satellites, 1959 α 1 (Vanguard II), 1959 η (Vanguard III), and 1960 ι2 (rocket of Echo I) are used for the long-range analysis.

2. Method of Analysis

The orbital elements every two days for the three satellites are derived by Smithsonian Astrophysical Observatory's Differential Orbit Improvement (DOI) program (GAPOSCHKIN, 1964) from precisely reduced BAKER-NUNN observations of four day interval, although eight day observations are used in one set when enough observations are not available in four days. To derive the orbital elements the first-order short-periodic perturbations due to the oblateness of the Earth and the perturbations due to tesseral harmonics in the Earth potential are taken into account.

The adopted tesseral harmonic coefficients through $C_{4,4}$ and $S_{4,4}$ and BAKER-NUNN camera station coordinates in the DOI are those derived by IZSAK (1964) from satellite observations. This first step of analysis was performed by Miss P. STERN of the Smithsonian Astrophysical Observatory.

In Table 2 the adopted time interval, variation ranges of mean orbital elements and the area-to-mass ratio (A/M) are given for each

Table 2. *Orbital Data for the Three Satellites*

	1959 α 1	1959 η	1960 ι2
Periods	Feb. 18, 59 Aug. 24, 63	Sept. 18, 59 July 1, 63	Sept. 14, 60 July 1, 63
e	0.1659–0.1642	0.1900 – 0.1886	0.0114
n	11.442 –11.480	11.061 – 11.089	12.197
a (km)	8 317 – 8 299	8 507 – 8 493	7 971
A/M (cgs)	0.21	0.27	0.21

satellite. In computing the solar radiation pressure perturbations the constant values of (A/M) given in Table 1 are used, although the shapes of 1959η and 1960 ι2 are not spherical.

The orbital elements for each epoch are corrected by subtracting the luni-solar periodic and solar radiation pressure perturbations from them. And the amplitudes of long-periodic perturbations with arguments ω, 2ω and 3ω are evaluated by using odd zonal harmonic coefficients given in Table 1, and after subtracting these long-periodic perturbations from the orbital elements we have mean orbital elements at each epoch.

Then the secular motions of the longitude of the ascending node and of the argument of perigee can be computed from the mean orbital elements thus derived with the even zonal harmonic coefficients in Table 1. Using these secular motions the residuals for the longitude of the ascending node and the argument of perigee are computed at each epoch.

3. Analysis of Residuals

During the time intervals adopted in this analysis, the semi-major axes of 1959 α1, 1959 η and 1960 ι2 are diminished, respectively, by 18 kilometers, 14 kilometers and 20 meters by air-drag effects. Since the air-drag effects to the motion of 1960 ι2 are very small because of large perigee height, the variations of the semi-major axis and the mean motion are extremely small for this satellite.

The mean perigee distance, q, and the mean inclination, i, for 1959 α1 and 1959 η are solved as functions of the semi-major axis, a, (in kilometers) by the method of least squares as follows;

for 1959 α1,

$$\begin{aligned} q &= \underset{\pm 7}{6936.400} + \underset{\pm 14}{0.0337}\,(a - 8300), \\ i &= \underset{\pm 5}{32^\circ\!.87945} + \underset{\pm 7}{0^\circ\!.39} \times \frac{a - 8300}{a}, \end{aligned} \qquad (1)$$

for 1959 η,

$$\begin{aligned} q &= \underset{\pm 19}{6890.935} + \underset{\pm 23}{0.0089}\,(a - 8490), \\ i &= \underset{\pm 11}{33^\circ\!.35528} - \underset{\pm 12}{0^\circ\!.52} \times \frac{a - 8490}{a}. \end{aligned} \qquad (2)$$

Since the inclination equation for 1959 η shows that the mean inclination increases with time, I am afraid that the Eqs. (1) and (2) are still affected by some systematic errors.

For 1960 ι2 the mean values of the eccentricity, the inclination, the daily anomalistic mean motion in revolutions, the semi-major axis and the perigee distance are solved as linear functions of time as follows;

$$\left.\begin{aligned}
e &= 0.011421 - 0.18 \times 10^{-7}(t - t_0), \\
&\quad\ \pm 1 \qquad\ \pm 2 \\
i &= 47^\circ 23246 - 0^\circ 26 \times 10^{-6}(t - t_0), \\
&\quad\ \pm 3 \qquad\ \pm 12 \\
n &= 12.197037 + 0.44 \times 10^{-7}(t - t_0), \\
&\quad\ \pm 1 \qquad\ \pm 3 \\
a &= 7971.457 - 0.20 \times 10^{-4}(t - t_0), \\
&\quad\ \pm 1 \qquad\ \pm 3 \\
q &= 7880.417 + 0.128 \times 10^{-3}(t - t_0), \\
&\quad\ \pm 4 \qquad\ \pm 15
\end{aligned}\right\} \tag{3}$$

where

$$t_0 = \text{April } 5.0,\ 1962.$$

The mean perigee distance is increased with time by one of the above equations.

The residuals for the node and the argument of perigee are solved by the method of least squares as linear functions of time, and the coefficients of time t in the solutions give $(O - C)'$ s based on J_n in Table 2 for the secular motions. The amplitudes of trigonometric terms with argument ω are also computed from the residuals of four orbital elements, and are regarded as $(O - C)'$ s based on J_n in Table 2. $(O - C)'$ s thus derived are given in Table 3 together with the residuals given in the previous paper (Kozai, 1964). It must be noticed that the standard

Table 3. *$(O - C)$ for Secular Motions and Amplitudes of Long-Periodic Terms*

	1959 α 1		1959 η		1960 ι 2	
	New	1964	New	1964	New	1964
	Secular motions (degrees per day)					
$\dot{\omega} \times 10^6$	45.4	5	69.3	18	205	230
	±0.4	±20	±1.5	±30	±7	±300
$\dot{\Omega} \times 10^6$	−1.3	0	−5.2	1	−4.6	−4
	±0.2	±6	±0.3	±9	±0.2	±15
	Amplitudes of long-periodic terms with argument ω					
$\omega \times 10^4$	−19	−7	−25	6		
	±5	±8	±8	±5		
$\Omega \times 10^4$	1.7	−1	7	−1	−1.6	−0.1
	±1.3	±5	±2	±2	±0.5	±1.3
$i \times 10^5$	12	1	−8	−1	−9	−8
	±6	±8	±9	±9	±4	±8
$e \times 10^6$	−1.5	−2	−5	0	−1.6	−0.1
	±1.2	±2	±2	±2	±0.5	±1.3

deviations assigned to $(O-C)'$ s are estimated by least square method, and that the errors in the computed values from the theory should be much larger. Therefore, the agreement between the present $(O-C)'$s and the previous ones may be satisfactory.

4. Periodic Fluctuations

After correcting secular changes to the orbital elements, second residuals are obtained. In order to find out periods of periodic fluctuations in the orbital elements due to unknown sources, auto-correlation functions are computed from the second residuals. Figs. 1, 2 and 3 give the auto-correlation functions for the argument of perigee, the longitude of the ascending node, the inclination, and the perigee distance of 1959 α 1 and 1959 η, respectively, and in Figs. 4 and 5 those for the mean motion, the eccentricity, the inclination and the longitude of the ascending node of 1960 ι2 are given. On the horizontal axes of these figures time lags are given in days.

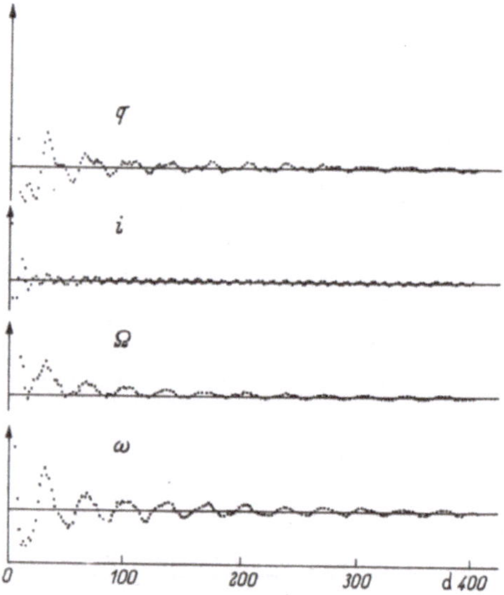

Fig. 1. Auto-correlation functions for 1959 α1. Abscissa represents time lags in days.

The auto-correlation functions for 1959 α1 in Fig. 1 are similar to curves of damped oscillations, therefore, show that in the four orbital elements there are fluctuations expressed by sine curves with variable phase angles, that is, by sine curves with acceleration terms of time in the arguments.

The estimated periods of fluctuations are 32 days for the node and the argument of perigee and 11 days for the inclination. For the perigee distance two fluctuations with 15 day and 32 day periods are super-posed. It is known that the lunar parallactic term with argument $☾ - \omega - \Omega$ produces inequalities with 32 day period and has been neglected in computing the lunar perturbations. However, the observed amplitudes of 32 day period fluctuations are too large to be identified with one of the lunar parallactic terms.

Figs. 2 and 3 show similar situations for 1959 η, although the auto-correlation functions have more complicated form than those in Fig. 1, since the constant value of A/M was adopted to compute the solar radiation pressure perturbations whereas the shape of this satellite is not spherical.

The estimated periods of fluctuations appearing in the orbital elements of 1959 η are exactly same as those for 1959 α 1; that is, 32 days for the node and the argument of perigee, 11 days for the inclination, and 32 days and 15 days for the perigee distance. For this satellite it

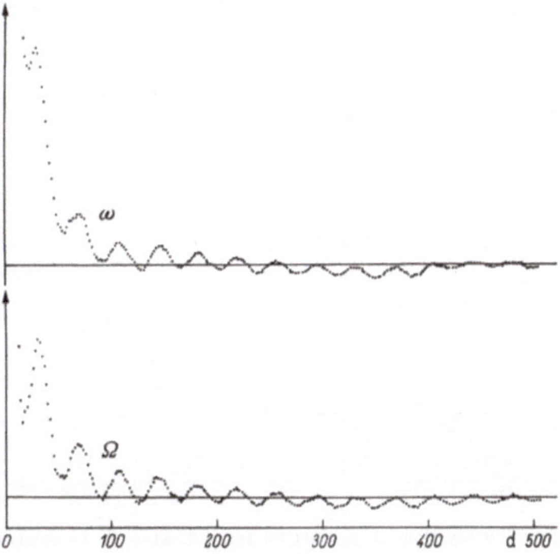

Fig. 2. Auto-correlations in ω and Ω for 1959 η. Abscissa represents time lags in days.

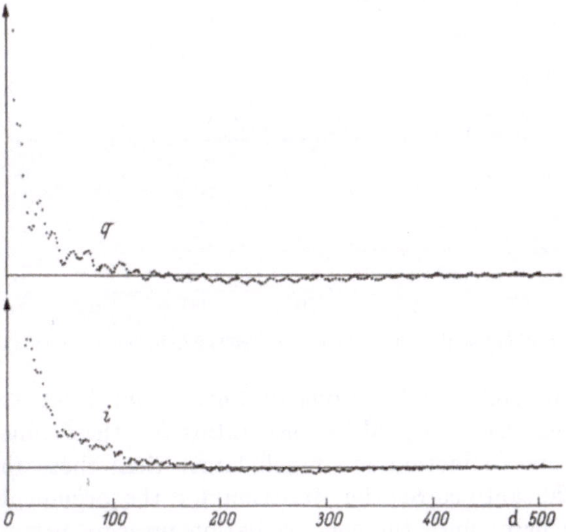

Fig, 3. Auto-correlation functions in q and i for 1959 η. Abscissa represents time lags in days.

is expected that inequalities with 15 day period of argument $nt - 11 \times \times (\mu - \Omega)$ (μ being the Greenwich siderial time) appear in the perigee distance and the argument of perigee by resonance from the sectorial harmonics $J_{11,11}$ in the gravitational potential.

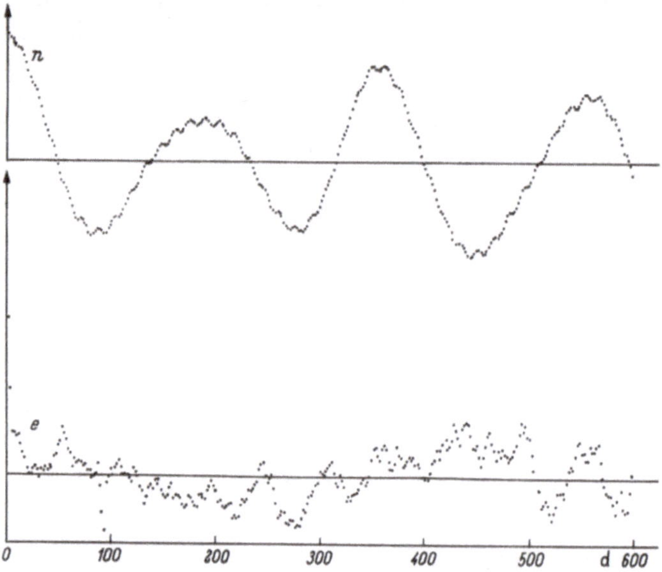

Fig. 4. Auto-correlation functions in n and e for 1960 ι 2. Abscissa represents time lags in days.

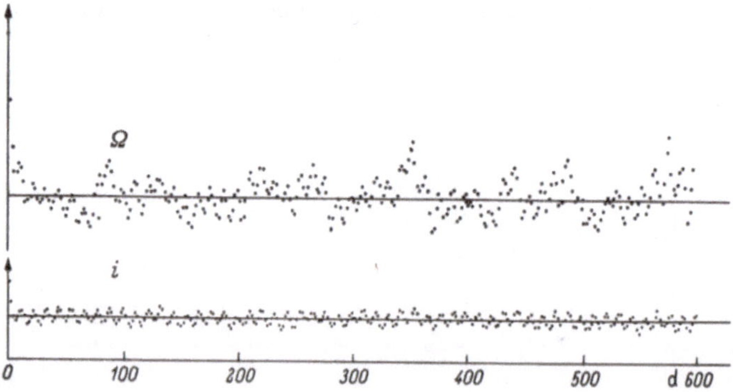

Fig. 5. Auto-correlation functions in Ω and ι for 1960 ι 2. Abscissa represents time lags in days

The auto-correlation functions in Figs. 4 and 5 for 1960 ι 2 have regular forms, that is signal-to-noise ratios for these functions, especially of the mean motion, are much larger than those for the other satellites. The auto-correlation function for the argument of perigee is not given here, since the solar radiation pressure perturbations are too large because of small eccentricity to be computed with enough accuracy in this element and any significant scientific results might not be extracted from the argument of perigee.

The estimated period of the fluctuations in the node and the inclination is 11.1 days, however, it seems that the phase angles of these fluctuations are shifting from time to time. It should be noticed that the 11 day period is common to the fluctuations in the inclination of the three satellites, and it might be suspected that these fluctuations have geometrical background, for example, we might suspect that they are due to the motion of the equator.

However, solving these inequalities for 1960 ι 2 by assuming that the arguments are linear functions of time, we have

$$\delta\Omega = \underset{\pm 2}{1''7} \sin\left\{\frac{360^\circ}{11.1}(t - t_0) + \underset{\pm 9}{44^\circ}\right\}, \quad \delta i = \underset{\pm 1}{1''3} \cos\left\{\frac{360^\circ}{11.1}(t - t_0) + \underset{\pm 3}{32}\right\}, \tag{4}$$

where

$$t_0 = \text{April 5.0, 1962.}$$

Since we can regard that the phase angles in the node and inclination inequalities are same, these fluctuation may be supposed to have dynamical origins.

The daily anomalistic mean motion for 1960 ι 2 is expressed by the following formula quite well;

$$\left.\begin{aligned}
&(n - 12.197038) \times 10^5 = 0.58 \times 10^{-2} t + \\
&+ 0.4 \cos\left(\frac{360^\circ}{14.5} t + \underset{\pm 15}{184^\circ}\right) + 0.9 \cos\left(\frac{360^\circ}{117.0} t - \underset{\pm 4}{9^\circ}\right) \\
&+ 1.9 \cos\left(\frac{360^\circ}{182.63} t + \underset{\pm 3}{187^\circ}\right) + 0.5 \cos\left(\frac{360^\circ}{292.00} t + \underset{\pm 5}{132^\circ}\right) \\
&+ 0.7 \cos\left(\frac{360^\circ}{365.25} t + \underset{\pm 4}{160^\circ}\right) + 0.7 \cos\left(\frac{360^\circ}{900} t + \underset{\pm 5}{63^\circ}\right),
\end{aligned}\right\} \tag{5}$$

where the initial epoch is April 5.0, 1962.

The first periodic term on the right-hand side of (5) has 14.5 day period and is identified as the term with argument $\omega + n\,t - 12(\mu - \Omega)$ due to $J_{12,12}$ harmonics in the gravitational potential. Fluctuations with the same period appear also in the node, inclination and the eccentricity. The third and fifth periodic terms are a half year and one year period fluctuations, respectively.

5. Fluctuation of J_2

When I started this analysis, I expected that I could detect fluctuations of J_2 from the residuals of the nodes for the three satellites. Since the rotation velocity of the earth is changing with one year period, we can expect that the principal moment of inertia, C, as well as J_2 is also fluctuating with the same period.

However, I have found from the auto-correlation functions of the nodes that 1960 ι 2 is the only one satellite that has large signal-to-noise ratio for the node. For the longitude of the node I derive the following fluctuation with 354 day period;

$$1''.2 \sin\left(\frac{360^\circ}{354} t - \underset{\pm 10}{13^\circ}\right). \tag{6}$$

If this fluctuation is entirely due to the variation of J_2, we can write

$$\delta J_2 = 2.0 \times 10^{-9} \cos\left(\frac{360^\circ}{354} t + 167^\circ\right), \tag{7}$$

where the initial epoch is April 5.0, 1962. Although the period of this term is not exactly one year, the trigonometric term in (6) has a good correlation with a curve representing (Atomic Time 1—Universal Time 1).

Since I could not check this result by the other two satellites, I cannot insist that J_2 is really changing.

This study is partly supported financially by the Asahi Academic Fund in 1963.

References

Gaposchkin, E. M.: Smithsonian Astrophysical Observatory, Special Report No. 161 (1964).
Izsak, I. G.: unpublished data 1964.
Kozai, Y.: Publ. A. S. Japan **16**, 263—284 (1964).

Discussion

Dr. Kaula stresses the following points.

(1) The terms of argument $\omega + n\,t - 12(\mu - \Omega)$ appearing on the right of the mean anomaly Eq. (5) is unlikely to be due to $J_{12,12}$ since a near resonant term from $J_{l,12}$ for l even has a coefficient of order of e, and the eccentricity is small. It is more likely to be due to the coefficients $J_{l,12}$ with l odd.

(2) It would be worth while to investigate whether any of the semi-monthly, etc. . . fluctuations are due to ocean tide effects. R. R. Newton (1965) has found a variation in inclination of one satellite which would correspond to a perturbation by a tidal component leading the sun by 1/2 hour.

(3) The magnitude of the δJ_2 in Eq. (7) is physically plausible.

References

Newton, R. R.: An observation of the satellite perturbation produced by the solar tide, J. Geophys. Res. **70**, in press, 1965.

Orbit Determination Using Linearized Differential Equations

By

Robert S. Long

University of Canterbury, Christchurch, New Zealand

Abstract. This paper explains the use of quasilinearization (the generalized Newton-Raphson process) as applied to the problem of determining orbits from various types of observational data, i.e. angular data, range, and range-rate. The non-linear differential equations controlling the orbit are replaced by a set of linear equations whose solutions, obtained numerically, converge to the orbit, at least if the orbit is well-determined by the data. Numerical examples show that the convergence is rapid.

Résumé. Ce travail explique l'utilisation de la quasilinéarisation (procédé de Newton-Raphson généralisé) pour le problème de la détermination des orbites à partir de diverses sources d'observations: observations d'angles, de distances, de variations de distances. Les équations différentielles non linéaires qui définissent l'orbite sont remplacées par un ensemble d'équations linéaires dont les solutions obtenues numériquement, convergent vers l'orbite à condition que celle-ci soit bien déterminée par les données. Des exemples numériques montrent que la convergence est rapide.

Note. This work was begun while the author was on sabbatical leave at the Astrodynamics Department, University of California, Los Angeles, and was supported by the Air Force Office of Scientific Research under Grant No. USAF AFOSR 241-63. The author wishes to express his sincere thanks to the Council of the University of Canterbury for granting sabbatical leave, to Dr. Samuel Herrick for generous provision of facilities at the University of California, to Dr. E. T. Pitkin for willing and valuable instruction in the art of programming, and to Dr. J. C. Butcher for comments and discussion on the numerical work involved.

Introduction

We consider the problem of determining the orbit of a satellite or planet (a point mass), given the forces acting on it, and given also the minimum amount of observational data sufficient to fix the orbit.

This is a problem of great analytical difficulty. The usual astronomical approach is to determine a preliminary orbit, as accurately as possible, by means of various specially developed formulae [*1*], and then to improve this by the process of differential correction, (the usual Newton-Raphson process), until the observational data are satisfied to the required degree of accuracy. The process of differential correction requires that the preliminary orbit must be sufficiently good for the process to converge. Also the process requires the calculation of a set of partial derivatives, and only in the case of the inverse-square law field can these be found from analytical formulae. In all other cases they must be calculated by a numerical perturbation scheme, and numerical differentiation can sometimes prove troublesome.

The purpose of the present paper is to show how the method of quasilinearization, which is a generalized NEWTON-RAPHSON process, can be used to provide a strongly convergent process without the need for numerical differentiation, or even, in some cases, the need for determining a preliminary orbit. The method is an iterative one, and has three useful characteristics,

(1) the initial approximation need not be very good,

(2) no numerical differentiation is required, and

(3) the convergence is quadratic i.e. the number of correct digits is approximately doubled at each iteration. The use of quasilinearization in this connection was first demonstrated for the case of two-dimensional motion, with angular observations [2]. In this paper we extend the method to three dimensions and also consider radar observations of range and range-rate.

Whether the method will be found to be of practical value is a question which can be decided only by practising and experienced orbital analysts. The results reported here, although not based on "real" data, and referring to comparatively simple situations, would appear to indicate that the method may indeed be of more than theoretical interest.

Basic Formulae

The equations of motion of the satellite, under any forces, may be written

$$\ddot{\boldsymbol{r}} = \boldsymbol{f}(\boldsymbol{r}, \dot{\boldsymbol{r}}, t), \tag{1}$$

where $\boldsymbol{r} = (x, y, z)$ is the position vector of the satellite relative to rectangular Cartesian axes, and $\boldsymbol{f} = (f_1, f_2, f_3)$ is the resultant force per unit mass on it. The explicit dependence of $\boldsymbol{f}$ on the time t allows for such cases as the field due to moving planets. If $\boldsymbol{f}$ does not depend on $\dot{\boldsymbol{r}}$ we write

$$\ddot{\boldsymbol{r}} = \boldsymbol{f}(\boldsymbol{r}, t). \tag{2}$$

In the general case, where $\dot{\boldsymbol{r}}$ enters explicitly, it is convenient to write

$$\boldsymbol{r} = (x_1, x_2, x_3),$$
$$\dot{\boldsymbol{r}} = (x_4, x_5, x_6),$$

so that

$$\dot{x}_1 = x_4, \quad \dot{x}_2 = x_5, \quad \dot{x}_3 = x_6. \tag{3}$$

Putting $\boldsymbol{x} = (x_1, x_2, x_3, x_4, x_5, x_6)$, Eqs. (1) and (3) may be written

$$\dot{x}_i = F_i(\boldsymbol{x}, t), \tag{4}$$

where $F_1 = x_4$, $F_2 = x_5$, $F_3 = x_6$ and $F_{i+3} = f_i$ for $i = 1, 2, 3$. The method of quasilinearization may be applied to any of the sets of Eqs. (1), (2) or (4), but (1) is not suitable for the numerical integration involved. We shall develop the theory in terms of the most general form (4).

It is clear that in order to determine a solution of these equations a minimum of six items of information is needed. We shall consider five types of observation (I) measurement of the angles of right ascension and declination at three times, (II) measurement of position at two times, (III) measurement of range (distance from the observer) at six times, (IV) measurement of range at three times and simultaneous measurement of range-rate (time rate-of-change of range) at these times, (V) measurement of range-rate at six times. We now formulate the equations (boundary conditions) for each case.

Case (I): Let $\boldsymbol{R} = (R_1, R_2, R_3)$ be the position vector of the origin relative to the observer. We suppose that $\boldsymbol{R}$ is known at all times. Let ϱ be the magnitude of $\boldsymbol{r} + \boldsymbol{R}$, and let $\boldsymbol{L} = (L_1, L_2, L_3)$ be the unit vector in the direction of the vector $\boldsymbol{r} + \boldsymbol{R}$. Then

$$\boldsymbol{r} + \boldsymbol{R} = \varrho \, \boldsymbol{L}. \tag{5}$$

If now α, δ are the right ascension and declination of the satellite as viewed from the observer, then

$$L_1 = \cos\alpha \cos\delta, \quad L_2 = \sin\alpha \cos\delta, \quad L_3 = \sin\delta, \tag{6}$$

so that $\boldsymbol{L}$ is known at each observation point. Eq. (5) gives

$$(\boldsymbol{r} + \boldsymbol{R}) \times \boldsymbol{L} = 0, \tag{7}$$

thus eliminating ϱ. This is equivalent to the three scalar equations

$$\left.\begin{aligned} L_2(x_3 + R_3) - L_3(x_2 + R_2) &= 0, \\ L_3(x_1 + R_1) - L_1(x_3 + R_3) &= 0, \\ L_1(x_2 + R_2) - L_2(x_1 + R_1) &= 0. \end{aligned}\right\} \tag{8}$$

Only two of these equations are independent, so that we have two equations at each of the three observation times, yielding six boundary conditions.

Case (II): Here $\boldsymbol{r}$ is known at two times, giving six equations.

Case (III): From Eq. (5)

$$(\boldsymbol{r} + \boldsymbol{R})^2 = \varrho^2 \tag{9}$$

and this holds at six observation times, with ϱ known at each time.

Case (IV): Differentiating Eq. (5) we obtain

$$\dot{\boldsymbol{r}} + \dot{\boldsymbol{R}} = \dot{\varrho}\,\boldsymbol{L} + \varrho\,\dot{\boldsymbol{L}}. \tag{10}$$

Now $\boldsymbol{L}$ is a unit vector, so that $\boldsymbol{L}^2 = 1$ and so $\boldsymbol{L}\,.\,\dot{\boldsymbol{L}} = 0$. Hence, from Eqs. (5) and (10),

$$(\boldsymbol{r} + \boldsymbol{R}) \cdot (\dot{\boldsymbol{r}} + \dot{\boldsymbol{R}}) = \varrho\,\dot{\varrho}. \tag{11}$$

If ϱ and $\dot{\varrho}$ are known at three observation times, then Eqs. (9) and (11) hold at these times, again yielding six boundary conditions.

Case (V): Substituting the value of ϱ from Eq. (9) into Eq. (11) we obtain

$$(\boldsymbol{r} + \boldsymbol{R}) \cdot (\dot{\boldsymbol{r}} + \dot{\boldsymbol{R}}) = |\boldsymbol{r} + \boldsymbol{R}|\,\dot{\varrho}, \tag{12}$$

with $\dot{\varrho}$ known at six times. Here $|\boldsymbol{r} + \boldsymbol{R}|$ denotes the magnitude of $\boldsymbol{r} + \boldsymbol{R}$.

Summarizing, we have in all cases a set of six boundary conditions which we write

$$Q_j\big(\boldsymbol{x}(t_j)\big) = 0, \quad j = 1, 2, \ldots, 6. \tag{13}$$

Note that some of the t_j may coincide.

Our objective is now to find a set of functions $\boldsymbol{x}(t)$, satisfying Eqs. (4) and (13). The general principle of solution using quasilinearization may be found explained in references [*2*] and [*3*], but for the sake of completeness we shall explain it briefly here. The method is an iterative one, based on the linearization of Eqs. (4), in the sense that, if $\boldsymbol{x}^*$, $\boldsymbol{x}$ are two successive iterates, then we write

$$\boldsymbol{x} = \boldsymbol{x}^* + (\boldsymbol{x} - \boldsymbol{x}^*),$$

insert this expression in the right-hand side of Eq. (4), and expand the functions F_i in powers of $\boldsymbol{x} - \boldsymbol{x}^*$, retaining only the linear terms. We obtain

$$\dot{\boldsymbol{x}} = \boldsymbol{F}(\boldsymbol{x}^*, t) + \boldsymbol{B}(\boldsymbol{x} - \boldsymbol{x}^*) \tag{14}$$

where the matrix $\boldsymbol{B}$ has elements

$$B_{ij} = \frac{\partial}{\partial x_j} F_i(\boldsymbol{x}^*, t), \tag{15}$$

and i, j take the values $1, 2, \ldots, 6$. If $\boldsymbol{x}^*$ is a known function of t (even if only in numerical terms), then Eq. (14) is a set of linear differential equations in $\boldsymbol{x}$, and may be written

$$\dot{\boldsymbol{x}} = \boldsymbol{B}\,\boldsymbol{x} + \boldsymbol{D} \tag{16}$$

where the vector $\boldsymbol{D}$ is given by

$$\boldsymbol{D} = \boldsymbol{F} - \boldsymbol{B}\,\boldsymbol{x}^*. \tag{17}$$

The partial differentiations in Eq. (15) are performed analytically, and the B_{ij} calculated from the resulting formulae. Thus $\boldsymbol{B}$ and $\boldsymbol{D}$ are independent of $\boldsymbol{x}$ and are calculated from $\boldsymbol{x}^*$ and t only, and so are known functions of t.

Now let $\boldsymbol{p}$ be any particular solution of Eq. (16), and let $\boldsymbol{w}_1, \boldsymbol{w}_2, \ldots, \boldsymbol{w}_6$ be six linearly independent vector solutions of the associated homogeneous equations

$$\dot{\boldsymbol{x}} = \boldsymbol{B}\,\boldsymbol{x}. \tag{18}$$

It follows from the theory of linear differential equations that, if $c_1, c_2, \ldots, c_6$ are any constants,

$$\boldsymbol{x} = \sum_{k=1}^{6} c_k\, \boldsymbol{w}_k + \boldsymbol{p} \tag{19}$$

is a solution of Eq. (16), and conversely any solution of (16) can be expressed in the form (19) by suitable choice of the constants c_k.

General Discussion

The $\boldsymbol{w}_k$ and $\boldsymbol{p}$ are all computed by numerical integration of the appropriate Eqs. (18) or (16), starting from some arbitrary set of initial conditions. However, the vectors $\boldsymbol{w}_k$ must be linearly independent. The values of $\boldsymbol{w}_k$ and $\boldsymbol{p}$ at the observation times t_j are noted, and using these values, the expressions (19) are substituted into Eqs. (13), to give a set of algebraic equations for the c_k. These are solved numerically, and then again using (19), the value of $\boldsymbol{x}$ at say t_1 is computed. Then starting from $\boldsymbol{x}(t_1)$, the *exact Eqs.* (*4*) are integrated numerically to produce the values $\boldsymbol{x}(t)$ at any time i.e. a current approximate orbit. If the values $\boldsymbol{x}$ so obtained satisfy the boundary conditions (13) to the required degree of accuracy, the solution is complete. If the required accuracy is not attained we return to the value $\boldsymbol{x}(t_1)$ computed as before from Eq. (19). Starting from $\boldsymbol{x}(t_1)$ we integrate the *linear Eqs.* (16), compute B_{ij} and D_i for the current iterate, and so begin the next iteration. Thus for *testing* purposes (agreement with observation) we must use the exact Eqs. (4) (since these are the dynamical equations controlling the system), while for *iterative* purpose we use the linear Eqs. (16).

To summarize, each iteration involves four main steps, (1) the computation of B_{ij} and D_i from the previous iteration:

(2) the determination of the $\boldsymbol{w}_k$ and $\boldsymbol{p}$ by numerical integration of Eqs. (16) and (18),

(3) the numerical solution of the algebraic Eqs. (13) for c_k,

(4) testing the current approximate orbit (derived from the exact Eqs. (4) for agreement with the boundary conditions (13).

To start the iterative process, the initial approximation may consist of *any* suitable set of values, an essential requirement being that B_{ij} and D_i can be computed, ready for the first iteration. Thus we could use either a dynamical orbit derived from Eqs. (4) and an initial estimate of $\boldsymbol{x}(t)$, or else use linear interpolation based on estimates of $\boldsymbol{r}$ at two or more times. If the initial guess is not good enough, it may be expected that the process will either fail to converge or will converge to an incorrect solution. For some cases, however, the convergence is so strong, that almost any initial guess, however, poor, gives convergence to the correct solution.

One feature to be noted, which helps to explain the strong convergence, is the fact that every iterate, except the initial approximation, satisfies the boundary conditions *exactly*, (except for truncation error in the integration formulae, etc.). Also if convergence does occur, then the converged functions $\boldsymbol{x}$ necessarily satisfy the dynamical Eqs. (4). For in Eq. (14) the differences $\boldsymbol{x} - \boldsymbol{x}^*$ will be very small, and (14) reduces to (4).

In cases (I) and (II), when Eqs. (19) are substituted into Eqs. (7) and (8), the resulting equations are linear in the c_k, and so are readily solvable. In cases (III), (IV) and (V) the equations are quadratic in the c_k, and so must be solved by some iterative method. Thus within each main iteration we have a sequence of subsidiary iterations to determine the c_k. In the present work an ordinary NEWTON-RHAPHSON process was used. This involves the calculation of a set of partial derivatives, and these are found analytically, using the boundary conditions (9), (11) or (12). Also in these last three cases we require a set of starting values for the c_k. In cases (III) and (IV), where ϱ is known, such starting values may be found if approximate values of right ascension and declination are known, so that $\boldsymbol{r}$ is known approximately at two or more times. Then the linear epuations of case (II) become available to provide the required initial values of c_k. In case (V), where $\dot{\varrho}$ is given, if the angles are again known approxymately, but if ϱ is not known, then from Eq. (10) we have that

$$(\dot{\boldsymbol{r}} + \dot{\boldsymbol{R}}) \cdot \boldsymbol{L} = \dot{\varrho}\,.$$

These equations are also linear in the c_k, and so could provide approximate values of the c_k.

It is clear that if we are dealing with Eqs. (2) then exactly the same process can be applied, the resulting linear system being

$$\ddot{x}_i = \sum_{j=1}^{3} B_{ij}\, x_j + D_i\,, \qquad i = 1, 2, 3, \tag{20}$$

where the B_{ij} and D_i are defined in a similar manner to that used in the previous case.

There remains the question of a suitable integration process. If we are dealing with Eqs. (2) and (20), then we may use the GAUSS-JACKSON, or second-sum, method, which has good accuracy and usually requires only one evaluation of the functions at each step. It suffers from the disadvantages that a special starting procedure is needed, and also that $\dot{\boldsymbol{r}}$ is not computed, while $\dot{\boldsymbol{r}}$ is needed in the case of range-rate observations. However, $\boldsymbol{r}$ could be found from $\boldsymbol{r}$ by suitable quadrature formulae. In the case of Eqs. (4) and (16) any standard method may be used e.g. Adams-Moulton or Runge-Kutta. In the present investigation a fourth-order Runge-Kutta method was used. No special starting procedure is needed, but it has one disadvantage, in that it is necessary to calculate the B_{ij} and the D_i at the mid-point of each integration interval, and so $\boldsymbol{x}^*$ (the previous iterate) must be known at these mid-points. But the integration scheme has determined the $\boldsymbol{x}^*$ only at the end-points of each interval, and so the required values must be found by interpolation.

Numerical Results and Discussion

For simplicity, the orbits studied were those of a small satellite moving under the inverse-square law field of a massive body. The results quoted here are for a nearly circular orbit generated from the initial values $\boldsymbol{r} = (1 \cdot 1, 0, 0)$, $\dot{\boldsymbol{r}} = 0{,}0 \cdot 95{,}0)$, with the units chosen so that the force is unity at unit distance from the centre. The three observing stations all lie approximately on a spherical surface of unit radius, and the observations all lie within a total time interval of 0.5 of a time unit. If this situation is applied to an earth satellite, the time interval is approximately 6 minutes, which is reasonable. The satellite would be at a height of approximately 400 miles.

The three observation stations have coordinates

$$S_1(0.97, 0.24, 0.0), \quad S_2(0.88, 0.44, -0.22), \quad S_3(0.95, 0.22, 0.22),$$

and the six observations at times $t = 0, 0.1, 0.2, 0.3, 0.4, 0.5$, were taken in the order S_1, S_2, S_3, S_1, S_2, S_3. The computation was done on an IBM 1620. With an integration step-size of $h = 0.1$, the fourth-order Runge-Kutta integration process was found to provide an accuracy of six significant digits.

In each case the initial path was taken to consist of a straight line determined by estimates of its end-points.

Case (I): The values of right ascension and declination were taken at times $t = 0.1, 0.3, 0.5$. To demonstrate the effectiveness of the method in this case a deliberately poor initial path was selected, by assuming that at times $t = 0.1$, $t = 0.5$, the value of ϱ was nearly 5, whereas the values of ϱ were close to 0.4. The initial path then consisted of the line joining the points $(3.5, -3.5, 2.5)$, $(1.6, 3.5, -3.0)$. Estimates of velocity were not required, since in our examples the coefficients B_{ij} and D_i depend on position only. However, the initial velocity at $t = 0.1$ may be taken from the above estimates to be approximately $(-4.0, 14.0, -11.0)$.

In the following tables the first line shows the initial estimated values, the last line shows the true values, and the intermediate lines are the results of successive iterations, all calculated at the first observation time. The value of N is the number of the iteration. For clarity, all irrelevant digits are omitted.

N	x	y	z
	3.5	−3.5	2.5
1	1.097	0.092	0.001
2	1.09587	0.09487	0.00000
3	1.095870	0.094881	0.000000
	1.095870	0.094881	0.0

N	$\dot{x}$	$\dot{y}$	$\dot{z}$
	−4.0	14.0	−11.0
1	−0.210	0.915	−0.002
2	−0.08257	0.94645	0.00001
3	0.082542	0.946430	0.000003
	0.082543	0.946433	0.0

If the third iterate is used to generate an orbit, the residuals are found to be zero, to six significant digits.

Case (II)[1]: Here the given data consist of the coordinates of the path at times $t = 0$, $t = 0.5$, and one would naturally choose the straight line joining these as a first approximation. But again in order to demonstrate the strong convergence of the method, the initial path was chosen to be that joining the points (2.0, 0, 0) and (2.0, −0.5, 0). This is, almost, a path of twice the radius of the actual path, and in the opposite sense.

In this case the boundary conditions cause the values x, y, z at $t = 0$ to be correct for each iteration, so no table is given for these.

N	$\dot{x}$	$\dot{y}$	$\dot{z}$
	0.0	−1.0	0.0
1	−0.082	0.917	0.000
2	0.00003	0.95001	0.00000
3	0.000000	0.950000	0.000000
	0.0	0.95	0.0

[1] An example of case (II) may also be found in [*4*], which was published while this work was in progress.

In the above cases the convergence is very satisfactory, and the linear boundary conditions enable full advantage to be taken of this. In the following cases, where the boundary conditions are quadratic, the convergence is still excellent, but the starting values of the c_k have to be chosen with more care.

Case (III): For testing purposes, the initial path was taken to be the straight line joining the points (1.4, 0.4, 0.3), (1.3, 0.8, 0.3). These points correspond to values of right ascension and declination which differ from their actual values by over 20°. The results follow.

N	x	y	z
	1.4	0.4	0.3
1	1.110	0.001	−0.001
2	1.100001	0.000001	0.000001
3	1.100000	0.000000	0.000000
	1.1	0.0	0.0

N	$\dot{x}$	$\dot{y}$	$\dot{z}$
	−0.2	0.8	0.
1	−0.04	0.96	0.01
2	−0.000019	0.949998	−0.000005
3	−0.000002	0.950001	0.000002
	0.0	0.95	0.0

The residuals for the third iterate are zero to six significant digits.

As explained earlier, within each major iteration there is a sequence of minor iterations to determine the c_k, using the method of case (II) to find starting values for the c_k. The number of minor iterations required in each of the three major iterations was 6, 4, and 2, respectively. Compared with the numerical integration required for each major iteration, each minor iteration requires relatively little time, the main operation being the inversion of the 6×6 matrix. It is clear that the overall convergence is as effective as in cases (I) and (II).

If a rather worse initial approximation is taken for the c_k, it is found that the minor iterations no longer converge, but usually oscillate violently. In one example, however, convergence to an incorrect orbit was found to occur. This incorrect, neighbouring orbit satisfied the observational data.

The preceding comments also apply to cases (IV) and (V), except that for case (V) (where range-rate only is given), the starting values for c_k need to be much more accurate, errors of more than about one degree in right ascension and declination giving rise to divergence. But where convergence did occur, it was as effective as in the earlier cases.

Conclusions and Comments

Apart from the strong convergence, the most striking feature of the method is that the problem of determining the orbit is reduced in all cases to the same analytical form, namely the solution of six algebraic equations in six unknowns c_k. This is true for all fields of force, for all types of orbit, and all types of observational data.

If the algebraic equation are linear, then the solution of the equations is unique, readily determined, and the convergence to the orbit is very rapid, from almost any initial path. In such cases, we appear to have a true method of orbit determination. However, some qualification is necessary. The author considered the case of angular data separated in time by approximately one period of revolution of the satellite (the two-pass case). Here the initial approximation needs to be fairly good, or else convergence to an incorrect orbit is found to occur. This is due to the fact that the linear Eqs. (8) are unchanged if $\boldsymbol{L}$ is replaced by $-\boldsymbol{L}$, and so convergence can occur to an orbit determined by reversing the sign of one or more of the observed vectors $\boldsymbol{L}$.

If the equations are non-linear, then their solution can only be found by iteration from an initial approximation, and this initial approximation must be sufficiently good. The method then provides an orbit correction technique, and the convergence is rapid. No numerical differentiation is required at any stage. If a rapid, effective method could be found for providing good approximations to the roots of the algebraic equations, then the method would provide an orbit determination technique in these cases also. Ambiguities such as that noted in discussing the numerical results of case (III) could then be eliminated by taking extra observational data into account.

In general, if we have more than the minimum set of six observations they can be combined by the usual least squares technique. The boundary conditions are then more complicated, but the principles remain exactly the same.

Additional features can be incorporated. For example, suppose that the gravitational constant μ of the attracting body is unknown. Then we treat μ as a dependent variable, just like x, and add the equation $\dot{\mu} = 0$ to the list of differential equations defining the orbit. Thus μ remains constant in each iteration, but may vary from one iteration to the next.

In the first iteration great accuracy is not required, and a large integration interval could be used, and then reduced as convergence is approached. Also, from Eq. (14), it is clear that the B_{ij} need not be calculated accurately, since they are multiplied by a factor which tends to zero. However, inaccurate values of the B_{ij} may slow the convergence, especially in the early stages.

References

[1] BAKER, R. M. L., and M. W. MAKEMSON: An Introduction to Astrodynamics, New York/London: Academic Press 1962, pp. 240/41.

[2] BELLMAN, R., H. KAGIWADA and R. KALABA: Orbit Determination as a Multipoint Boundary-Value Problem and Quasilinearization, Proc. Nat. Acad. Sci. 48, 1327 (1962).

[3] KALABA, R.: On Nonlinear Differential Equations, the Maximum Operation, and Monotone Convergence, J. Math. Mech. 8, 519–574 (1959).

[4] MCGILL, R., and P. KENNETH: Solution of Variational Problems by Means of a Generalized Newton-Rhaphson Operator, AIAA J. 2, 1761–1773 (1964).

Discussion

Prof. HERRICK indicates that he and Mr. KENNETH FORD have undertaken a comparison of this method with that of differential correction, both analytically and computationally. They expect to employ somewhat more sober examples that those in which Mr. LONG has found such spectacular convergence, and which may be looked upon as more in the realm of preliminary orbit determination than in that of orbit correction.

Although Prof. HERRICK was first skeptical toward a process that does not integrate the initial approximation accurately to a representation of the observational data, it appears now clearly to him that, as the term $\boldsymbol{B}(\boldsymbol{x} - \boldsymbol{x}^*)$ in Eq. (14) converges to zero in successive approximations, this equation forces the solution in the end to agree with the differential Eq. (4) or $\boldsymbol{x} = \boldsymbol{F}(\boldsymbol{x}, \boldsymbol{t})$.

Mr. LONG states that these developments are a part of the general method as originally set down by Prof. HESTENES.

Prof. HERRICK believes that the lack of accurately integrated representation of purely observational data, and of the resulting residuals between observation and computation, has a disadvantage in that one cannot tell certainly when to stop, and is very likely to go through more approximations that are necessary. This fact is notable, for example, in a comparison of the classical Laplacian and Gaussian orbit methods. The former method, at least in the LEUSCHNER form, calculates accurate observation residuals and bases a differential correction upon them. The corresponding residuals in the latter method are not purely observational, but are in a sense corrupted by computational data. The result, in the specific example investigated by S. HERRICK, was that three unnecessary corrections were made: purely observational residuals were found to be negligible in the first approximation—and also, of course, in the three additional ones.

To obtain such residuals in Mr. LONG's method, one would have to perform non-linear integrations now and then, in addition to the seven quasilinearized integration for each step.

Dr. BAKER and his team has been working on range and (or) range-rate systems for many years. One problem that is frequently encountered is that of ambiguous solutions. Their experience has also shown that, as Mr. LONG has found, when only range or range-rate are available upon which to establish a preliminary orbit, one is ultimately faced with a relaxation problem and, because of near singularities over short arcs, this problem is always a most difficult one.

Universal Variables[1]
(Abstract)

By

Samuel Herrick

University of California, Los Angeles, California, U. S. A.

§ 1. **„Universal" variables, parameters, and formulae** are those that can be used with any of the two-body conic-section orbits, including the circular and rectilinear extremes of the ellipse, and the rectilinear extremes of the hyperbola and parabola. Their primary contribution is non-singular transition between orbits of differing type, in perturbations by variation-of-parameters, or in either observational or thrust correction. They are also advantageous, for example, for "on-board" computers that will encounter a variety of orbit types.

The simplest of the universal formulations is that in which the constant of gravitation, μ, is absorbed into the independent variable by such equations as

$$\tau = \sqrt{\mu}(t - t_0), \qquad \dot{\boldsymbol{r}} = \frac{d\boldsymbol{r}}{d\tau}, \qquad \ddot{\boldsymbol{r}} = \frac{d^2\boldsymbol{r}}{d\tau^2}. \tag{1}$$

Then, for example, the determination of $\boldsymbol{r}$, $\dot{\boldsymbol{r}}$ position and velocity at time t, from $\boldsymbol{r}_0$, $\dot{\boldsymbol{r}}_0$ at time t_0 is accomplished by:

$$r_2^2 = \boldsymbol{r}_0 \cdot \boldsymbol{r}_0, \qquad \dot{s}_0^2 = \dot{\boldsymbol{r}}_0 \cdot \dot{\boldsymbol{r}}_0, \qquad D_0 = r_0 \dot{r}_0 = \boldsymbol{r}_0 \cdot \dot{\boldsymbol{r}}_0, \tag{2}$$

$$\alpha = -\frac{1}{a} = \dot{s}_0^2 - \frac{2}{r_0}, \qquad c_0 = 1 + \alpha r_0. \tag{3}$$

For comparison with the ellipse: $c_0 = e \cos E_0$, $D_0 = \sqrt{a}\, e \sin E_0$. These quantities are entered in their unspecialized or "universal" forms, however, into the equivalent of KEPLER's equation:

$$\tau = r_0 \hat{X} + D_0 \hat{C} + c_0 \hat{U} \tag{4}$$

[1] Complete version and references in *Astronomical Journal* **70**, No. 4, 309 (1965). This work supported by: (a) NASA Goddard Space Flight Center, (b) USAF Office of Scientific Research, (c) North American Aviation Space & Information Systems Division.

to be solved for the universal variables $\hat{X}$, $\hat{C}$, $\hat{U}$ in conjunction with

$$\hat{U} = \frac{\hat{X}^3}{3!} + \alpha \frac{\hat{X}^5}{5!} + \alpha^2 \frac{\hat{X}^7}{7!} + \cdots,$$
$$\hat{C} = \frac{\hat{X}^2}{2!} + \alpha \frac{\hat{X}^4}{4!} + \alpha^2 \frac{\hat{X}^6}{6!} + \cdots. \tag{5}$$

Then r may be obtained from the companion formulae

$$r = r_0 = D_0 \hat{S} + c_0 \hat{C}, \qquad \hat{S} = \hat{X} + \alpha \hat{U}. \tag{6}$$

Again for comparison we note the elliptic equivalents:

$$\hat{X} = \sqrt{a}(E - E_0), \qquad \hat{U} = a^{3/2}[(E - E_0) - \sin(E - E_0)],$$
$$\hat{S} = \sqrt{a}\sin(E - E_0), \qquad \hat{C} = a[1 - \cos(E - E_0)].$$

From these we may infer Eqs. (4), (5), (6), and that $\hat{U}$, $\hat{C}$, and $\hat{S}$ may be determined in other ways equivalent to special machine devices for determining sines and cosines. We may also see that Eqs. (4), (5), (6) apply equally to the hyperbola, for which

$$\hat{X} = \sqrt{-a}(F - F_0), \qquad \hat{S} = \sqrt{-a}\sinh(F - F_0)$$

and to the parabola, for which

$$\hat{X} = \hat{S} = D - D_0, \qquad D = r\dot{r} = \sqrt{2q}\tan\frac{v}{2}.$$

Finally the well-known expressions for f, g, $\dot{f}$, $\dot{g}$, $\boldsymbol{r}$, $\dot{\boldsymbol{r}}$ become

$$f = 1 - \frac{\hat{C}}{r_0}, \qquad \dot{f} = -\frac{\hat{S}}{r r_0}, \qquad g = \tau - \hat{U}, \qquad \dot{g} = 1 - \frac{\hat{C}}{r_0}, \tag{7}$$

$$\boldsymbol{r} = f\boldsymbol{r}_0 + g\dot{\boldsymbol{r}}_0, \qquad \dot{\boldsymbol{r}} = \dot{f}\boldsymbol{r}_0 + \dot{g}\dot{\boldsymbol{r}}_0. \tag{8}$$

§ 2. Alternative sets of universal variables form a doubly infinite family with the set introduced by Eqs. (4), (5), (6) since there are two places in which we may introduce arbitrary coefficients. Designating these β and $\bar{n}$, we may rewrite Eqs. (4), (5), (6):

Replacing	*By*
$\alpha = -1/a$	$\bar{\alpha} = \beta\alpha = -\beta/a$,
τ	$\bar{M} = \bar{n}\tau$,
r	$\bar{r} = \bar{n}\beta^{1/2} r$,
r_0	$\bar{r}_0 = \bar{n}\beta^{1/2} r_0$,
D_0	$\bar{D}_0 = \bar{n}\beta D_0$,
c_0	$\bar{c}_0 = \bar{n}\beta^{3/2} c_0$,
$\hat{X}$	$\bar{X} = \beta^{-1/2}\hat{X}$,
$\hat{S}$	$\bar{S} = \beta^{-1/2}\hat{S}$,
$\hat{C}$	$\bar{C} = \beta^{-1}\hat{C}$,
$\hat{U}$	$\bar{U} = \beta^{-3/2}\hat{U}$.

After $\bar{U}, \bar{C}, \bar{S}, \bar{r}$ have been determined from the modified Eqs. (4), (5), (6), we must in effect determine, by direct calculation or by an equivalent procedure,

$$\hat{U} = \beta^{3/2}\,\bar{U}, \qquad \hat{C} = \beta\,\bar{C}, \qquad \hat{S} = \beta^{1/2}\,\bar{S}, \qquad r = \bar{n}^{-1}\,\beta^{-1/2}\,\bar{r}$$

in order to proceed with Eqs. (7) for $f, g, \dot{f}, \dot{g}$.

Several possible sets of β and $\bar{n}$ are discussed in the complete paper. These include the set $\beta = \tau^2/r_0^2$, $\bar{n} = 1/\tau$ which leads to Stumpff's form of the universal variables, for which comparative notation is supplied.

§ 3. Formulae for variation-of-parameters and for partial differential coefficients are developed in the complete paper, both as a basis of comparison of the various possible sets of universal variables, parameters, and formulae, and for the practical use of the set defined by Eqs. (1) through (8) in perturbation integrations and in differential corrections and related statistical or maneuvering problems. The paper ends with a discussion of some universal parameters or elements alternative to $\boldsymbol{r}_0$ and $\dot{\boldsymbol{r}}_0$.

§ 4. Addendum to Astronomical Journal article. It appears that some further generalization of the alternative sets of universal variables and parameters (§ 2 above) may be achieved by setting

$$\boldsymbol{r} = f\,\boldsymbol{r}_0 + \bar{g}\,\bar{\dot{\boldsymbol{r}}}_0, \qquad \bar{g} = \bar{n}\,g, \qquad \bar{\dot{\boldsymbol{r}}} = \bar{n}^{-1}\,\dot{\boldsymbol{r}}_0, \tag{8}$$

$$\bar{\dot{\boldsymbol{r}}} = \frac{1}{\bar{n}}\,\frac{d\boldsymbol{r}}{d\tau} = \frac{1}{\bar{n}\sqrt{\mu}}\,\frac{d\boldsymbol{r}}{dt} = \bar{\dot{f}}\,\boldsymbol{r}_0 + \dot{g}\,\bar{\dot{\boldsymbol{r}}}_0, \qquad \bar{\dot{f}} = \frac{1}{\bar{n}}\,\dot{f}. \tag{9}$$

Thus with Stumpff's form ($\bar{n} = 1/\tau$) the modified $f, \bar{g}, \bar{\dot{f}}, \dot{g}$ become more symmetrical:

$$\begin{aligned} f &= 1 - \frac{\tau^2\,\bar{C}}{r_0^3}, & \bar{\dot{f}} &= -\frac{\tau^2\,\bar{S}}{r\,r_0^2}, \\ \bar{g} &= 1 - \frac{\tau^2\,\bar{U}}{r_0^3}, & \dot{g} &= 1 - \frac{\tau^2\,\bar{C}}{r\,r_0^2} \end{aligned} \tag{10}$$

and

$$\bar{\dot{\boldsymbol{r}}} = \tau\,\dot{\boldsymbol{r}} = \tau\,\frac{d\boldsymbol{r}}{d\tau} = (t - t_0)\,\frac{d\boldsymbol{r}}{dt}, \tag{11}$$

[We correct an error in sign in the $\dot{f}$ Eqs. (36) and (42) of the A.J. article. — Cf. also errata in A.J. 70, No. 6, 447 (1965).]

Sur le Développement du Potentiel de la Terre par les Fonctions de Lamé

Par

G. N. Doubochine

Université de Moscou, U.R.S.S.

Résumé. L'auteur propose de développer le potentiel terrestre en coordonnées ellipsoïdales. L'ellipsoïde de référence peut être choisi aussi voisin que possible de l'ellipsoïde représentant la surface de la Terre. Dans ce cas, une des coordonnées ellipsoïdales λ, est la distance à la surface de la Terre. Les développements en puissances de λ convergeraient rapidement dans la théorie du mouvement d'un satellite artificiel.

Abstract. A development of the potential of the Earth in ellipsoidal coordinates is proposed. The reference ellipsoid may be chosen as close as possible from the ellipsoid representing the surface of the Earth. In this case, one of the ellipsoidal coordinates, λ, represents the distance to the surface of the Earth. Developments in powers of λ would be rapidly converging in the theory of the motion of an artificial satellite.

C'est la force d'attraction de la Terre qui est, comme on le sait, la force principale, gouvernant les mouvements des corps célestes artificiels.

On sait de même qu'on ne peut pas en maints cas envisager la Terre, comme un point matériel, mais qu'il faut prendre en considération ses dimensions, sa forme et sa structure intérieure.

C'est pourquoi, dans l'astrodynamique, un rôle considérable est joué par le champ de gravitation de la Terre, défini par son potentiel en un point extérieur $\mathcal{P}$ de masse unité:

$$U(\mathcal{P}) = f \int\limits_{(T)} \frac{dm}{\Delta}, \tag{1}$$

où f est la constante d'attraction, dm la masse élémentaire concentrée au point $\mathcal{P}'$, $\Delta = \overline{\mathcal{P}\mathcal{P}'}$ et l'intégrale est étendue sur toute la masse de la Terre.

Il est impossible d'exprimer la fonction $U(\mathcal{P})$ par une formule finie et il est nécessaire de la représenter par quelque série infinie.

Les représentations classiques du potentiel nous donnent, comme on le sait, des développements par les polynômes harmoniques ou par les fonctions sphériques de LAPLACE.

Soit $0\,x\,y\,z$ le système des coordonnées, ayant pour origine le centre de la Terre et dont les axes sont dirigés suivant les axes principaux d'inertie de la Terre. Alors, nous avons,

$$U(\mathscr{P}) = U(x, y, z) = f \sum_{n=0}^{\infty} \frac{U_n(x, y, z)}{r^{2n+1}}, \tag{2}$$

où $r^2 = x^2 + y^3 + z^2$, $U_n(x, y, z)$ est un polynôme harmonique d'ordre n, dont les coefficients sont définis uniformément par la forme et la structure de la Terre.

La série (2) converge absolument dans le domaine $r > a$, où a est le rayon équatorial de la Terre.

En employant les coordonnées sphériques r, θ, ψ, liées aux coordonnées rectangulaires par les relations

$$x = r \sin\theta \cos\psi, \qquad y = r \sin\theta \sin\psi, \qquad z = r \cos\theta, \tag{3}$$

nous aurons le développement suivant

$$U(\mathscr{P}) = U(r, \theta, \psi) = f \sum_{n=0}^{\infty} \frac{Y_n(\theta, \psi)}{r^{n+1}}, \tag{4}$$

où $Y_n(\theta, \psi)$ représente la fonction sphérique d'ordre n définie par la formule

$$Y_n(\theta, \psi) = \sum_{k=0}^{n} P_n^k(\cos\theta)\,[A_{nk} \cos k\psi + B_n r_k \sin k\,\psi], \tag{5}$$

dont les coefficients sont définis également par la forme et la structure de la Terre.

On peut calculer ces coefficients par des mesures gravimétriques directes ou par les mouvements observés des satellites artificiels de la Terre.

On ne détermine pratiquement assez exactement que deux ou trois des premiers coefficients zonaux, tandis qu'on ne connait encore que très grossièrement même les premiers coefficients tesseraux.

Cette circonstance, ainsi que la très lente convergence (pour les satellites voisins) des séries, rend l'application des développements classiques peu commode et pousse à chercher pour la représentation de la fonction $U(\mathscr{P})$ des formes nouvelles.

Pendant le Symposium «La dynamique des satellites» (28—30 Mai 1962, Paris) j'ai remarqué qu'on peut employer pour le développement du potentiel terrestre les fonctions ellipsoïdales, comme avait fait A. LIAPOUNOFF dans sa théorie célèbre sur les figures d'équilibre des fluides en rotation.

Cette communication est consacrée à un certain développement de remarque indiquée.

Introduisons d'abord, au lieu des coordonnées rectangulaires, ou sphériques, les coordonnées sphéro-coniques par les formules:

$$\left.\begin{aligned} x &= r\sin\theta\cos\psi = r\,\frac{\sqrt{\mu^2-h^2}\sqrt{h^2-\nu^2}}{h\sqrt{k^2-h^2}},\\ y &= r\sin\theta\sin\psi = r\,\frac{\sqrt{k^2-h^2}\sqrt{k^2-\nu^2}}{k\sqrt{k^2-h^2}},\\ z &= r\cos\theta = r\,\frac{\mu\nu}{hk}, \end{aligned}\right\} \tag{6}$$

où $h>0$ et $k>h$ sont des constantes convenablement choisies et tels que $h\leqq|\mu|\leqq k$, $|\nu|\leqq h$.

Avec $\gamma = <\mathscr{P}\,O\,\mathscr{P}'$, nous avons

$$\cos\gamma = \cos\theta\cos\theta' + \sin\theta\sin\theta'\cos(\psi-\psi') \tag{7}$$

et alors nous pouvons écrire la formule connue de la théorie des fonctions de Lamé:

$$P_n(\cos\gamma) = \frac{\pi}{2(2n+1)}\sum_{s=0}^{2n+1} E_n^s(\mu)\,E_n^s(\nu)\,E_n^s(\mu')\,E_n^s(\nu'), \tag{8}$$

où $E_n^s(\xi)$ est la fonction de Lamé du premier ordre et de degré n:

$$E_n^s(\xi) = \begin{cases} \Pi(\xi) & s\equiv 0 \pmod 4,\\ \Pi(\xi)\sqrt{\xi^2-h^2} & s\equiv 1 \pmod 4,\\ \Pi(\xi)\sqrt{\xi^2-k^2} & s\equiv 2 \pmod 4,\\ \Pi(\xi)\sqrt{\xi^2-h^2}\,\sqrt{\xi^2-k^2} & s\equiv 3 \pmod 4, \end{cases}$$

où $\Pi(\xi)$ est un polynôme du degré respectivement égal à n, $n-1$, $n-1$, $n-2$. Selon la parité ou imparité de n, le nombre des fonctions de première espèce (où les fonctions $K(\xi)$) est égal à $1+\frac{1}{2}n$ ou à $\frac{1}{2}(n+1)$, le nombre des fonctions de deuxième et de troisième espèce (les fonctions $L(\xi)$ et $M(\xi)$) est égal à $\frac{1}{2}n$ ou à $\frac{1}{2}(n+1)$, le nombre des fonctions de quatrième espèce (les fonctions $N(\xi)$) est égal à $\frac{1}{2}n$ ou à $\frac{1}{2}(n-1)$.

Les fonctions sphériques et les fonctions de Lamé sont liées par les relations suivantes:

$$\begin{aligned} P_n^{2m}(\cos\theta)\cos 2m\psi &= \sum \alpha\,K(\mu)\,K(\nu);\\ P_n^{2m+1}(\cos\theta)\cos(2m+1)\psi &= \sum \alpha\,L(\mu)\,L(\nu);\\ P_n^{2m}(\cos\theta)\sin 2m\psi &= \sum \alpha\,N(\mu)\,N(\nu);\\ P_n^{2m+1}(\cos\theta)\sin(2m+1)\psi &= \sum \alpha\,M(\mu)\,M(\nu) \end{aligned}$$

où α sont des coefficients constants et le nombre des divers termes de la somme est égal au nombre des fonctions de la classe correspondante.

Pour obtenir l'expression de $U(\mathscr{P})$ en coordonnées sphéro-coniques, il est plus simple de procéder de la manière suivante: dans le développement de l'inverse de la distance Δ^{-1} suivant les polynômes de LEGENDRE

$$\frac{1}{\Delta} = \sum_{n=0}^{\infty} \frac{r'^n P_n(\cos\gamma)}{r^{n+1}} \qquad (r' < r), \tag{9}$$

changeons $P_n(\cos\gamma)$ par son expression (8) et substituons après l'expression (9) pour Δ^{-1} dans la formule (1).

Il n'est pas difficile de présenter le résultat sous la forme:

$$U(\mathscr{P}) = U(r, \mu, \nu) = f \sum_{n=0}^{\infty} \frac{U_n(\mu, \nu)}{r^{n+1}}, \tag{10}$$

où

$$U_n = \sum_{s=1}^{2n+1} U_n^s E_n^s(\mu) E_n^s(\nu), \tag{11}$$

et U_n^s sont des coefficients constants, définis par la formule suivante:

$$U_n^s = \frac{\pi}{2(2n+1)} \int_{(T)} r'^n E_n^s(\mu') E_n^s(\nu') \, dm.$$

Il est clair, que, pour la Terre, les U_n^s sont des constantes caractéristiques, qu'on peut déduire des constantes A_{ns} et B_{ns} et qu'on peut déterminer également par les observations directes des satellites artificiels. Mais dans ce dernier cas, il est nécessaire, bien entendu, d'avoir une théorie analytique du mouvement du satellite, donnant les coordonnées sphéro-coniques en fonction du temps.

On obtient ces fonctions par intégration analytique des équations différentielles du mouvement, qu'on peut écrire sans difficulté ayant l'expression (10) pour la fonction U, et compte tenu du fait que la force vive est définie en coordonnées nouvelles par la formule:

$$T = \frac{1}{2} \left\{ \dot{r}^2 + \frac{r^2(\mu^2 - \nu^2)\dot{\mu}^2}{(\mu^2 - h^2)(k^2 - \mu^2)} + \frac{r^2(\mu^2 - \dot{\nu}^2)\nu^2}{(h^2 - \nu^2)(k^2 - \nu^2)} \right\},$$

qu'on peut déduire facilement à l'aide des formules (6).

Envisageons maintenant le cas général des coordonnées ellipsoïdales. Soit

$$\frac{X^2}{a^2} + \frac{Y^2}{b^2} + \frac{Z^2}{c^2} = 1 \quad (a \geqq b > c) \tag{12}$$

l'ellipsoïde donné. Alors, pour les valeurs fixes de coordonnées rectangulaires x, y, z, l'équation

$$\frac{x^2}{a^2 + \theta} + \frac{y^2}{b^2 + \theta} + \frac{z^2}{c^2 + \theta} = 1 \quad (a \geqq b > c) \tag{13}$$

admet toujours les trois racines réelles λ, μ, ν, de sorte que

$$-a^2 < \nu < -b^2 < \mu < -c^2 < \lambda < +\infty$$

En remplaçant successivement dans (13) la quantité θ par λ, μ, ν et en résolvant les équations obtenues, nous trouvons

$$\left.\begin{aligned} x^2 &= \frac{(a^2+\lambda)(a^2+\mu)(a^2+\nu)}{(a^2-b^2)(a^2-c^2)}, \\ y^2 &= \frac{(b^2+\lambda)(b^2+\mu)(b^2+\nu)}{(a^2-b^2)(c^2-b^2)}, \\ z^2 &= \frac{(c^2+\lambda)(c^2+\mu)(c^2+\nu)}{(a^2-c^2)(b^2-c^2)}. \end{aligned}\right\} \tag{14}$$

Notons qu'en vertu du théorème de Vietà nous aurons,

$$r^2 = x^2 + y^2 + z^2 = a^2 + b^2 + c^2 + \lambda + \mu + \nu. \tag{15}$$

Pour $\lambda = 0$ le point est situé sur l'ellipsoïde (12). En désignant ce point par $\mathcal{P}_0$ et ces coordonnées par x_0, y_0, z_0, nous avons

$$r_0^2 = x_0^2 + y_0^2 + z_0^2 = a^2 + b^2 + c^2 + \mu + \nu. \tag{16}$$

et nous pouvons écrire

$$r^2 = r_0^2 + \lambda, \tag{17}$$

d'où on voit immédiatement que la variable λ caractérise aussi la distance du point $\mathcal{P}$ à la surface de l'ellipsoide fondamental (12).

On sait d'après la théorie des fonctions ellipsoïdales, que le produit

$$V_n^s = E_n^s(\lambda)\, E_n^s(\mu)\, E_n^s(\nu) \tag{18}$$

satisfait à l'équation de Laplace correspondante, d'où il s'ensuit que chaque polynôme harmonique $U_n(x, y, z)$, transformé en fonction des variables λ, μ, ν, se présente sous la forme d'une combinaison linéaire des fonctions V_n^s.

Alors, le potentiel de la Terre, défini par la formule (1), s'écrit aussi sous la forme:

$$U(\mathcal{P}) = U(\lambda, \mu, \nu) = f \sum_{n=0}^{\infty} \frac{V_n(\lambda, \mu, \nu,)}{(r_0^2+\lambda)^{n+\frac{1}{2}}}, \tag{19}$$

où

$$V_n(\lambda, \mu, \nu) = \sum_{s=0}^{2n} C_n^s\, E_n^s(\lambda)\, E_n^s(\mu)\, E_n^s(\nu). \tag{20}$$

Les coefficients constants C_n^s, qu'on peut exprimer par les coefficients des polynômes U_n, sont définis entièrement par la forme et la structure de la Terre.

Si nous construisons la théorie du mouvement des satellites, alors les coefficients C_n^s pourront être déterminés par les observations des satellites (aussi comme les coefficients des fonctions sphériques).

Mais, le mouvement d'un satellite est déterminé par les équations différentielles, qu'il est facile de former.

En effet, en coordonnées ellipsoïdales la force vive est définie par la formule:

$$T = \tfrac{1}{2}[H_1^2 \dot{\lambda}^2 + H_2^2 \dot{\mu}^2 + H_3^2 \dot{\nu}^2], \tag{21}$$

où

$$4H_1^2 = \frac{(\lambda - \mu)(\lambda - \nu)}{(a^2 + \lambda)(b^2 + \lambda)(c^2 + \lambda)},$$

$$4H_2^2 = \frac{(\mu - \nu)(\mu - \lambda)}{(a^2 + \mu)(b^2 + \mu)(c^2 + \mu)},$$

$$4H_3^2 = \frac{(\nu - \lambda)(\nu - \mu)}{(a^2 + \nu)(b^2 + \nu)(c^2 + \nu)}.$$

En introduisant maintenant les impulsions généralisées

$$\Lambda = \frac{\partial T}{\partial \dot{\lambda}} = H_1^2 \dot{\lambda}, \qquad M = \frac{\partial T}{\partial \dot{\mu}} = H_2^2 \dot{\mu}, \qquad N = \frac{\partial T}{\partial \dot{\nu}} = H_3^2 \dot{\nu}$$

nous aurons encore

$$T = \frac{1}{2}\left[\frac{1}{H_1^2} \Lambda^2 + \frac{1}{H_2^2} M^2 + \frac{1}{H_3^2} N^2\right],$$

alors l'équation de Hamilton-Jacobi s'écrit sous la forme

$$\frac{1}{H_1^2}\left(\frac{\partial W}{\partial \lambda}\right)^2 + \frac{1}{H_2^2}\left(\frac{\partial W}{\partial \mu}\right)^2 + \frac{1}{H_3^2}\left(\frac{\partial W}{\partial \nu}\right)^2 = 2U + 2C, \tag{22}$$

et la résolution du problème se réduit à la recherche d'une intégrale complète de cette équation.

Pour conclure remarquons, que pour les satellites très voisins, la quantité λ est très petite et qu'il est commode de développer la fonction des forces en série de puissances de λ.

Car, en outre, le choix de l'ellipsoïde fondamental est tout-à-fait arbitraire, on peut choisir pour cet ellipsoïde un certain ellipsoïde très voisin de la surface de la Terre (par exemple, l'ellipsoïde de Krassovsky):

Soit $\mathscr{P}(\lambda, \mu, \nu)$ un point de l'espace, où se trouve à un moment donné notre satellite; soit $\mathscr{P}_0(0, \mu, \nu)$ le point de l'ellipsoïde (12), correspondant (au sens d'Ivory) au point $\mathscr{P}$.

Alors $d = \overline{\mathscr{P}\mathscr{P}_0}$ est défini par la formule:

$$d^2 = \left(\frac{\sqrt{a^2+\lambda}}{a} - 1\right)^2 x_0^2 + \left(\frac{\sqrt{b^2+\lambda}}{b} - 1\right)^2 y_0^2 + \left(\frac{\sqrt{c^2+\lambda}}{c} - 1\right)^2 z_0^2, \tag{23}$$

d'où on voit immédiatement que, pour les petites valeurs de λ, la distance d sera aussi très petite et inversement.

Pour $\lambda < c^2$, on peut développer la distance d en série absolument convergente des puissances de λ, de la forme:

$$d = \sum_{k=1}^{\infty} D_k \lambda^k, \tag{24}$$

où

$$D_1 = \frac{1}{2}\sqrt{\frac{x_0^2}{a^4} + \frac{y_0^2}{b^4} + \frac{z_0^2}{c^4}} < \frac{r_0}{2c^2} \leqq \frac{a}{2c^2},$$

$$D_2 = -\frac{1}{8}\left(\frac{x_0^2}{a^4} + \frac{y_0^2}{b^4} + \frac{z_0^2}{c^4}\right)^{-1/2}\left(\frac{x_0^2}{a^6} + \frac{y_0^2}{b^6} + \frac{z_0^2}{c^6}\right), \quad D_3 = \cdots.$$

Avec les mêmes conditions, la fonction $E_n^s(\lambda)\,(r_0^2 + \lambda)^{-n-1/2}$ peut être développée aussi en puissances de λ, de sorte que nous aurons pour le potentiel terrestre la représentation suivante:

$$U(\mathscr{P}) = f\sum_{n=0}^{\infty} U_n(\mu, \nu)\,\lambda^n, \tag{25}$$

dont les coefficients dépendent de μ et de ν uniquement:

Il est aisé de voir, que nous avons:

$$f\,U_0(\mu, \nu) = U(x_0, y_0, z_0) = U(\mathscr{P}_0),$$

$$f\,U_1(\mu, \nu) = \frac{x_0}{2a^2}\,U'_x(\mathscr{P}_0) + \frac{y_0}{2b^2}\,U'_y(\mathscr{P}_0) + \frac{z_0}{2c^2}\,U'_z(\mathscr{P}_0).$$

Discussion

Dr. Cook points out that, in spheroidal coordinates, Vinti showed that one can determine an exact solution if the potential is such that $J_{2n+1} = 0$, $J_{2n} = (-1)^n (J_2)^{2n-2}$. The condition of separability, whatever the coordinates used, is that the potential should be of the form:

$$\Sigma\frac{V_i}{h_i^2},$$

where h_i is the coefficient of the differential dx_i in the metric of the coordinate system and V_i is a function of x_i only.

Professor Duboshin agrees with this of course; he says that developing in ellipsoidal coordinates will lead to a certain new type of intermediate orbits that will have to be treated according to the perturbation theory based on approximations. One of the main interests is the fact that one of the variables will be the distance between the satellite and the Earth and as such may lead to rapidly convergent series for close satellites. This possibility should be checked, since, as Dr. Kozai pointed out, the usual development in spherical harmonics is not rapidly converging.

The Effect of Precession of the Earth's Axis of Rotation on the Motion of an Artificial Satellite

By

Dirk Brouwer

Yale University Observatory, New Haven, Connecticut, U.S.A.

Abstract. The problem of making allowance for the motion of the equatorial plane on the motion of an artificial satellite is treated by formulating the equations for the variation of elements in an inertial system, in which the ecliptic at an arbitrary date is used as $x\ y$ plane. The result obtained by Goldreich (1965) is confirmed. Inclinations near 0° or 90° require special consideration, but it is shown that they are not exceptional.

Résumé. Pour tenir compte du mouvement du plan de l'équateur dans le problème du mouvement d'un satellite artificiel, on écrit les équations aux variations des éléments dans un système d'inertie dans lequel le plan de l'écliptique à une date donnée arbitraire est pris pour plan des $x\ y$. Le résultat obtenu par Goldreich (1965) est confirmé. Les inclinaisons voisines de 0° ou 90° doivent être examinées avec soin mais l'on montre qu'elles ne posent pas de problèmes exceptionnels.

Introduction

The equations of motion of an artificial satellite are commonly formulated on the tacit assumption that a coordinate system with Z axis along the axis of rotation of the earth and the X and Y axes in the Earth's equatorial plane may be treated as an inertial coordinate system. It is of interest to inquire how the solution is affected if the precessional motion of the earth's axis is taken into account.

The problem was treated by Kozai (1960), who was primarily interested in obtaining explicit corrections to the elements of an artificial satellite. Kozai related the instantaneous equator of date to a fixed equator of a standard date. He included both the precession and nutation.

More recently, Goldreich (1965) considered the problem from a more general dynamical point of view. He starts from the equations for the variations of the elements in an instantaneous equatorial system, adding to the oblateness perturbations the corrections that must be introduced to allow for the Coriolis and centripetal accelerations that arise on account of the motion of the coordinate system. He concludes that if the motion of the satellite's ascending node on the equatorial plane has a period which is short compared with the planet's precession period, then the satellite orbit will maintain a constant inclination to the planet's moving equator.

A procedure different from both treatments quoted will be followed in this article. The equations for the variations of the elements are initially stated for an inertial coordinate system and an appropriate transformation is introduced to obtain the equations for the variations of the elements referred to the equatorial system of date.

Although in principle any arbitrary inertial coordinate system would serve, it appears that a coordinate system with the $x\,y$ plane coinciding with the fixed ecliptic for a standard date, say 1950.0, has a special advantage. For such a coordinate system the obliquity of the moving equator at T relative to the fixed ecliptic at T_0 has the expression (NEWCOMB 1906)

$$\varepsilon_1(T) = \varepsilon_0(T_0) + 0''\!.06\,(T - T_0)^2 - 0''\!.008\,(T - T_0)^3,$$

$T - T_0$ being expressed in centuries. Only approximate values of the coefficients of the quadratic and cubic terms are given. The important thing is that the term proportional to $T - T_0$ is absent. Hence, except for the periodic terms known as the nutation in the obliquity, $\varepsilon_1 = \text{const}$ is an excellent approximation.

The Equations of Motion

Let I be the inclination and Ω the longitude of the ascending node (Q) of a satellite orbit referred to the fixed ecliptic for a standard date, say, 1950.0.

Let ε_1 be the inclination and Φ the longitude of the ascending node (N) of the moving equator on the fixed ecliptic plane.

Let γ be the inclination of the satellite orbit relative to the moving equator and ψ arc $N\,Q'$, Q' being the ascending node of the satellite orbit on the moving equator.

If terms of short period, having the mean anomaly of the satellite in their arguments are ignored, the time derivatives of I and Ω are given by

$$\frac{dI}{dt} = -\frac{1}{n\,a^2\,(1 - e^2)^{1/2}\sin I}\,\frac{\partial R}{\partial \Omega}, \qquad \frac{d\Omega}{dt} = +\frac{1}{n\,a^2\,(1 - e^2)^{1/2}\sin I}\,\frac{\partial R}{\partial I}$$

with

$$R = \frac{3}{4}\,J_2\,n^2\,\frac{a_e^2\cos^2\gamma}{(1 - e)^{3/2}},$$

n being the mean motion of the satellite, a_e the Earth's equatorial radius.

From the spherical triangle formed by the three planes (the fixed ecliptic, the equator, and the orbital plane) it is evident that γ may be considered to be a function of ε_1, I, and $\Omega - \Phi$. Then if

$$\varepsilon_1 = \text{constant},$$
$$\Phi = \Phi_0 - p\,t$$

are introduced, a simple reduction yields

$$\frac{d\gamma}{dt} = p\,\frac{\partial\gamma}{\partial(\Omega - \Phi)},$$

$$\frac{d\psi}{dt} = \frac{1}{n\,a^2\,(1 - e^2)^{\frac{1}{2}}\sin I}\,\frac{\partial(\psi,\gamma)}{\partial(\Omega - \Phi, I)}\,\frac{\partial R}{\partial\gamma} + p\,\frac{\partial\psi}{\partial(\Omega - \Phi)}.$$

The following relations may be obtained from the spherical triangle:

$$\frac{\partial\gamma}{\partial(\Omega - \Phi)} = \sin\varepsilon_1,\ \sin\psi,$$

$$\frac{\partial\psi}{\partial(\Omega - \Phi)} = \cos\varepsilon_1 + \sin\varepsilon_1\cot\gamma\cos\psi,$$

$$\frac{\partial(\psi,\gamma)}{\partial(\Omega - \Phi, I)} = \frac{\sin I}{\sin\gamma}.$$

If further

$$b_1 = \frac{3}{2}\,n\,J_2\left(\frac{a_e}{a}\right)^2\frac{1}{(1 - e^2)^2},$$

the equations reduce to

$$\frac{d\gamma}{dt} = p\sin\varepsilon_1\sin\psi,$$

$$\frac{d\psi}{dt} = -b_1\cos\gamma + p\cos\varepsilon_1 + p\sin\varepsilon_1\cot\gamma\cos\psi.$$

The first term in the expression for $d\psi/dt$ is evidently the motion of the node obtained if the equations referred to the equatorial coordinate system treated with neglect of the procession.

Special Cases

Three cases are distinguished.

Case I.

$$p \ll b_1,$$
$$p \ll \sin\gamma$$

A solution of the form

$$\gamma = \gamma_0 + \sum C_j\cos j\,(b\,t - \beta),$$
$$\psi = -b\,t + \beta + \sum D_j\sin j\,(b\,t - \beta),$$
$$b \approx b_1\cos\gamma_0 - p\cos\varepsilon_1$$

may be obtained by successive approximations. The solution shows that γ oscillates about a mean value γ_0 with a period close to the period of revolution of the node in the equatorial plane.

Case II. For sufficiently small values of γ the last term of the expression for $d\psi/dt$ becomes singular. In this case the introduction of

$$X = \sin\gamma \cos\psi, \qquad Y = \sin\gamma \sin\psi$$

leads to the equations

$$\dot{X} - (b_1 \cos\gamma - p \cos\varepsilon_1)\, Y = 0,$$
$$\dot{Y} + (b_1 \cos\gamma - p \cos\varepsilon_1)\, X = p \sin\varepsilon_1 \cos\gamma.$$

But $\cos\gamma = (1 - X^2 - Y^2)^{1/2}$ may be expanded in powers of $X^2 + Y^2$, and the solution may be obtained by successive approximations. In the first approximation the linear equations

$$\dot{X} - (b_1 - p \cos\varepsilon_1)\, Y = 0,$$
$$\dot{Y} + (b_1 - p \cos\varepsilon_1)\, X = p \sin\varepsilon_1$$

are obtained, with the solution

$$X = \frac{p \sin\varepsilon_1}{b_1 - p\cos\varepsilon} + C\cos(b\,t - \beta),$$
$$Y = -C\sin(b\,t - \beta)$$

with

$$b = b_1 - p\cos\varepsilon_1 .$$

Again, it is seen that γ oscillates about a mean value γ_0.

Case III. If γ is close to 90°, let $\cos\gamma = k$, k being a small positive or negative quantity. If $\sin\gamma$ is put equal to unity, the equations may be simplified to

$$\frac{dk}{dt} = -p \sin\varepsilon_1 \sin\psi,$$

$$\frac{d\psi}{dt} = p\cos\varepsilon_1 - k(b_1 - p\sin\varepsilon_1 \cos\psi).$$

If $b_1 \gg p$, the second equation may be approximated by

$$\frac{d\psi}{dt} = p\cos\varepsilon_1 - k\,b_1$$

from which

$$\frac{d^2\psi}{dt^2} = b_1\, p \sin\varepsilon_1 \sin\psi .$$

This is the equation of the mathematical pendulum. In general, $\sin\psi$ will be developable in the form of a harmonic series. For small oscillations in ψ the mean motion is $(b_1\, p \sin\varepsilon_1)^{1/2}$, so that in obtaining k from the equation for dk/dt this expression will appear as divisor. In general, therefore, k (and therefore γ) will be oscillatory in character, although the solution is different from that of Case I. Solutions in the vicinity of the unstable equilibrium solution of the pendulum will require a special discussion.

After the solution for $\varepsilon_1 = \text{const}$, $p = \text{const}$ has been obtained it remains to evaluate the effect on the argument of the perigee and to introduce the changes with time in ε, and in Φ_0 that have been ignored in this discussion. Over an interval of the order of the principal nutation period the latter effects should correspond to the nutation terms obtained by KOZAI.

References

GOLDREICH, P.: Astronomical J. **70**, 5 (1965).
KOZAI, Y.: Astronomical J. **65**, 621 (1960).
NEWCOMB, S.: 1906, Compendium of Spherical Astronomy, (New York: The Macmillan Company, p. 241); reprinted by Dover Publications, 1960.

Discussion

It would be of interest, according to Mr. KAULA, to adapt the treatment in Dr. BROUWER's paper to the change of inclination in the evolution of the moon's orbit due to tidal friction where the inclination of the equator on the ecliptic, ε, also changes with time. This change is either ignored or treated in an unsatisfactory manner in most solutions of the tidal friction problem.

Professor DUBOSHIN suggests to compare this effect of precession to the effect due to the rotation of the satellite around its center of gravity.

The Perturbations of the Orbital Elements Caused by the Pressure of the Radiation Reflected from the Earth

By

L. Sehnal

Astronomical Institute of the Czechoslovak Academy of Sciences, Ondřejov, ČSSR

Abstract. The formulae for the changes of the elements of the orbits of artificial satellites of the Earth are given. The perturbative force considered here is the pressure of the solar radiation, reflected from the Earth. The effect of the Earth's shadow is investigated using an approximative theory.

Résumé. L'auteur donne les formules pour la variation des éléments des satellites artificiels de la Terre. La force perturbatrice considérée est la pression de la radiation solaire réfléchie par la Terre. L'auteur étudie l'effet de l'ombre de la Terre à l'aide d'une théorie approchée.

In a previous paper [*1*], I investigated the components of the disturbing function, which described the effect of the reflected solar radiation pressure. The changes of the elements can be then computed, using the Lagrangian equations in the GAUSSian form.

We can get by this substitution the changes of all orbital elements. However, the most important changes will be here the secular perturbations. But the mutual position of the orbital plane and the illuminated hemisphere of the Earth change very rapidly and the physical conditions of reflection are also in no case constant. The secularity of the perturbations is then very limited and so we shall consider the secular changes only during one revolution of the satellite. The perturbations of the elements will be of course very small and so we shall use the transformation between the increments of the true anomaly (v) and time (t) in the following form:

$$r^2\,dv = k\,\sqrt{M}\,\sqrt{p}\,dt\,.$$

Here, k is the gravitational constant, M the mass of the Earth, p is the orbital parameter and r is the position vector of the satellite. Then we compute the change of an arbitrary element σ as the integral

$$\Delta\sigma = \frac{1}{2\pi}\int\limits_0^{2\pi} \frac{d\sigma}{dt}\,\frac{1}{k\sqrt{M}}\,\frac{r^2}{\sqrt{p}}\,dv\,. \tag{1}$$

If the satellite enters the Earth's shadow, we shall have two integrals. We denote as ω the angle between perigee and the perpendicular projection of the Earth-Sun direction on the orbital plane, and $u = \omega + v$. The first integral will then be computed from $v = 0$ (if $u = 0$) to $v = v_1$, corresponding to the point where the satellite enters into the shadow. If the satellite goes out from the shadow at $v = v_2$, the second integral is computed between the limits $v = v_2$ and $v = 2\pi$ (if $u = 2\pi$). Then we have

$$\Delta\sigma = \int_0^{\nu_1} \frac{d\sigma}{d\nu} d\nu + \int_{\nu_2}^{2\pi} \frac{d\sigma}{d\nu} d\nu . \tag{2}$$

This case with these two integrals is, however, much more complicated. So I tried to simplify this case by an approximation. We do not consider the value of the reflected radiation presure as vanishing in the points, where the orbit goes into and out from the Earth's shadow. We consider it only to be very small and to be vanishing only in one point, which lies on the intersection of the perpendicular projection of the Earth-Sun line into the orbital plane, with the orbit. This projection has two intersections with the orbit. The first lies above the illuminated hemisphere and the value of the reflected radiation has here its maximum. The other point of intersection lies in the shadow and the reflected radiation pressure here vanishes. Between these two limiting points, we suppose the disturbing radiation pressure to have a continuous course.

Now we can compute the perturbations by the formula (1). Some of the formulae for the changes of elements are then obtained in a simple closed form, but others (for $\Delta\omega$, Δe) contain the infinite series with Bessel functions. The terms of those series were obtained by the evaluation of the integrals

$$\int_0^{2\pi} \frac{r}{a} \cos m v \, dv =$$

$$2\pi(-1)^m \left(\frac{e}{2}\right)^m (1-e^2)^{(2-m)/4} .$$

Now we can give all the formulae for the changes of the elements. The expressions of the constants z and q, given below, are valid if the satellite does not enter the Earth's shadow. In other case, we shall have these expression for z and q:

$$z = \frac{\pi/2 - j}{2} \frac{\pi}{4} ; \qquad q = 2 . \tag{3}$$

The constants are given explicitly with their derivation in a paper mentioned [1]. The resulting formulae are:

$$\Delta a = a_K\{-\tfrac{1}{2}\, C_0\, I_0(z)\, e \sin\omega + 2 I_1(z)\, [(q-1)\,(C_0\, e + C_1) \sin\omega + \\ + \tfrac{1}{4}\, C_1\, e \sin 2\omega] + \\ + I_2(z)\, [C_0\, e \sin\omega - C_1\, e(q-1) \sin 2\omega] + \\ + \tfrac{1}{2}\, I_3(z)\, C_1\, e \sin 2\omega\},$$

$$\Delta e = e_K \Big\{-\tfrac{1}{4}\, C_0\, I_0 \sin\omega + I_2[C_0 \sin\omega - C_1(q-1) \sin 2\omega] + \\ + \frac{1}{e} I_1\, C_3\, (q-1) \sin\omega - \\ - \frac{1}{e}\left(\frac{R}{p}\right)^2 \Big[2\,(q-1) \sum_{p=1}^{\infty} [I_{2p-1}\left(\frac{e}{2}\right)^{2p-1} (1-e^2)^{(1-2p)/4} \sin(2p-1)\omega\Big] + \\ + \sum_{p=1}^{\infty} (I_{2p-1} + I_{2p+1}) \left(\frac{e}{2}\right)^{2p} (1-e^2)^{(1-p)/2} \sin 2p\omega\Big]\Big\},$$

$$\Delta\Omega = \Omega_K \Big\{\frac{1}{2}\, C_1(q-1) \sin\omega \sqrt{1-e^2} - \\ - \Big[C_0(q-1) + \frac{1}{2}\, C_1 \cos\omega\Big] \sin\omega\, \frac{e}{2}\, (1-e^2)^{3/2} + \\ + [C_0 \sin 2\omega + C_1(q-1) \sin\omega] \left(\frac{e}{2}\right)^2 \sqrt{1-e^2} - \\ - \frac{1}{4}\, C_1 \sin 2\omega \left(\frac{e}{2}\right)^3 (1-e^2)^{-1/2}\Big\},$$

$$\Delta i = i_K \Big\{\frac{1}{2}\, [C_0 + C_1(q-1) \cos\omega] \sqrt{1-e^2} - \\ - \Big[C_0(q-1) \cos\omega + \frac{1}{2}\, C_1 + \frac{1}{4}\, C_1 \cos 2\omega\Big] \frac{e}{2}\, (1-e^2)^{3/2} + \\ + \Big[\frac{1}{2}\, C_0 \cos 2\omega + \frac{1}{2}\, C_1(q-1) \cos\omega\Big] \left(\frac{e}{2}\right)^2 \sqrt{1-e^2} - \qquad (4) \\ - \frac{1}{4}\, C_1 \sin 2\omega \left(\frac{e}{2}\right)^3 (1-e^2)^{-1/2}\Big\},$$

$$\Delta\omega = \omega_K \Big\{-\Big[\frac{1}{2}\, C_0 \cos\omega + \frac{1}{2}\, C_1(q-1)\Big] I_0 - \\ - \frac{1}{2} [C_0 \cos\omega + C_1(q-1) \cos 2\omega]\, 2 I_2 + \\ + \Big[C_0(q-1) \cos\omega + \frac{1}{4}\, C_1 \cos 2\omega\Big] I_1 + \frac{1}{4}\, C_1 \cos 2\omega\, I_3 + \\ + \frac{3}{2} \left(\frac{R}{p}\right)^3 [(q-1) \cos\omega\, I_1] + \\ + \frac{1}{2}\, \frac{1}{1-e^2} \left(\frac{R}{p}\right)^2 \Big[\frac{1}{2}\, I_1 \sqrt{1-e^2} - \\ - \sum_{p=1}^{\infty} 2 I_{2p-1} \sin\omega \sin(2p-1)\, (-1)^{2p-1} \left(\frac{e}{2}\right)^{2p-1} (1-e^2)^{\frac{2}{3}-2p}\Big]\Big\}.$$

$I_i = I_i(z)$ are the BESSEL functions of the first kind, j is the mutual inclination of the solar rays to the orbital plane, and

$$C_0 = \left(\frac{R}{p}\right)^2 + \frac{3}{4}\left(\frac{R}{p}\right)^3,$$

$$C_1 = e\left(\frac{R}{p}\right)^3,$$

$$q = \tan\left[\frac{\pi}{2}\left(1 + \frac{j - \frac{\pi}{2}}{j_0 - \frac{\pi}{2}}\right)\right] + 2,$$

$$j_0 = \arccos\frac{R}{r_0} + \frac{\pi}{2},$$

$$r_0 = \frac{p}{1 + e\cos\omega},$$

$$z = \frac{\frac{\pi}{2} - j}{j_0 - \frac{\pi}{2}}\,\frac{\pi}{2}.$$

The expressions a_K, e_K, Ω_K, i_K, and ω_K, depend on the physical constants and on the mutual position of the orbital plane and the illuminated hemisphere of the Earth. They are given as follows:

$$\left.\begin{aligned}
a_K &= \frac{2}{3}\,\frac{a^2}{k\sqrt{M}}\,\frac{p}{\pi}\,\Phi,\\
e_K &= \frac{1}{3}\,\frac{a^2(1-e^2)}{k\sqrt{M}}\,\frac{1}{\pi}\,\Phi,\\
\Omega_K &= \frac{1}{3}\,\frac{a^2(1-e^2)}{k\sqrt{M}}\,\frac{1}{\pi}\,\frac{1}{\sin i}\,\Phi,\\
i_K &= \frac{1}{3}\,\frac{a^2(1-e^2)}{k\sqrt{M}}\,\frac{1}{\pi}\,\Phi,\\
\omega_K &= \frac{1}{3}\,\frac{1}{e}\,\frac{a^2(1-e^2)}{k\sqrt{M}}\,\frac{1}{\pi}\,\Phi,
\end{aligned}\right\} \tag{5}$$

where

$$\Phi = \frac{\varkappa\,\alpha}{c}\,\frac{A}{m}\,\frac{1}{q}\cos k_0\cos k_1$$

and

$$k_0 = \frac{j}{j_0}\,\frac{\pi}{2} \qquad k_1 = \frac{j}{j_0}(j - \pi); \qquad j_0 = \arccos\frac{R}{r_0} + \frac{\pi}{2}$$

and A/m is the ration area/mass for a satellite. $\varkappa$ is the solar constant, α the Earth's albedo and c is the velocity of the light.

Now, in the case when the satellite does not enter the shadow, we compute the changes of elements using the formulae (4), with the constants given by (5). If the satellite penetrates the shadow, the values of constants are given by (3).

We can now compare the results, obtained in the case of shadowing with the use of accurate method, given by (2), with the results obtained by the formulae (4) and (5). Doing this, we can omit some of the physical constants, which are the same for both methods, since we are interested only in the relative results, obtained by the two methods. The formulae, obtained by the method given by (2), are, however, very complicated and I shall give here only the expression for the change Δa of the semi-major axis:

$$\Delta a = a_K (G_1 + G_2);$$

$$\begin{aligned} G_1 = {} & \tfrac{1}{2} C_0 \{\sin[(d_1 - c_1)\omega + (d_1 - c_1) v] + \sin[(d_1 + c_1)\omega + (d_1 + c_1) v]\} + \\ & + \tfrac{1}{2} C_0 e \{\sin[(d_1 - c_1)\omega + (1 + d_1 - c_1) v] + \sin[(d_1 + c_1)\omega + \\ & + (1 + d_1 + c_1) v]\} + \tfrac{1}{4} C_1 \{\sin[(d_1 - c_1)\omega + (d_1 - c_1 - 1) v] + \\ & + \sin[(d_1 - c_1)\omega + (1 + d_1 - c_1) v] + \sin[(d_1 + c_1)\omega + \\ & + (d_1 + c_1 - 1) v] + \sin[(d_1 + c_1)\omega + (1 + d_1 + c_1) v]\} + \\ & + \tfrac{1}{4} C_1 e \{\sin[(d_1 - c_1)\omega + (d_1 - c_1) v] + \sin[(c_1 + d_1)\omega + \\ & + (d_1 + c_1) v] + \sin[(c_1 + d_1)\omega + (2 + d_1 + c_1) v] + \\ & + \sin[(d_1 - c_1)\omega + (2 + d_1 - c_1) v]\}. \end{aligned} \tag{6}$$

The function G_2 can be obtained from the expression for G_1 through the change of sign and substitution c_2 for c_1 and d_2 for d_1. The meaning of the constants $c_{1,2}$ and $d_{1,2}$ depends on the value of the true anomaly for the entry and exit of the satellite from the shadow, and the constants are given in the mentioned paper [*1*].

It is not complicated to compute the numerical values from the expressions (3), (4) and (6). The results, obtained with the use of (3) and (4) does not differ more than 15% from the results, obtained by the accurate method and formulae. This precision, although not too great, can be considered as sufficient for our purposes, since the uncertainties in the physical description of the mechanism of the reflected radiation, the rapid changes in the Earth's albedo, the nonuniformity of the reflecting Earth's surface and other less known effects would cause greater deviation, beyond the precision of our formulae.

References

[*1*] SEHNAL, L.: U.A.I. Symposium No. 25 (under press).

Radiation on a Satellite in the Presence of Partly Diffuse and Partly Specular Reflecting Body

By

Robert M. L. Baker, Jr.

Computer Sciences Corporation and University of California, Los Angeles, California, U.S.A.

Abstract. The influence of radiation on a satellite in the presence of a partly diffuse and partly specular reflecting body is studied. Two important details of the problem, which, in general have not been studied, namely the values of the non-radial components of the radiation flux and the influence of satellite orientation in the resolution of the radiation flux into perturbative forces, are analyzed.

Résumé. On étudie l'influence de la radiation sur un corps à réflexion en partie diffuse en partie spéculaire. On analyse deux détails importants qui n'ont pas été encore étudiés en général, à savoir: les valeurs des composantes non radiales de la pression de radiation et l'influence de l'orientation de satellite sur la transformation du flux de radiation en forces perturbatrices.

1. Introduction

1.1 Related Research

The author's interest in the study of radiation pressure phenomena was stimulated by two papers presented at the May 1962 IUTAM Conference in Paris, France. At this earlier IUTAM Conference, it occured to the author that there exists an analogy between the distribution of emitted particles from a spherical satellite noted in a study of neutral free-molecular-flow drag (by the author in 1957—1958), and the distribution of photons reflected from a spherical reflecting body.

The study of radiation pressure and its influence on satellites has given rise to over forty scientific papers in the past five or six years. In 1960, the orbits of the Echo balloon and the Vanguard satellite were studied by a number of scientists, and the influence of both direct solar radiation pressure and, to a lesser degree, reflected Earth radiation were analyzed. This work has been described by Parkinson et al. (1960), Musen et al. (1960), Katz (1960), and Shapiro, et al. (1960).

Following this, a new rash of papers were written in 1961, 1962 (two of them were read at the IUTAM Conference), and 1963, presenting analyses of various aspects and influences of radiation pressure. Among these were papers by JAFFE (1961), study of Orbiting Astronomical Observatory with no Earth reflection; SHAPIRO and JONES (1961), a study of gas loss from the Echo balloon; BRYANT (1961), using the method of KRYLOFF-BOGOLIUBOFF to compute radiation pressure perturbations; ZADUNAISKY et al (1961), a study of Echo I; KOZAI (1961), including radiation pressure in the determination of the Earth's gravitational constants; KOSKELA (1962), an analysis of direct solar radiation pressure perturbations; SHAPIRO (1962), including Earth specularly reflected radiation together with a large number of other perturbation sources acting on objects orbiting about the Earth; SWALLEY (1962), obtains radiation pressure perturbations from a radiation flux for a spinning satellite; WYATT (1962), an important contribution to the solution of reflected radiation problems to be discussed in this paper; DENNISON (1962), a numerical study of total radiation flux[1]; HRYCAK (1962), analyzes total Earth-reflected radiation[1] and satellite temperature balance; LEVIN (1962), an important contribution to the study of Earth reflected radiation also to be discussed in this paper; CUNNINGHAM (1962), a numerical study of total radiation flux[1]; PIERCE (1963), a study of the passage of a satellite through the Earth's umbra; SHAPIRO (1963), a study of a more sophisticated satellite reflection model; and SEHNAL (1963), an important contribution to the understanding of reflected radiation phenomena also to be discussed presently. Of the foregoing papers, the author will single out three of them for more detailed consideration: WYATT (1962), LEVIN (1962), and SEHNAL (1963).

1.2 Requirements for Study

In view of the profusion of papers and supporting analyses dealing with the influence of radiation pressure on satellites, the reason why more study is needed may not be obvious. Before explaining this need, however, let us briefly examine the three major areas of influence of radiation flux upon satellites. First, and perhaps foremost, is the contribution of total radiation flux both from the Sun, and reflected from planets upon the thermal balance of a satellite. A consideration of CAMAC (1960), KREITH (1962), CUNNINGHAM (1962), DENNISON (1962), and HRYCAK (1962), indicates the importance of the total Earth-reflected radiation flux on the thermal control of Earth satellites. Second, is the dynamical effect of both the magnitude and direction of solar radiation on the orbit of a satellite or space vehicle. Here, except for CUNNINGHAM, general perturbations has been the tool for analysis.

[1] Near to the Earth, the total radiation flux is quite different from the radial component of radiation flux. Thus the total flux is not directly related to satellite dynamics.

Third, is the utilization of radiation pressure to achieve attitude stability of Earth satellites. Four of the earlier papers on this subject, FRYE et al (1959), SOHN (1959), NEWTON (1960), and IVES (1961), do not consider reflected solar radiation, nor do they consider radiation effects upon a spinning satellite. The paper by ULE (1963) considers the use of an array of spinning mirrors to achieve attitude stability using radiation pressure.

Because of the influence of satellite orientation, which cannot easily be expressed analytically, in all three of the aforementioned areas, it seems to the author that the use of special perturbations instead of general perturbations as a method of analyzing this influence is indicated. In addition, new advances in numerical procedures and computer techniques now allow for precise prediction up to one thousand revolutions. A generalized flow diagram of a special perturbations loop (with the generation of the radiation pressure perturbations shown in detail) is presented in Fig. 1. The areas of analysis that still warrant additional attention are shaded in this figure. The first of these shaded boxes indicates the generation of the transverse component to Earth reflected diffused radiation.

WYATT (1962) concentrated only upon the radial component of radiation flux. In his IR-1 to IR-3 models for infra-red Earth radiation, for example (although he considered the variation of infra-red radiation of differing latitudes) he set to zero the component normal to the equatorial plane. In his models AD-1 through AD-4, he considered albedo-diffuse Earth reflected radiation, but sets the transverse component to zero. In particular, in his model AD-3, he considered LAMBERT's Law of diffuse reflection but commented, "... the calculations, however, are formidable and it appears impossible without recourse to numerical techniques to obtain the components of the acceleration for a close satellite." An analytical solution to this problem (using a method developed by the author (1958)), will be given in the main body of this report, with supporting details in Appendix *A*. In his model AD-4, WYATT utilizes DANJON's diffuse reflection relationship. WYATT commented that the specular reflection is small, but developed models AS-1, and AS-2, for its analysis, in certain special cases. The consideration of the specular reflection flux is indicated in the second shaded box (Fig. 1) and is solved, in general form, in Appendix *B*.

LEVIN (1962) indicates one of the more disturbing features of the problem in the case of the Earth:

"Physically, the reflecting 'surface' is not the geometrical surface of a sphere but in large part is a reflection in depth through the atmosphere, and particular significance must be associated with cloud geometry. It is known that quite complex internal and mutual reflections and excitations occur within and between cloud

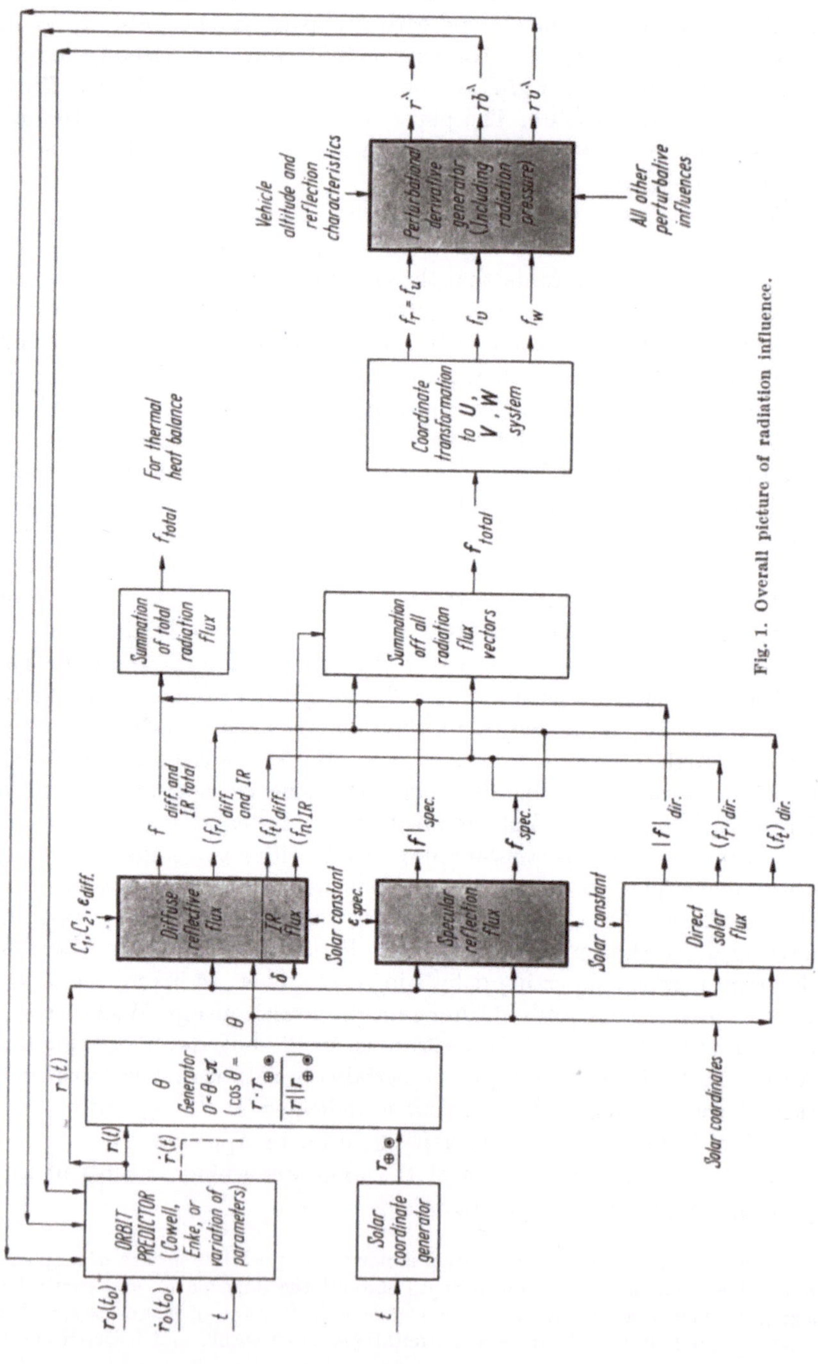

Fig. 1. Overall picture of radiation influence.

formations, and, consequently, the longer path length of rays through the atmosphere for elements inclined to the direction of the Sun compensates to some extent for their smaller projected area. The precise factor to employ is, of course, unknown (and variable both spatially and temporally) but will range between unity and $\sin\alpha \cos\beta$, yielding upper and lower limits, respectively. The value of unity has been used throughout the following to provide an upper bound, since for most space applications (e. g., temperature control, estimation of balloon satellite perturbations) it is the maximum value that must be considered for design and analysis purposes. Consequently, the reflected radiation emitted by each illuminated element is assumed to be independent of its orientation with respect to the Sun."

As indicated in this quotation, LEVIN does not consider the decrease in incident solar radiation near the terminator on the Earth. LEVIN also considers a two-dimensional problem as do most of the other investigators, and although he accounts for the transverse component, he concludes that it can be neglected. Levin does not consider specular reflection from the Earth nor does he carry out integration of the perturbative influences upon a satellite orbit.

SEHNAL (1963), similarly to LEVIN, assumes that the Sun is in the orbit plane, but neglects the transverse component. SEHNAL, like WYATT, actually integrates the perturbations acting on a satellite (using the general perturbations method) but in generating the perturbations resulting from a given solar flux (indicated by the third shaded box in Fig. 1), he considers only what amounts to a perfectly absorptive satellite; most of the other analysts e.g. KOSKELA (1962), makes this same assumption.

To the best of the author's knowledge, no comprehensive scheme exists for the prediction of spacecraft motion in three-dimensions which takes into account both diffused and specular reflection, and in which the transverse and normal components of radiation flux are considered. Furthermore, relaxation of the perfectly absorptive satellite body assumption has not been made. ACORD and NICKLAS (1963), as well as a few others, have at least considered it, but have not, at the same time, rigorously accounted for Earth reflected radiation. Finally, no analyst's have considered the influence of promptly emitted infrared radiation, a factor of profound influence in the analysis of a lunar orbiter.

The requirement for considering seemingly small effects such as the non-radial components of total radiation flux, Earth specular reflection, three-dimensional treatment of radiation-pressure perturbations, and the introduction of satellite orientation can, of course, be justified upon purely academic grounds. Nevertheless, these effects are also of practical importance, in that they represent perturbative accelerations acting on satellites that are comparable to those arising from higher order zonal, sectorial, and tesseral harmonics of the Earth's gravitational field, which are under intensive study at the present time, and are also comparable to certain electromagnetic perturbations and tidal influences which are currently undergoing careful study[1]. Thus, the following analyses represents the addition of one more essential detail to the vastly complex problem of the study of satellite dynamics and the prediction of the trajectories of artificial celestial bodies as determined from observations.

2. Analysis

2.1 Direct Solar Radiation

Of all of the radiation pressure problems, the one resulting from direct solar radiation has been studied the most thoroughly. In 1903, J. H. POYNTING considered the consequences of the absorption and re-emission of direct solar radiation on meteoritic dust. ROBERTSON (1937) introduced relativity into POYNTING's analysis. Further study of this effect was made by WYATT and WHIPPLE in 1950. It was over twenty years after the fundamental contributions of ROBERTSON that GARWIN (1958), first proposed the use of direct solar radiation as a motive force for interplanetary space vehicles, and coined the term "solar sailing." As noted in Section 1.1,the next few years saw a number of papers published analyzing the influence of direct solar radiation on geocentric orbits. In this Proceedings a paper by SEHNAL (p. 80) also presents an analysis of reflected solar radiation pressure perturbations.

Except during the passage of the spacecraft through the penumbra and umbra of a planet (usually the Earth), the radiation flux is from the direction of the Sun and has a magnitude equal to the local solar constant. Since the main analytical emphasis of this paper is upon reflected radiation, only an outline of the more important details of the direct solar-radiation problem, and the approach to its solution will be presented here.

[1] For a typical solar sailing communications satellite (under study by the Rome Air Development Command, Rome, New York, USA) which is composed of a 50 meter long, parallelogram-shaped sail 20 meters wide and weighing about 30 kg., the influence of direct solar radiation causes a perturbation of roughly 10^{-5} g's. Radiation reflected radially from the Earth gives it an additional acceleration of approximately 10^{-6} g's and the tangential acceleration component periodically amounts to about 10^{-7} g's. The higher-order zonal harmonics (J_3° and above) give rise to periodic accelerations amounting to about 10^{-6} g's. Depending upon satellite orientation, at a height of about 700 km the neutral drag amounts to between 10^{-5} g's and 10^{-7} g's, while the electrical effects are an order of magnitude or so smaller than this. The perturbative influence of the Moon on this spacecraft is about 10^{-7} g's. If a denser spacecraft were chosen, then the tangential component of Earth's reflection would drop to about 10^{-9} or 10^{-10} g's, but would still require consideration if an accuracy of $\pm 1\%$ or better were to be sought in higher-order zonal harmonics, or in air density.

The author is completing research on these problems now, but this topic does not fit within the scope of the present paper.

In this connection, however, it should be recognized that analyses of the details of reflected radiation phenomena, such as the computation of the tangential component of the reflected radiation, are worthless unless full account is taken of the detailed effects of direct solar radiation. The first of these is the solar constant itself; here lies a very important and fundamental constant. The solar constant is, to some extent, time variable; earlier estimates from a Smithsonian study (ABBOTT (1935)) indicated that a 3% variation over short time intervals was to be expected. More recently ALDRICH [(1954), p. 23] showed that the variation over 10 day averages was less than 2% and rms divation of monthly averages taken since 1923 is only 0.6%. Clearly, the actual extra-terrestial variation of the solar constant requires further investigation by means of scientific satellites. At the moment, it is premature to analyze the probable results of these experiments. The average value of the solar constant has been investigated for some time. Table 1 indicates some selected determinations of the solar constant since 1940.

Table 1. *Experimental Determinations of Solar Constant up to 1956 (Terrestrial Experiment)*

Experimeter	Solar constant
	($watts/cm^2$)
MOON (1940)	0.1322
ALLEN (1950)	0.1374
ALDRICH (1954)	0.1357
JOHNSON (1954)	0.1396
STAIR (1956)	0.1430

In Table 2 we find the values that have been used by various analysts in their studies of radiation pressure influences on satellite motion or on satellite thermal balance.

Table 2. *Values of Solar Constant Utilized by Orbital Analysts*

Analyst	Solar constant ($watts/cm^2$)
IVES (1961)	0.129
PARKINSON (1960)	0.135
CUNNINGHAM (1962)	0.1353
KRIETH (1962)	0.140
LEVIN (1962)	0.130
SHAPIRO (1963)	0.139

It is clear that a much more precise value of the solar constant should be sought through multiple scientific satellite experiments. A preliminary, reference value for the solar constant near the Earth might be 0.140 ± 0.005 $watts/cm^2$ (which corresponds to a solar pressure constant, $P_{\odot}$, near the Earth of 4.7×10^{-5} $dynes/cm^2$). It should, of course, be emphasized that this number should be regarded as only a very tentative value.

The second problem is the passage of the spacecraft through the penumbra. Most analyses of the penumbra, such as that of PIERCE

(1963), define it through use of simple geometry, based on the assumption of a spherical planet, that is devoid of atmosphere. Clearly most planets are slightly aspherical, and the sunlight is refracted somewhat in its passage through the planetary atmosphere. Furthermore, as KING-HELE pointed out, the clouds in the Earth's atmosphere sometimes obscure the Sun and cast shadow into space; thus causing the conical umbra to be irregular in cross section, not circular. These are important details to be dealt with in the refinement of the overall direct radiation-pressure problem's formulation.

The use of an aspherical planet, instead of a spherical one, presents no fundamental problem; it simply makes the geometry involved slightly more complicated. The refraction of sunlight through an atmosphere is, on the other hand, much more complicated and involves many of the problems to be considered in our analysis of the characteristics of the Earth's upper atmosphere, cloud cover, internal reflections, etc. The approach selected here is an analysis of ray tracing that has been applied by the author in the correction for atmospheric refraction for spacecraft photographs of ground points, in essence, the method involves an integration along the solar light ray, using the slope of the ray as an independent variable. The cloud obscuration problem, as well as the problems associated with the time-variable reflection properties of the Earth and other planets (to be discussed in Section 2.4), might be dealt with by telemetering data back to Earth on photon flux at the spacecraft and incorporating this information directly into the prediction computation.

2.2 Diffusely Reflected and IR Radiation from a Spherical Body

The Sun illuminates various areas on the surface of a spherical planet in proportion to the cosine of the Sun's zenith distance. As LEVIN (1962) points out the cosine relationship is not rigorous, but in order to avoid systematic errors involved in assuming that the area near the terminator (dawn-dusk line) is as brightly illuminated as the area in which the Sun is in the zenith, the cosine relationship will not be neglected. The diffuse reflection does not exactly follow the LAMBERT Law (cosine) relationship, according to WYATT (1962), but any other assumption (such as DANJON's) would probably be less realistic, so LAMBERT's diffuse reflection is assumed.

Since we are not restricted to simplified analytical forms (as are required by the general perturbations methods) and since some interesting effects may be observed when a satellite's orbital plane is nearly coincident with the plane of the terminator, a three-dimensional analysis is made. If dS' is a typical element of area on a spacecraft; $\left(\frac{dn}{dS\,dt}\right)_S$

is the flux of photons from the Sun incident on the reflecting planet (e.g., if dS is in cm², then at the Earth's distance from the Sun $\left(\frac{dn}{dS\,dt}\right)_S \cong$ $\cong 8 \times 10^{17}$ photons/cm²/sec assuming that the average photon wavelength is 5,560 Å); r is the radial distance of the spacecraft from the planet's center in terms of planetary radii, c is the cosine and s the sine of the angle (measured at the planet's center) between the direction of the Sun and the direction of the spacecraft, θ, then from the analysis of Appendix A, we find that:

Diffuse and IR total flux is given by

$$f_{\text{diff. and IR total}} = \frac{dn}{dt\,dS'} = \frac{2}{\pi r^2}\left(\frac{dn}{dS\,dt}\right)_S$$

$$\varepsilon'_{\text{diff.}}\{U(\theta_T-\theta)(I_1+I_2)+U(\theta-\theta_T)\,U(\theta'_T-\theta)\,I_3\}+$$

$$+\left\{(1-\varepsilon'_{\text{diff.}}-\varepsilon_{\text{spec.}})\left(\frac{dn}{dS\,dt}\right)_S \times \right.$$

$$\left.\times \frac{1}{2\pi(1-c_1/3)\,r^2}(C_1-C_2\,c_1\cos^2\delta_{\text{eq}}+C_3\,c_1\sin^2\delta_{\text{eq}}),\right\} \qquad (1)$$

where

$U(\theta_T-\theta)=1$ for $\theta<\theta_T$ (horizon is all illuminated),
$U(\theta_T-\theta)=0$ for $\theta\geqq\theta_T$ (terminator is in S_2),
$U(\theta-\theta_T)=0$ for $\theta<\theta_T$ (horizon is all illuminated),
$U(\theta-\theta_T)=1$ for $\theta\geqq\theta_T$ (terminator is in S_2),
$U(\theta'_T-\theta)=1$ for $\theta<\theta'_T$ (terminator is in S_1),
$U(\theta'_T-\theta)=0$ for $\theta\geqq\theta'_T$ (horizon is all dark),

θ_T is the value of θ when the terminator first enters the spacecraft's horizon (in S_2),

θ'_T is the value of θ when the terminator leaves the spacecraft's horizon (in S_1),

$\varepsilon'_{\text{diff.}}$ is the fraction of diffusely reflected visible light plus promptly emitted IR,

$\varepsilon_{\text{spec.}}$ is the fraction of specularly reflected light,

$$(I_1+I_2)=\left\{\left(\frac{\pi}{3}\right)+\frac{1}{r}\left(\frac{\pi}{4}\right)+\frac{1}{r^2}(0)+\frac{1}{r^3}\left(\frac{\pi}{6}-\frac{5}{18}\right)+\frac{1}{r^4}(0)+\cdots\right\}c,$$

$$(I_3)=\left\{\left(\frac{[\pi-\theta]}{3}c+\frac{s}{3}\right)+\frac{1}{r}\left(-\frac{\pi}{16}+\frac{\pi}{8}c+\frac{3\pi}{16}c^2\right)+\right.$$

$$+\frac{1}{r^2}\left(\frac{\pi}{4}-\frac{19}{24}+\frac{c^2}{2}\right)s+$$

$$+\frac{1}{r^3}\left(-\frac{\pi}{64}+\left[\frac{\pi}{12}-\frac{5}{36}\right]c+\frac{5\pi}{32}c^2-\frac{35\pi}{192}c^4\right)+$$

$$\left.+\frac{1}{r^4}\left(\frac{7}{4}-\frac{9\pi}{16}+\frac{11}{24}c^2-\frac{3}{4}c^4+\frac{c^2}{24s^2}\right)s+\cdots\right\}$$

$$c=\cos\theta \quad\text{and}\quad s=\sin\theta,$$

$$C_1 = \frac{\pi}{2} + \frac{1}{r}(0) + \frac{1}{r^2}\left(\frac{7}{3} - \frac{5\pi}{8}\right) + \frac{1}{r^3}(0) + \frac{1}{r^4}\left(\frac{11\pi}{8} - \frac{8}{3}\right),$$

$$C_2 = \frac{\pi}{8} + \frac{1}{r}\left(-\frac{2\pi}{15}\right) + \frac{1}{r^2}\left(1 - \frac{5\pi}{16}\right) + \frac{1}{r^3}(0) + \frac{1}{r^4}\left(\frac{24}{135} - \frac{167\pi}{184}\right),$$

$$C_3 = \frac{\pi}{4} + \frac{1}{r}\left(\frac{4\pi}{15}\right) + \frac{1}{r^2}\left(\frac{\pi}{8}\right) + \frac{1}{r^3}(0) + \frac{1}{r^4}\left(\frac{4}{9} - \frac{23\pi}{192}\right),$$

c_1 is the coefficient for delayed IR radiation in which the equator to pole variation in IR emission is given by $1 - c_1 \sin^2\varphi$ (φ being the latitude of the emitting incremental surface)

and

δ_{eq} is the declination of the spacecraft relative to the equator of the spherical reflecting body.

The foregoing diffuse and IR total flux represents a *scalar* summation of all of the diffuse and IR photon fluxes. Transverse diffuse flux (both visible and prompt IR) is given by

$$\frac{(f_t)_{\text{diff.}}}{\sin\theta} = \frac{1}{\sin\theta}\left(\frac{dn}{dt\,dS'}\right)_T = \frac{2}{\pi r^2}\left(\frac{dn}{dS\,dt}\right)_S \varepsilon'_{\text{diff.}}$$

$$\{U(\theta_T - \theta)\, I_4 + U(\theta - \theta_T)\, U(\theta'_T - \theta)\, I_5\}, \qquad (2)$$

where

$$I_4 = \left\{(0) + \frac{1}{r}(\pi/8) + \frac{1}{r^2}(-\pi/15) + \frac{1}{r^3}\left(\frac{4}{3} - \frac{7\pi}{16}\right) + \frac{1}{r^4}\left(\frac{-2\pi}{105}\right) + \cdots\right\}$$

and

$$I_5 = \left\{(0) + \frac{1}{r}\left(\frac{\pi c}{16} + \frac{\pi}{16}\right) + \frac{1}{r^2}\left(\frac{4}{15}\frac{c}{s} - \frac{4}{15}\frac{c^3}{s} - \frac{[\pi - \theta]}{15}\right)\right.$$

$$+ \frac{1}{r^3}\left(\frac{2}{3} - \frac{7\pi}{32} - \frac{\pi}{8}c^3\right)$$

$$\left. + \frac{1}{r^4}\left(\frac{-2[\pi - \theta]}{105} + \left[\frac{9}{35} - \frac{\pi}{16}\right]\frac{c}{s} - \frac{8}{15}\frac{c^3}{s} + \frac{64}{105}\frac{c^5}{s}\right) + \cdots\right\}.$$

(For $s = 0$ the U step functions multiplying I_5 are zero so that no indeterminancy in I_5 is present.)

Radial diffuse and IR flux is given by

$$(f_r)_{\text{diff. and IR}} = \frac{2}{\pi r^2}\left(\frac{dn}{dS\,dt}\right)_S \varepsilon'_{\text{diff.}} \{U(\theta'_T - \theta)\,[I_6 - U(\theta - \theta_T)\, I_7]\} +$$

$$+ \left\{(1 - \varepsilon'_{\text{diff.}} - \varepsilon_{\text{spec.}})\left(\frac{dn}{dS\,dt}\right)_S \times\right.$$

$$\times \frac{1}{2\pi(1 - c_1/3)\, r^2}\,(C_4 - C_5 c_1 \cos^2\delta_{eq} - C_6 c_1 \sin^2\delta_{eq}). \qquad (3)$$

where

$$I_6 \triangleq c\Big[(\pi/3) + \frac{1}{r}(\pi/4) + \frac{1}{r^2}(-\pi/15) + \\ + \frac{1}{r^3}(\pi/12 - 5/18) + \frac{1}{r^4}(-\pi/105)\Big] + \cdots,$$

$$I_7 \triangleq \frac{1}{3}[\theta\, c - s] + \frac{1}{r}(\pi/16)\,(1 + 2c - 3c^2) + \\ + \frac{1}{r^2}[(107/120 - \pi/4)\, s - (\theta/15)\, c \\ - (8/15)\, s\, c^2] + \frac{1}{r^3}[c(\pi/24 - 5/36) - (\pi/8)\, c^2 + (5\pi/24)\, c^4] \\ + \frac{1}{r^4} + [(11\pi/16 - 361/168)\, s - \theta\, c/105 - (8/21)\, s\, c^2 + (32/35)\, s\, c^4 \\ - (1/24)\, c^2/s] + \cdots,$$

$$C_4 = \frac{\pi}{2} + \frac{1}{r}(0) + \frac{1}{r^2}\left(\frac{1}{3} - \frac{3\pi}{4}\right) + \frac{1}{r^3}(0) + \frac{1}{r^4}\left(\frac{23\pi}{16} - \frac{23}{6}\right),$$

$$C_5 = -\frac{3\pi}{8} + \frac{1}{r}\left(-\frac{2\pi}{15}\right) + \frac{1}{r^2}\left(1 - \frac{17\pi}{48}\right) + \frac{1}{r^3}\left(\frac{4\pi}{105}\right) + \\ + \frac{1}{r^4}\left(-\frac{1}{18} + \frac{113\pi}{192}\right),$$

$$C_6 = \frac{\pi}{4} + \frac{1}{r}\left(\frac{4\pi}{15}\right) + \frac{1}{r^2}\left(\frac{\pi}{12}\right) + \frac{1}{r^3}\left(-\frac{8\pi}{105}\right) + \frac{1}{r^4}\left(\frac{4}{9} - \frac{11\pi}{96}\right).$$

Normal delayed IR flux is given by

$$(f_n)_{\mathrm{IR}} = \frac{1}{2\pi(1 - c_1/3)\, r^3}\left(\frac{dn}{dS\, dt}\right)_S (1 - \varepsilon'_{\mathrm{diff.}} - \varepsilon_{\mathrm{spec.}}) \\ c_1(C_7 \sin\delta_{\mathrm{eq.}} \cos\delta_{\mathrm{eq.}}), \tag{4}$$

where

$$C_7 = 4\left\{\left(\frac{\pi}{15}\right)\frac{1}{r} + \left(\frac{\pi}{24}\right)\frac{1}{r^2} + \left(-\frac{2\pi}{35}\right)\frac{1}{r^3} - \\ - \left(\frac{71}{360} + \frac{9\pi}{8}\right)\frac{1}{r^4} + \cdots.\right.$$

Fig. 2 exhibits the total diffuse radiation term (no specular or delayed IR),

$$\frac{2\varepsilon'}{\pi r^2}\{U(\theta_T - \theta)\,(I_1 + I_2) + U(\theta - \theta_T)\, U(\theta'_T - \theta)\, I_3\}$$

as a function of θ for values of r from 1.1 to 2.0 and $\varepsilon' = 0.3$. Fig. 4 exhibits the transverse component of diffuse radiation (no specular),

$$\frac{2\varepsilon'}{\pi r^2}\sin^2\theta\left(\frac{dn}{dS\, dt}\right)_S\{U(\theta_T - \theta)\, I_4 + U(\theta - \theta_T)\, U(\theta'_T - \theta)\, I_5\},$$

as a function of θ for the same range of r and $\varepsilon' = 0.3$. Fig. 3 exhibits the radial component of diffuse radiation (no specular or delayed IR),

$$\frac{2\varepsilon'}{\pi r^2}\left(\frac{dn}{dS\,dt}\right)_S \{U(\theta'_T - \theta)\,[I_6 - U(\theta - \theta_T)\,I_7]\}\,,$$

as a function of θ for the same range of r and $\varepsilon' = 0.3$. Note that each of these graphical values for flux need be multiplied by the appropriate $\varepsilon'_{\text{diff.}}$

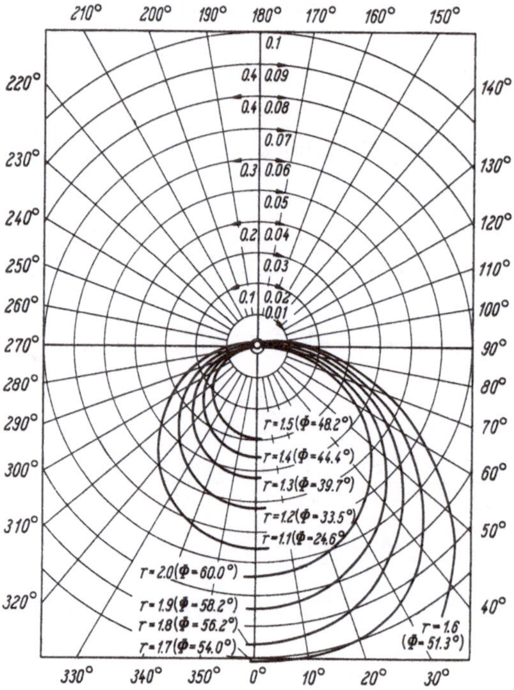

Fig. 2. Total diffuse flux term.

and divided by the assumed $\varepsilon' = 0.3$ in order to be employed in computations. As an example of the use of these graphs consider the determination of the transverse component of diffuse radiation flux for a satellite of Venus having a radial distance of 1.2 Venus radii and a Sun angle, θ, for 70°. From Fig. 3 we read the value 0.02 for $r = 1.2$ and $\theta = 70°$. Thus, since $\varepsilon'_{\text{diff.}}$ for Venus is 0.70, we find that (0.02) (0.7) $\div$ 0.3 = 0.044, i.e., the transverse diffuse flux is about 4.4% of the direct solar flux.

A multiplication factor is included in the term $(f_t)_{\text{diff.}}$ in order to resolve an indeterminacy in the coordinate transformation of subsection 2.5.1. The reflectivity factor, $\varepsilon'_{\text{diff.}}$, is given for various planets in

Section 2.4. Note that only in the calculation of the flux magnitude and the radial component, $(f_r)_{\text{diff.}}$, will the delayed infra-red albedo of the "hot" planet be introduced. Its value is most simply taken to be that of an isotropic source at the planet's center (distribution of relatively

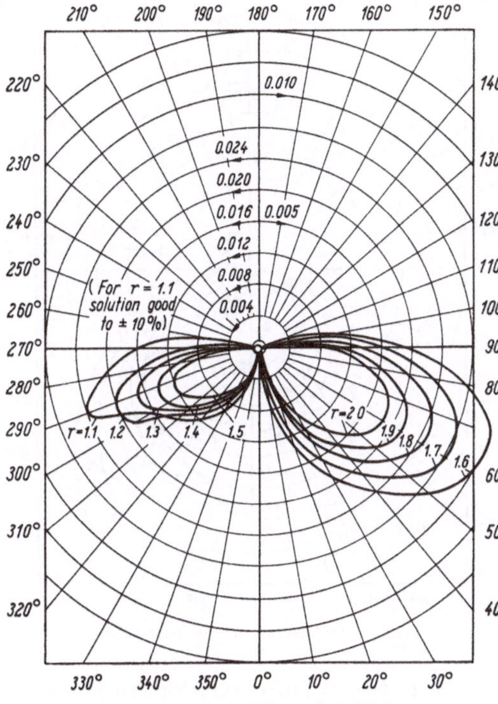

Fig. 3. Radial diffuse flux term.

"hot" areas of the Earth's surface are thus ignored) giving rise to a term of the form $(1 - \varepsilon'_{\text{diff.}} - \varepsilon_{\text{spec.}}) \frac{dn}{dt}\Big|_S \frac{dS'}{r^2}$. Instead, we adopt something similar to Wyatt's (1962) model IR-3, but consider that the actual delayed IR variation over the Earth follows the law $1 - c_1 \sin^2 \varphi$ where is φ the local reference latitude. This relationship, then, will give rise to certain constants (C_1 through C_7) found in the foregoing equations, which are similar in concept to Wyatt's "adjustable constants." A tentative value for c_1 for the Earth is given in Section 2.4.

The transverse component of diffusely reflected flux is the component in the Sun-planet-spacecraft plane that is perpendicular to the radius vector from the planet's center to the spacecraft. It arises both due to the terminator and due to the varying angle of illumination of the planetary surface by the Sun. The normal component of delayed IR flux is the component tangential to the local meridian of the spacecraft. It arises since, on the average, the planetary poles are colder and, hence, emit fewer delayed IR photons than the equatorial regions

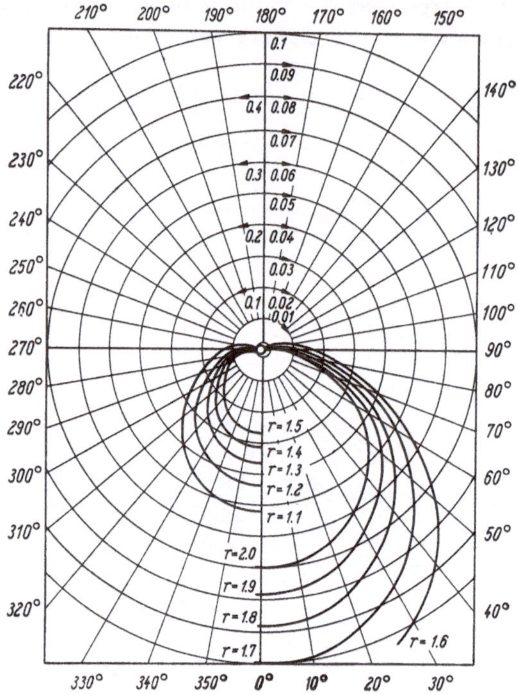

Fig. 4. Transverse diffuse flux term.

(this gives rise to the $1 - c_1 \sin^2 \varphi$ relationship). See Fig. 9, p. 113 for an illustration of these components.

2.3 Specularly-Reflected Radiation Emitted from a Spherical Body

Most analysts referenced in Section 1.1 have felt that the specularly reflected radiation from the Earth and other planets is, at most, 10% of the total reflection, with the exact percentage obviously depending upon the cloud cover of oceans and lakes. WYATT (1962) has investigated and discussed this in some detail, and it is not necessary to repeat his research. Consequently, assuming that a fraction $\varepsilon_{\text{spec.}}$ of the incident solar radiation is reflected specularly from a typical planar reflecting element on a spherical planet, then the radiation flux vector is, simply:

$$\boldsymbol{f}_{\text{spec.}} = k_3\, \varepsilon_{\text{spec.}} \left(\frac{dn}{dS\,dt}\right)_S \frac{\boldsymbol{\varrho}}{\varrho} \left(\frac{1}{1 + \varrho/\text{a. u.}}\right)^2, \tag{5}$$

where $\boldsymbol{\rho}$ is the vector from the point of specular reflection on the spherical planet to the spacecraft surface, dS', and k_3 is a constant of proportionality ($= 1/4$). Following the lead of WYATT (1962), the FRESNEL law of reflection is ignored and the Sun is assumed to be a point-source at infinity. The definition of the vector ρ is a bit involved. WYATT's Eq. (24), which involves an unknown angle "θ" (measured to the point on the Earth where the specular reflection accurs), is replaced by an iterative solution of:

$$(U_\odot \cdot L_{(i)}) - (U_\odot \cdot L)_{(i)} = g(\varrho_{(i)}) \tag{6}$$

for ρ and L, so that it is possible to form $\varrho = \rho L$. As defined in the appendices, $U_\odot$ is a unit vector directed to the Sun. A derivation and discussion of this solution as well as an alternative solution by J. ONDRASKI is given in Appendix *B*.

2.4 Reflection Properties of Earth, Moon, Venus, and Mars

There is a considerable diversity of opinion on the diffuse reflection albedo of the Earth and other planets for visible light. Table 3 summarizes the values used for the visible-light albedos (termed $\varepsilon_{\text{diff.}}$ in this report) for the Earth, Moon, Venus, and Mars. DANJON (1954, p. 737) feels that it is time variable between the limits of 0.32 and 0.52.

Table 3. *Albedo, $\varepsilon_{\text{diff.}}$, for Earth, Moon, Venus, and Mars*

Analyst	Earth	Moon	Venus	Mars
DANJON (1954)	0.32–0.52 (avg. of 0.40)	0.073	0.63	—
BALLINGER (1961)	0.40	0.07	0.76	0.15
WYATT (1962)	0.35–0.40 (chooses 0.40)	—	—	—
CUNNINGHAM (1962)	0.34	—	—	—
KRIETH (1962)	0.40	—	0.61	0.15
LEVIN (1962)	0.40	—	—	—
SHAPIRO (1963)	0.40	—	—	—
WATSON (1965) (personal communication)	—	0.085 (0.05 Maria and 0.18 bright uplands, average 0.085)	—	—
Adopted reference values (1965)	0.39	0.085	0.70	0.15

The most popular value for the $\varepsilon_{\text{diff.}}$ for the Earth 0.40. Unfortunately, this quantity is related to cloud cover etc., and is, no doubt, both time and geographically variable. The reflection properties for Venus and, especially, for the Moon seem to be much less time variable than for the Earth. Thus a more definitive radiation-pressure analysis for spacecraft in the vicinity of these bodies may be much more fruitful than it would be for the Earth. As a preliminary measure the reference values shown in Table 3 are adopted.

It should be emphasized that in the case of dynamical analyses no distinction need be drawn between visible light diffuse reflection and prompt IR emission. This is particularly true in the case of the Moon where $\varepsilon_{\text{diff.}} = 0.085$ but, according to a personal communications with KEN WATSON and B. C. MURRAY (1965), $\varepsilon'_{\text{diff.}} = 1.00$ i.e., the dark portions of the Moon essentially emit no photons visible or IR; and the illuminated portions of the Moon promptly emit most of the received solar flux in the IR.

In a sense this is fortunate for the reflective properties of the Moon for visible light do not follow the conventional LAMBERT Law. The Moon is a strong backscatterer (similar to road sign reflectors) and a very complex relationship should be used instead of $\cos\gamma$ (B. HAPKE (1963)). We shall neglect this complication in light of the small influence of lunar diffuse radiation for visible light (a better approximation to the true phenomena could be made by employing a fictitous specular component to lunar reflection).

The adjustable constant c_1 can be approximately developed by using a T^4 emission law. As an example, if one assumes for the Earth that the average polar temperature is 240 °K and the average equatorial temperature is 300 °K, then, $1/(1-c_1) = T^4$ equat.$/T^4$ polar so that $c_1 \cong 0.40$. In the case of the Mars the diffuse reflection may not be as significant as is the IR.

As was discussed by WYATT (1962), the question of specularly reflected light depends upon the calmness of ocean waters, the relative prevalence of small lakes in any given area, etc. It seems reasonable to set $\varepsilon_{\text{spec.}}$ to zero for all of the planets (noted in Table 3) except for the Earth. Prior to the analysis of the scientific satellite experiments suggested earlier, it is possible only to offer a pure guess i.e., that $\varepsilon_{\text{spec.}}$ will average out to about 0.01.

The uncertainty and time variability of both the solar constant and the Earth's reflection properties, discussed above, must be reckoned with if the rigorous analysis of reflected solar radiation that has been accomplished in the foregoing sections of this report is to be justified. In view of this uncertainty about the properties of specularly-reflected emission from a planet, it is necessary to agree that a great deal of effort would be wasted in the labor of accounting precisely for the specular photon flux. But what about the influence of the tangential component of diffuse reflected radiation? First, let it be postulated that the basic uncertainty in the solar constant can be reduced to 0.1% (i.e., assuming that it is possible either to define for past times or (hopefully) predict for future times the time-variable solar constant to within 0.1%). If $\varepsilon'_{\text{diff.}}$. can be defined to a accuracy of, say, $\pm 5\%$, then the transverse component of diffusely reflected radiation can be computed to $\pm 5\%$. Since this transverse component amounts conservatively speaking to about 10% of the total diffuse reflection vector's magnitude, then it will amount to about 2 to 3% of the maximum transverse component of the direct solar radiation[1]. Consequently, in order to estimate radiation influences to within 0.1%, it is necessary to define the transverse diffusely reflected component to about 3%. Since $\varepsilon'_{\text{diff.}}$ is uncertain to 5% it is reasonable to regard as justified an analysis of the transverse component that is accurate within 3 to 5%. Such a criterion has been utilized in selecting the point at which the series in Appendix *A* have been truncated.

Other constants, associated with the reflection properties, geometry, and attitude of the spacecraft itself, must also be assumed to be definable to the same degree of precision (i.e., to about 1 to 1/10th of a percent), but a demand for this degree of an accuracy in spacecraft characteristics is not felt to be unreasonable.

2.5 Synthesis of Radiation-Pressure Perturbations on a Spacecraft

2.5.1 Coordinate transformations from Sun-line system. If $\boldsymbol{U}_\odot$ is a unit vector directed to the Sun, and $\boldsymbol{U}$ is a unit vector from the planet's center to the spacecraft, then the perpendicular to the $X - Y$ plane of Fig. 6 is $\boldsymbol{U}_\odot \times \boldsymbol{U}$ and the unit vector in the transverse direction is

$$\frac{(\boldsymbol{U}_\odot \times \boldsymbol{U}) \times \boldsymbol{U}}{|(\boldsymbol{U}_\odot \times \boldsymbol{U}) \times \boldsymbol{U}|}.$$

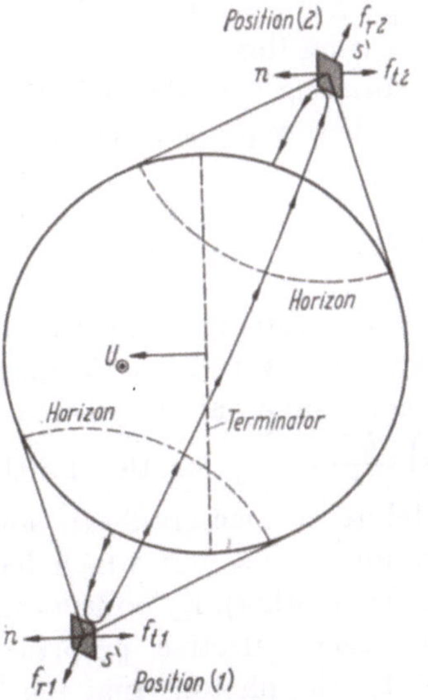

Fig. 5. Special orbit chosen for qualitative analysis.

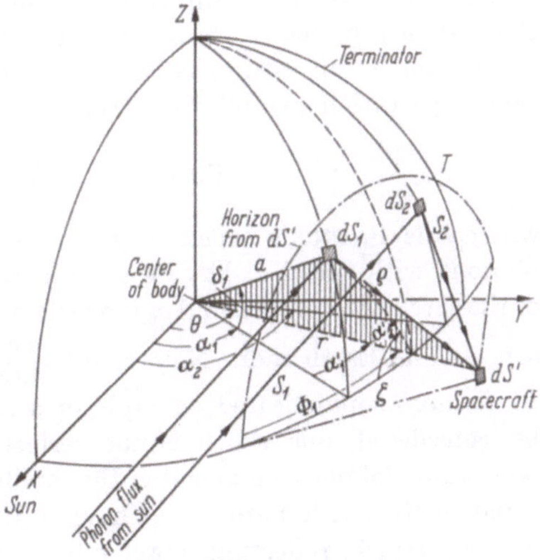

Fig. 6. Reflection geometry.

[1] These figures are confirmed by LEVIN's (1962) less generalized analysis. See Figs. 3 and 4 of his article.

By the triple vector product identity (noting that $\boldsymbol{U} \cdot \boldsymbol{U}_{\odot} = \cos\theta$),

$$(\boldsymbol{U}_{\odot} \times \boldsymbol{U}) \times \boldsymbol{U} = \boldsymbol{U} \cos\theta - \boldsymbol{U}_{\odot}.$$

The absolute value $|\boldsymbol{U}\cos\theta - \boldsymbol{U}_{\odot}|$ is $\sin\theta$. Thus, the unit vector in the transverse direction $\boldsymbol{T}$ is:

$$\boldsymbol{T} = \frac{\boldsymbol{U}\cos\theta - \boldsymbol{U}_{\odot}}{\sin\theta}. \tag{7}$$

Since θ can become zero and the unit vector can become indeterminate, it is clear why the f_t was divided by $\sin\theta$. Thus the perturbative radiation pressure perturbations become:

$$\boldsymbol{f} = \boldsymbol{f}_{\text{total}} = \Big[\boldsymbol{f}_{\text{dir.}} + \boldsymbol{f}_{\text{spec.}} + (f_n)_{\text{IR}}\, \boldsymbol{I}^{\backslash} \times \boldsymbol{U} + (f_r)_{\text{diff. and IR}}\, \boldsymbol{U} + \\ + \left\{\frac{(f_t)_{\text{diff.}}}{\sin\theta}\right\} (\boldsymbol{U}\cos\theta - \boldsymbol{U}_{\odot})\Big] \tag{8}$$

where $f_{\text{dir.}}$ is the direct component of solar radiation; the other quantities are defined previously.

($\boldsymbol{I}^{\backslash} = \boldsymbol{I}\sin\alpha_{\text{eq.}} - \boldsymbol{J}\cos\alpha_{\text{eq.}}$ with $\boldsymbol{I}$ being a unit vector to the reference equinox, $\boldsymbol{J}$ a unit vector in the reference equator perpendicular to $\boldsymbol{I}$, and $\alpha_{\text{eq.}}$ is the local right ascention of the spacecraft.)

2.5.2 Development of perturbative derivatives. Another detail of the satellite radiation effects problem that is usually overlooked or discounted is the fact that the radiation pressure acceleration is *not* necessarily in the same direction as the radiation flux vector. Only in the case of a perfectly absorptive (strictly speaking with its surface all at the same temperature) satellite does the radiation pressure perturbation acceleration exhibit the form

$$\dot{\boldsymbol{r}}^{\backslash}_{\text{rad.}} = \left(\frac{dS'}{M}\right) \boldsymbol{f}\, h\, \bar{\nu}/c\, g_0, \tag{9}$$

where h is PLANCK's constant, $\boldsymbol{f}$ is defined, as before, to be the flux of photons with speed c, having an average frequency $\bar{\nu}$, M is the mass of the satellite in grams, and g_0 is acceleration of gravity at unit distance, e.g., at one Earth radii. Note that $\left(\frac{dn}{dt\,dS}\right)_S \frac{h\bar{\nu}}{c} = P_{\odot}$ for the Earth.

A more sophisticated example of a satellite or spacecraft can now be considered, one which is not perfectly absorptive, but which has no shading (of one component of the satellite by another). By postulating a flat plate (with both sides having the same reflective proprties) having area A, reflecting power $\varepsilon\,(0 \leqq \varepsilon \leqq 1)$, the photons that reach the plane surface would give rise to the following $\boldsymbol{f}$—directed acceleration:

$$\dot{\boldsymbol{r}}^{\backslash}_{\text{rad.}} = \left(\frac{A\,h\,\bar{\nu}}{g_0 M}\right) (\boldsymbol{n} \cdot \boldsymbol{f})\, \frac{\boldsymbol{f}}{|\boldsymbol{f}|}, \tag{10}$$

where $\boldsymbol{n}$ is a unit vector normal to the plate. Because of the re-emitted photons, there will be an $\boldsymbol{n}$-directed force resulting from their diffuse reflection amounting to:

$$-\left(\frac{2}{3}\right)\left(\frac{A\,h\,\bar{\nu}}{c g_0 M}\right)(\boldsymbol{n}\cdot\boldsymbol{f})(1-\sigma)\,\varepsilon\,\boldsymbol{n}\,, \tag{11}$$

where $(1-\sigma)$ is the fraction of the photons that are diffusely reflected. The (2/3) term appears exactly as it did in the case of free molecule flow as noted by the author in (1958); in this analog the diffusely emitted molecules are replaced by diffusely emitted photons. In addition to this, there will be a force component *opposite* to the specularly reflected photon beam (i.e., the fraction σ of the photons being specularly reflected). A unit vector in the direction of this reflected beam, $\boldsymbol{B}$, is given a by solution of the three components of the equation

$$\boldsymbol{B} = \frac{\boldsymbol{f}}{|\boldsymbol{f}|} - \boldsymbol{n}\left(2\,\frac{\boldsymbol{f}}{|\boldsymbol{f}|}\cdot\boldsymbol{n}\right) \tag{12}$$

for B_x, B_y, and B_z. This equation requires that the incident beam (along $\boldsymbol{f}$), the reflected beam ($\boldsymbol{B}$) and the surface normal ($\boldsymbol{n}$) be located in the same plane, and that the angle of incidence be equal to the angle of reflection. Thus, the total radiation pressure perturbation is given by:

$$\dot{\boldsymbol{r}}'_{\substack{\text{rad.}\\ \text{total}}} = \left(\frac{A\,h\,\bar{\nu}}{c g_0 M}\right)(\boldsymbol{n}\cdot\boldsymbol{f})\left\{\frac{\boldsymbol{f}}{|\boldsymbol{f}|} - \frac{2(1-\sigma)\varepsilon}{3}\,\boldsymbol{n} - \sigma\,\varepsilon\,\boldsymbol{B}\right\}. \tag{13}$$

An alternative way to account for specular radiation pressure has been utilized by Acord and Nicklas (1963). Essentially, they recognize the fact that a portion of the incident photon radiation and the specularly reflected radiation combine to yield a force normal to the planar surface i.e., expressed in the notation used in this paper and allowing for reflected radiation, the Acord and Nicklas results are:

$$\dot{\boldsymbol{r}}'_{\substack{\text{rad.}\\ \text{total}}} = \left(\frac{A\,h\,\bar{\nu}}{c g_0 M}\right)\boldsymbol{n}\cdot\boldsymbol{f}\left\{2\left(\frac{\boldsymbol{f}}{|\boldsymbol{f}|}\right)\cdot\boldsymbol{n}\,\varepsilon\,\sigma\,\boldsymbol{n} - \frac{2}{3}\,\varepsilon(1-\sigma)\,\boldsymbol{n} + (1-\sigma\,\varepsilon)\left(\frac{\boldsymbol{f}}{|\boldsymbol{f}|}\right)\right\}. \tag{14}$$

Either result could be utilized.

The perturbative in the $\boldsymbol{F}$, $\boldsymbol{G}$, $\boldsymbol{W}$ system which remains defined for $e = 0$ and $i = 0$ [Baker (1960)] are

$$\boldsymbol{a}' = \frac{\dot{D}'\,\boldsymbol{r} - D'\dot{\boldsymbol{r}} - D\,\dot{\boldsymbol{r}}'}{\sqrt{\mu}}. \tag{15}$$

in which $\dot{\boldsymbol{r}}'$ is the *total* perturbative acceleration (i.e., occasioned by radiation pressure, drag, asphericity of the planets, other bodies, etc.):

$$D' = \boldsymbol{r}\cdot\dot{\boldsymbol{r}}'/\sqrt{\mu}\,,$$

and

$$\dot{D}' = \left(\frac{2}{\sqrt{\mu}}\right) \boldsymbol{r} \cdot \dot{\boldsymbol{r}}',$$

$$h' = \boldsymbol{r} \times \dot{\boldsymbol{r}}'/\sqrt{\mu},$$

and the mean longitude is given by an integration of:

$$L = L_0 + n_0 (t - t_0) + k \int_{t_0}^{t} \int_{t_0}^{t} n' \, dt^2 + k \int_{t_0}^{t} L' \, dt \tag{16}$$

in which

$$n' = -\tfrac{3}{2} n \, a \, \dot{D}'/\sqrt{\mu},$$

$$L' = l' - \frac{2 \, D'}{\sqrt{a}} - \frac{e^2 \, v'}{1 + \sqrt{1 - e^2}},$$

where:

$$l' = \frac{z (r \, \dot{b}')}{(1 + W_z) \sqrt{\mu \, p}} \quad (r \, \dot{b}' = \boldsymbol{W} \cdot \dot{\boldsymbol{r}}')$$

and

$$- e^2 \, v' = (W_y \, a_z - W_z \, a_y) \, a_x' + (W_z \, a_x - W_x \, a_z) \, a_y' + \\ + (W_x \, a_y - W_y \, a_x) \, a_z'.$$

3. Conclusion

3.1 Qualitative Analysis of Special Case

Qualitative conclusions regarding the magnitude of reflected Earth radiation (both radial and transverse components) at a spacecraft whose orbital plane is in the ecliptic, and which receives light from a uniformly emitting hemisphere, are given by LEVIN (1962). SEHNAL (1963) considers a simplified Earth reflection model, and assumes that the Sun is in the orbit plane and that the spacecraft is perfectly absorptive (i.e., the transverse component is not considered). He then applies the general perturbations method to define the secular changes in period and eccentricity. SHAPIRO (1962) also develops secular variations in the orbital elements (a, e, ω, Ω, and i), but assumes that *all* radiation from the Earth is specularly reflected (as noted in WYATT (1962) and in Section 2.4 of this report, such an assumption is not an especially valid one). Finally, WYATT (1962) gives a very thorough analysis of secular changes in period and eccentricity for all of his various models, but assumes a perfectly absorptive satellite and neglects the transverse (orthogonal) component of reflected radiation.

In view of these prior studies, a rather special satellite orbit has been singled out for qualitative analysis. It will be assumed to have an inclination near 80° (and, therefore, nearly synchronous with the

Sun if it is a geocentric satellite); an altitude of about 0.2 planetary radii (thus emphasizing the influence of reflected radiation), and a near-zero eccentricity. It will also be assumed that the spacecraft can be represented by a diffuse-reflecting plane, S', that is oriented with its normal to the Sun. The special orbit is not the "satellite in the ecliptic plane" model orbit usually chosen nor is the spacecraft assumed to be perfectly absorptive. Since the influences of direct solar radiation and the radial component of reflected radiation have been dealt with in the papers already cited, this discussion will concentrate on the transverse component.

For the sake of this qualitative analysis, the orbit should be established at an inclination of about 20° with respect to the plane of the terminator. Fig. 5 gives a picture of the chosen orbit. Note that, in this case, the transverse component, as a result of reflected radiation, is relatively large at position (1) and relatively small at position (2) (see Fig. 3 for $\theta = 70°$ and $\theta = 110°$ at $r = 1.2$). A semi-quantitative analysis shows that at position (1) the perturbative acceleration normal to the orbit plane at position (1) is about 3% of the direct solar perturbative acceleration whereas at position (2) it is less than 1/10% of direct solar perturbative acceleration. Obviously, a relative difference of about 3% of total solar perturbative acceleration exists between the two positions. It would be possible to establish the exact difference, thus permitting entry into the "general" perturbations equations of, say, KOSKELA (1962, pp. 78—80) or the "special perturbations" equations of Section 2.5.2 and thereby define accurately the variations in the orbital elements for a given satellite area/mass quotient. Instead of this detailed approach, it might be more informative to estimate qualitatively the variations in inclination, i, and longitude of the ascending node, Ω. From pp. 174 and 175 of BAKER (1960) the perturbative derivatives are

$$i' = (r^2\,\dot{b}')\cos u/\sqrt{\mu\,p}$$

and

$$\Omega' = (r^2\,\dot{b}')\sin u/\sqrt{\mu\,p}\,\sin i, \tag{17}$$

where, for geocentric orbits, i' and Ω' are expressed in units of radians per k_e^{-1} min. (i.e., radians per 13.6 minutes); $(r^2\,\dot{b}')$ is the perturbative acceleration normal to the orbit plane in g's, u is the argument of latitude $\triangleq v + \omega$; μ equals one Earth mass, and p is the semi-latus rectum in Earth radii. An area/mass quotient of about 100 cm²/gm (approximately that of the Echo I) and $P_{\odot} = 4.7 \times 10^{-5}$ dynes/cm² are assumed. At position (2), the additional normal component of perturbative acceleration amounts to about 3% of the direct solar; thus:

$$(r^2\,\dot{b}') = \frac{(1 + 2/3)\,(100)\,(4.7 \times 10^{-5})}{(981)}\,(0.03)\ \text{g's} = 2.7 \times 10^{-7}\ \text{g's},$$

where the 2/3 approximates the diffuse reflection of the plane S'. At position (1) $u \cong -90°$ and at position (2) $u \cong +90°$. Thus the secular variation in i is not very large.

In the case of Ω, however, there is a definite secular variation. At position (2) the instantaneous variation is about 2.5×10^{-7} radians/k_e^{-1}min. An average over the "upper" portion of this orbit gives a secular change in Ω amounting to about 5×10^{-7} radians per revolution. If the satellite orbit plane is not perturbed from its position relative to the plane of the terminator (remember it is nearly synchronous with the Sun) for about a month, then its node will be moved about 2′ of arc during this time. Clearly, this is not a large effect, but as mentioned in the introduction, it must be taken into account in any precise analysis. In fact, if observational data are accurate to 0″.1 or 0″.01 of arc or even better, then the transverse component should be computed even for Earth satellites having a mass /area quotient of, say, 10 gms/cm².

As a qualitative example of the influence of IR radiation, let us consider a satellite near Mars. The influence of radiation pressure for a Mars orbiter would be ten percent greater, relative to the Earth, than the central force term. Thus radiation pressure perturbations would play a slightly more significant role in the neighborhood of Mars than they do near the Earth. Furthermore, since Mars emits about 50% of its radiation in the delayed IR, the influence of delayed IR radiation will probably predominate. Because Mars emits more IR in its "equatorial" zones, one can imagine Mars as girdled by an IR source. The resolution of the normal, $(f_n)_{IR}$, and the radial, (f_r), will give rise to a resultant total flux that has a significant component perpendicular to the martian equator. Thus the radiation pressure perturbation will act *oppositely*, to the second martian zonal harmonic. Because the second harmonic is poorly defined at present, it is difficult to access the effect quantitatively, nevertheless it should be significant for a Mars orbiter having area to mass quotients of about 10 gm/cm².

3.2 Computer Program Discussion

The substantive results of a special perturbations analysis such as this is the preparation of a computer algorithm to be applied by other analysts to special problems. The rudiments of this program are to be found in the equations of Sections 2.2, 2.3 and 2.5, and in the flow diagram for this process is essentially presented in Fig. 1. The development of a complete computer program covering all aspects of radiation pressure is anticipated at Computer Sciences Corporation, El Segundo, California, USA. The basic special perturbations computer program makes use of HERRICK's universal variables; a Fortran II listing of it is to be found in Appendix E of PITKEN (1965),

3.3 Acknowledgements

The author would like to acknowledge the extensive assistance of his graduate class in advanced astrodynamics at the University of California, Los Angeles, U.S.A. Especially helpful were the contributions of Col. R. R. Lochry, K. W. Behnke, J. Y. Miyamoto, E. F. Stoops, B. Smith, B. D. Ward, T. Bullock, and R. G. Totten.

Appendix A

Analytical Integrations of Diffusely Reflected Photon Fluxes and IR Photon Fluxes

A.1 Total Diffuse Flux. Consider the photons diffusely reflected from "patches" dS_1 and dS_2 on the spherical body's surface, and the incremental area of the spacecraft, dS' with a location specified by vector $\boldsymbol{r}$. The coordinates of dS_1 are α_1, δ_1, with $\alpha_1 + \alpha_2' = \theta$. The coordinates of dS_2 are α_2, δ_2, with $\theta + \alpha_2' = \alpha_2$. Let θ be the angle between $\boldsymbol{r}$ and the vector between the planetocenter and the Sun. For ease of presentation dS_1 and dS_2 are defined as symmetrically located about the "meridian" of the spacecraft in the Sun—oriented $X - Y$ or $\alpha - \delta$ system shown in Fig. 6. The unit of distance is the radius of the spherical, planetary, reflecting body. Therefore,

$$\delta_1 = \delta_2 \triangleq \delta .$$

Note that on the horizon from the spacecraft (dS')

$$\cos\varphi = 1/r = \cos\varphi_2 .$$

On the surface S_1 and S_2 (the areas of the planet that are observable from the spacecraft and symmetrically located about its meridian)

$$\cos\beta_{1,2} = \cos\alpha_{1,2}' \cos\delta ,$$

where the angle β is defined in Fig. 7.

On the horizon

$$\beta_{1,2} = \varphi ;$$

therefore,

$$\cos\alpha_{1,2}' \cos\delta = 1/r \quad \text{(A-1)}$$

on the horizon.

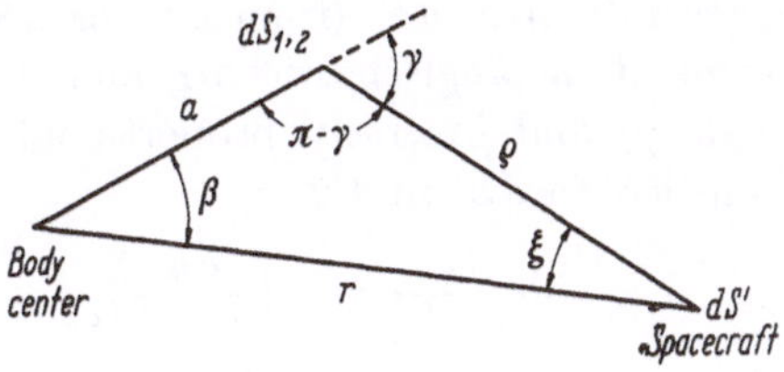

Fig. 7. Triangle $\varrho\ r\ a$.

At T (the intersection of the horizon line and the terminator) $\alpha_{1,2} = \pi/2$, if, the limit horizon line actually crosses the terminator; therefore,

$$\cos\delta_{1,2\,T} = \frac{1}{r\cos\alpha_{1,2\,T}'} = \frac{1}{r\cos(\pi/2 - \theta)} = \frac{1}{r\sin\theta} ;$$

i.e., since

$$\theta - \alpha'_{1T} = \alpha_{1T} = \pi/2 \therefore \alpha'_{1T} = \theta - \pi/2$$

and

$$\theta + \alpha'_{2T} = \alpha_{2T} = \pi/2 \therefore \alpha'_{2T} = \pi/2 - \theta .$$

However, the horizon from dS' only crosses the terminator if

$$\pi/2 - \varphi \leqq \theta \leqq \pi/2 + \varphi \therefore \theta_T \triangleq \pi/2 - \varphi ,$$

for θ_T in the first quadrant, i.e., for the terminator in S_2, or

$$\theta_T = \pi/2 - \cos^{-1}(1/r) = \sin^{-1}(1/r) . \tag{A-2}$$

Thus, if $\theta \geqq \theta_T$ the horizon from the spacecraft and the terminator will intersect (or, at least, oscullate). If θ_T is in the second quadrant, call it θ'_T, then the terminator will be in S_1 and $\theta'_T \triangleq \pi/2 + \varphi$.

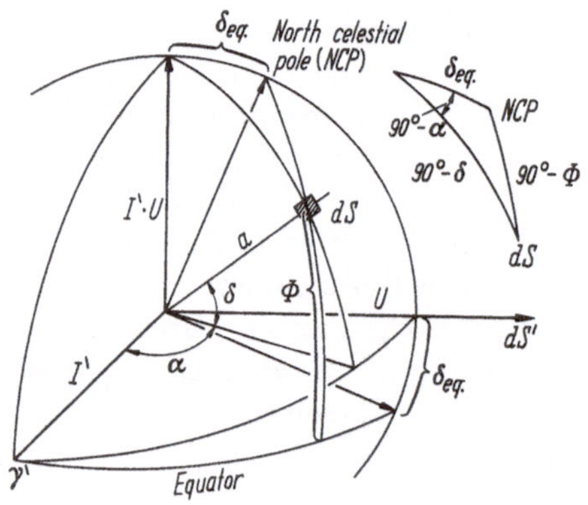

Fig. 8. Coordinate system for IR component.

The number of photons per unit time reaching dS_1 or dS_2 of the sphere; i.e., $[dn/dt]$ is proportional to

$$\left(\frac{dn}{dS\,dt}\right)_S \cos\alpha_{1,2} \cos\delta \, dS_{1,2} ,$$

where $\left(\frac{dn}{dS\,dt}\right)_S =$ the total number of photons reaching the surface of the spherical planet per unit time per unit area; i.e., the local planetary solar constant where $\cos\alpha_{1,2}\cos\delta$ is included to account for the cosine decrease of incident photons per unit area radiated from the front of the spherical body.

The number of photons per unit time that reach an element of spacecraft area, dS' (from dS_1 or dS_2) is directly proportional to the cosine of the angle between $\boldsymbol{\varrho}$ and the surface normal at $dS_{1,2}$ (i.e., the angle γ) and inversely proportional to the square of the distance ϱ from dS to dS' in Fig. 8

$$\therefore \frac{dn}{dt\,dS'} = k_1 \left(\frac{dn}{dS\,dt}\right)_S \varepsilon'_{\text{diff.}} \frac{\cos\alpha_{1,2}\cos\delta\cos\gamma\, dS_{1,2}}{\varrho^2} ,$$

where k_1 is a constant of proportionality.

From Fig. 7

$$1 = |\boldsymbol{a}| = \varrho\cos(\pi - \gamma) + r\cos\beta = -\varrho\cos\gamma + r\cos\beta ,$$

$$\therefore \cos\gamma = \frac{r}{\varrho}\left(\cos\beta - \frac{1}{r}\right) = \frac{r}{\varrho}\left\{\cos\beta - \frac{1}{r}\right\}$$

(since r is measured in units of a),

where β is the angle between the vector $\boldsymbol{r}$ from the center of the sphere to the spacecraft incremental area dS' (in units of sphere radii) and $\boldsymbol{a}$ is a vector from the sphere center to the patch, dS, on the sphere's diffusely reflecting surface.

Indeed $\cos\beta = \cos\alpha' \cos\delta$, and

$$\therefore \cos\gamma = \frac{r}{\varrho}\left\{\cos\alpha' \cos\delta - \frac{1}{r}\right\}.$$

Of course, $dS_{1,2} = a^2 \cos\delta\, d\alpha_{1,2}\, d\delta = (1)^2 \cos\delta\, d\alpha_{1,2}\, d\delta$.

Therefore, the number of photons per unit time per unit area striking the spacecraft from the patches dS_1 or dS_2 (located at $\alpha_{1,2}$, δ on the spherical body) is provided by

$$\frac{dn}{dt\,dS'} = k_1 \left(\frac{dn}{dS\,dt}\right)_S \varepsilon'_{\text{diff.}} \frac{\cos\alpha_{1,2}\cos\delta}{\varrho^2}\left(\frac{r}{\varrho}\right) \times$$
$$\times \left\{\cos\alpha'_{1,2}\cos\delta - \frac{1}{r}\right\} \cos\delta\, d\alpha_{1,2}\, d\delta, \tag{A-3}$$

where all distances are measured in terms of units of a.

Consider the magnitude of the flux. The only differences between the surfaces S_1 and S_2 is that on S_1

$$\alpha_1 + \alpha'_1 = \theta \mathrel{\therefore} \alpha_1 = \theta - \alpha'_1 \quad \text{while on} \quad S_2$$
$$\theta + \alpha'_2 = \alpha_2$$

$$\therefore \cos\alpha_1 = \cos(\theta - \alpha'_1) = \cos\theta \cos\alpha'_1 + \sin\alpha'_1$$

and

$$d\alpha_1 = -d\alpha'_1;$$

while

$$\cos\alpha_2 = \cos(\theta + \alpha'_2) = \cos\theta\cos\alpha'_2 - \sin\theta \sin\alpha'_2$$

and

$$d\alpha_2 = d\alpha'_2.$$

Therefore, the contribution of $\sin\theta \sin\alpha'$ cancels after integrating over S_1 and S_2.

Consider the factor $\frac{r}{\varrho^3}$ in Eq. (A-3),

$$\varrho^2 = r^2 + a^2 - 2\,\text{ar}\cos\beta$$

or

$$\varrho = r\left[1 + 1/r^2 - \frac{2}{r}\cos\alpha'\cos\delta\right]^{1/2},$$

$$\therefore \frac{r}{\varrho} = P_0 + \frac{1}{r}P_1 + \frac{1}{r^2}P_2 + \frac{1}{r^3}P_3 + \cdots,$$

where the LEGENDRE polynomials[1] are given by

$$P_0 = 1,$$
$$P_1 = \cos\beta = \cos\alpha' \cos\delta,$$
$$P_2 = \frac{1}{2}(3\cos^2\beta - 1),$$
$$P_3 = \frac{1}{2}(5\cos^3\beta - 3\cos\beta),$$
$$P_4 = \frac{1}{8}(35\cos^4\beta - 30\cos^2\beta + 3)$$

and

$$\therefore \frac{1}{\varrho^3} = \frac{1}{r^3}\Big\{1 + \frac{1}{r}[3\cos\alpha'\cos\delta] + \frac{1}{r^2}\frac{3}{2}[5\cos^2\alpha'\cos^2\delta - 1] + \frac{1}{r^3}\frac{5}{2}[7\cos^3\alpha'\cos^3\delta - 3\cos\alpha'\cos\delta] + \frac{1}{r^4}\frac{15}{8}[21\cos^4\alpha'\cos^4\delta - 14\cos^2\alpha'\cos^2\delta + 1] + \cdots\Big\}$$

$$\frac{r}{\varrho^3} = \frac{1}{r^2}\Big\{1 + \frac{1}{r}[3\cos\alpha'\cos\delta] + \cdots\Big\}.$$

Considering first $\theta < \pi/2$ and integrating over S_1 and S_2^*, first with δ then with α', we find that for S_1

$$\left(\frac{dn}{dt\,dS'}\right) = 2\frac{1}{r^2}k_1\left(\frac{dn}{dS\,dt}\right)_S \varepsilon'_{\text{diff.}} \int_{\delta_1=0}^{\delta_1=\cos^{-1}(1/r\cos\alpha_1')} (d\delta_2 = d\delta) \int_{\alpha_1'=\varphi_1=\cos^{-1}(1/r)}^{\alpha_1'=0\,(\text{over } S_1)} [1]\,(d\alpha_1 = -d\alpha_1')$$

and for S_2

$$\left(\frac{dn}{dt\,dS'}\right) = 2\frac{1}{r^2}k_1\left(\frac{dn}{dS\,dt}\right)_S \varepsilon'_{\text{diff.}} \int_{\delta_2=0}^{\delta_2=\cos^{-1}(1/r\cos\alpha_2')} (d\delta_2 = d\delta) \int_{\alpha_2'=0}^{\alpha_2'=\varphi_2=\cos^{-1}(1/r)\,(\text{over } S_2)} [2]\,(d\alpha_2 = d\alpha_2'),$$

where $\varphi_2 = \pi/2 - \theta$ and

$$[1] \triangleq [\cos\theta\cos\alpha_1' + \sin\theta\sin\alpha_1'][\cos^2\delta]\left[\cos\alpha_1'\cos\delta - \frac{1}{r}\right]$$
$$\Big\{\Big[1 + \frac{3}{r}\cos\alpha_1'\cos\delta\Big] + \frac{1}{r}\frac{3}{2}[5\cos^2\alpha_1'\cos^2\delta - 1] + \frac{1}{r^3}\frac{5}{2}(7\cos^3\alpha_1'\cos^3\delta - 3\cos\alpha_1'\cos\delta] + \frac{1}{r^4}\frac{15}{8}[21\cos^4\alpha_1'\cos^4\delta - 14\cos^2\alpha_1'\cos^2\delta + 1]\Big\} + \cdots.$$

$$[2] \triangleq [\cos\theta\cos\alpha_2' - \sin\theta\sin\alpha_2'][\;][\ldots].$$

---

[1] By carrying up to P_4 we achieve an accuracy of better than 5% in our subsequent calculation for $r > 1.2$.

* Note that HRYCAK (1962) carries out a similar analytical integration, but uses an approximate equation [Eq. (6), p. 1295] that he indicates is 60% in error.

Let

$$I_1 + I_2 \triangleq \int [1] + \int [2] = 2 \int [\cos\theta \cos\alpha_2'] [\cos^2\delta] .$$

$$\left[\cos\alpha_2' \cos\delta - \frac{1}{r}\right] \left[1 + \frac{3}{r} \{\cos\alpha_2' \cos\delta\} + \cdots\right] d\alpha_2' \, d\delta .$$

In addition, we must subtract the portion behind the terminator—if the terminator actually cuts the cup; i.e., when $\alpha_{2T}' = \frac{\pi}{2} - \theta$, see Eq. (A-2); therefore, we subtract

$$\underbrace{-2U(\theta - \theta_T)}_{\text{step } fc} \int_{\delta_2 = 0}^{\delta_2 = \delta_{2T} = \cos^{-1}(1/r \cos\alpha_2')} (d\delta_2 = d\delta) \int_{\alpha_2' = \pi/2 - \theta}^{\alpha_2' = \cos^{-1}(1/r) = \Phi_2} (d\alpha_2 = d\alpha') \, [2] \frac{k_1}{r^2} \left(\frac{dn}{dS\,dt}\right)_S$$

for

$$\frac{\pi}{2} - \varphi < \theta < \pi/2 .$$

Let

$$\frac{dn}{dt\,dS'} = \frac{2}{r^2} k_1 \left(\frac{dn}{dS\,dt}\right)_S \varepsilon'_{\text{diff.}} \{I_1 + I_2 - U(\theta - \theta_T) I_3'\} ,$$

where

$U(\theta - \theta_T) = 0$ for $\theta < \theta_T$ (horizon is all illuminated),

$U(\theta - \theta_T) = 1$ for $\theta \geqq \theta_T$ (terminator is in S_2).

For $\pi > \theta > \pi/2$

$$\frac{dn}{dt\,dS'} = \frac{2}{r^2} k_1 \varepsilon'_{\text{diff.}} \left(\frac{dn}{dS\,dt}\right)_S \{I_3\} \, U(\theta_T' - \theta) ,$$

where

$U(\theta_T' - \theta) = 0$ for $\theta > \theta_T'$ (horizon is all dark),

$U(\theta_T' - \theta) = 1$ for $\theta \leqq \theta_T'$ (terminator is in S_1).

Following in detail the analysis of R. R. Lochry: (verified by B. Smith), we have

$$\left(\frac{r}{\varrho}\right)^3 \left(\cos\beta - \frac{1}{r}\right) = \cos\beta + \frac{1}{r} (3\cos^2\beta - 1) + \frac{3}{2r^2} (5\cos^3\beta - 3\cos\beta) +$$
$$+ \frac{1}{2r^3} (35\cos^4\beta - 30\cos^2\beta + 3) +$$
$$+ \frac{5}{8r^4} (63\cos^5\beta - 70\cos^3\beta + 15\cos\beta) + \cdots$$

or

$$\left(\frac{r}{\varrho}\right)^3 \left(\cos\beta - \frac{1}{r}\right) = \lambda_0 + \lambda_1 \cos\beta + \lambda_2 \cos^2\beta + \lambda_3 \cos^3\beta +$$
$$+ \lambda_4 \cos^4\beta + \lambda_5 \cos^5\beta + \cdots ,$$

where

$$\lambda_0 = \frac{3}{2r^3} - \frac{1}{r}, \qquad \lambda_3 = \frac{15}{2r^2} - \frac{175}{4r^4},$$

$$\lambda_1 = \frac{75}{8r^4} - \frac{9}{2r^2} + 1, \qquad \lambda_4 = \frac{35}{2r^3},$$

$$\lambda_2 = \frac{3}{r} - \frac{15}{r^3}, \qquad \lambda_5 = \frac{315}{8r^4},$$

$$\cos\beta = \cos\alpha' \cos\delta \leqq \frac{1}{r}.$$

Substituting into the total flux integral, we find

$$f_{\text{diff. and IR total}} = \frac{k_1}{r^2} \varepsilon'_{\text{diff.}} \frac{dn}{dS\,dt} \int\int (\cos\alpha \cos^2\delta)\,(\lambda_0 + \lambda_1 \cos\alpha' \cos\delta + \lambda_2 \cos^2\alpha' \cos^2\delta + \lambda_3 \cos^3\alpha' \cos^3\delta + \lambda_4 \cos^4\alpha' \cos^4\delta + \lambda_5 \cos^5\alpha' \cos^5\delta)\, d\alpha\, d\delta,$$

$$f_{\text{diff. and IR total}} = \frac{k_1}{r^2} \varepsilon'_{\text{diff.}} \frac{dn}{dS\,dt} \int\int \cos\alpha\,(\lambda_0 \cos^2\delta + \lambda_1 \cos\alpha' \cos^3\delta + \lambda_2 \cos^2\alpha' \cos^4\delta + \lambda_3 \cos^3\alpha' \cos^5\delta + \lambda_4 \cos^4\alpha' \cos^6\delta + \lambda_5 \cos^5\alpha' \cos^7\delta)\, d\alpha\, d\delta.$$

By taking advantage of the symmetry with respect to the $r - \theta(X\,Y)$ plane, we have

$$f_{\text{diff. and IR total}} = \frac{2k_1}{r^2} \varepsilon'_{\text{diff.}} \frac{dn}{dS\,dt} \int_{\alpha'=\alpha'_1}^{\alpha'=\Phi} \cos(\theta - \alpha') \int_{\delta=0}^{\delta=\omega} \left(\sum_{i=0}^{5} \lambda_1 \cos^i\alpha' \cos^{(i+2)}\delta\right) d\alpha\, d\delta + $$

$$+ \frac{2k_1}{r^2} \varepsilon'_{\text{diff.}} \frac{dn}{dS\,dt} \int_{\alpha'=\alpha'_2}^{\alpha'=\Phi} \cos(\theta + \alpha') \int_{\delta=0}^{\delta=\omega} \left(\sum_{i=0}^{5} \lambda_1 \cos^i\alpha' \cos^{(i+2)}\delta\right) d\alpha\, d\delta.$$

For "no terminator" case the foregoing equals

$$\frac{2k_1}{r^2} \varepsilon'_{\text{diff.}} \frac{dn}{dS\,dt} [I_1 + I_2],$$

where

$$I_1 = \int_{\alpha'=0}^{\alpha'=\Phi} (\cos\theta \cos\alpha' + \sin\alpha') \int_{\delta=0}^{\delta=\omega} \left(\sum_{i=0}^{5} \lambda_1 \cos^i\alpha' \cos^{(i+2)}\delta\right) d\,\alpha' d\,\delta,$$

$$I_2 = \int_{\alpha'=0}^{\alpha'=\Phi} (\cos\theta \cos\alpha' - \sin\theta \sin\alpha') \int_{\delta=0}^{\delta=\omega} \left(\sum_{i=0}^{5} \lambda_1 \cos^i\alpha' \cos^{(i+2)}\delta\right) d\alpha'\, d\delta$$

so that

$$I_1 + I_2 = 2\cos\theta\cos\alpha' \int\limits_{\alpha'=0}^{\alpha'=\Phi} d\alpha' \int\limits_{\delta=0}^{\delta=\omega} \left(\sum_{i=0}^{5} \lambda_1 \cos^i\alpha' \cos^{i+2}\delta\right) d\delta,$$

where

$$\omega \triangleq \cos^{-1}\left(\frac{1}{r\cos\alpha'}\right).$$

For the δ-integration we can use

$$\int \cos^n dx = \frac{1}{n}\cos^{n-1}x\sin x + \frac{n-1}{n}\int \cos^{n-2} dx$$

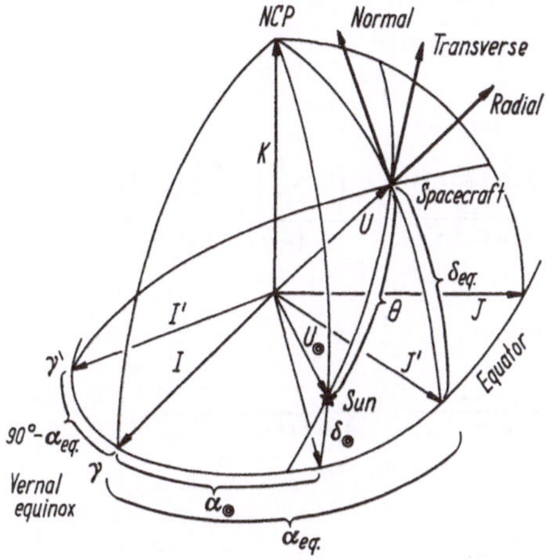

Fig. 9. Illustration of photon flux component directions.

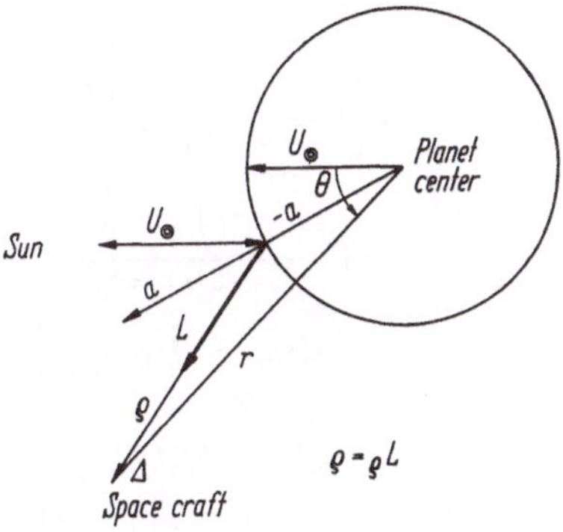

Fig. 10. Specular reflection geometry.

which yields (see Fig. 11)

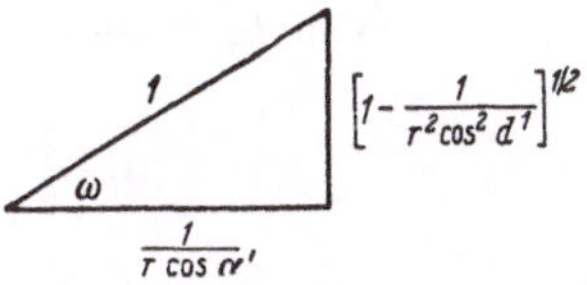

Fig. 11

$$\int \cos x\, dx = \sin x,$$

$$\int \cos^2 x\, dx = \frac{x}{2} + \frac{1}{2}\sin x(\cos x),$$

$$\int \cos^3 x\, dx = \frac{1}{3}\sin x(2 + \cos^2 x),$$

$$\int \cos^4 x\, dx = \frac{3x}{8} + \frac{1}{8}\sin x(3\cos x + 2\cos^3 x),$$

$$\int \cos^5 x\, dx = \frac{1}{15}\sin x(8 + 4\cos^2 x + 3\cos^4 x),$$

$$\int \cos^6 x\, dx = \frac{15}{48}x + \frac{\sin x}{48}(15\cos x + 10\cos^3 x + 8\cos^5 x),$$

$$\int \cos^7 x\, dx = \frac{1}{35}\sin x(16 + 8\cos^2 x + 6\cos^4 x + 5\cos^6 x),$$

$$I_1 + I_2 = 2\cos\theta\cos\alpha' \int\limits_{\alpha'=0}^{\alpha=\Phi} \lambda_0 \left[\frac{\delta}{2} + \frac{1}{2}\sin\delta\cos\delta\right] +$$

$$+ \lambda_1 \cos\alpha' \left[\frac{1}{3}\sin\delta(2+\cos^2\delta)\right] +$$

$$+ \lambda_2 \cos^2\alpha' \left[\frac{3\delta}{8} + \frac{\sin\delta}{8}(3\cos\delta + 2\cos^3\delta)\right] +$$

$$+ \lambda_3 \cos^3\alpha' \left[\frac{\sin\delta}{15}(8 + 4\cos^2\delta + 3\cos^4\delta)\right] +$$

$$+ \lambda_4 \cos^4\alpha' \left[\frac{5}{16}\delta + \frac{\sin\delta}{48}(15\cos\delta + 10\cos^3\delta + 8\cos^5\delta)\right] +$$

$$+ \lambda_5 \cos^5\alpha' \left[\frac{\sin\delta}{35}(16 + 8\cos^2\delta + 6\cos^4\delta +$$

$$+ 5\cos^6\delta)\right]\Bigg|_{\delta=0}^{\delta=\cos^{-1}\left(\frac{1}{r\cos\alpha'}\right)}.$$

Expanding:

$$\sin\omega = \left[1 - \frac{1}{r^2\cos^2\alpha'}\right]^{1/2} = 1 + \frac{\left(\frac{1}{2}\right)(-1)}{r^2\cos^2\alpha'} + \frac{\left(-\frac{1}{2}\right)\left(\frac{1}{2}\right)(-1)^2}{(2!)\, r^4\cos^4\alpha'} +$$

$$+ \frac{\left(-\frac{3}{2}\right)\left(-\frac{1}{2}\right)(-1)^3}{(3!)\, r^6\cos^6\alpha'} + \cdots,$$

$$\sin\omega = 1 - \frac{1}{2r^2c^2\alpha'} - \frac{1}{8r^4c^4\alpha'} - \frac{1}{16r^6c^6\alpha'} - \cdots,$$

$$\cos\omega = \frac{1}{r\,c'\,\alpha},$$

$$\omega = \frac{\pi}{2} - \frac{1}{r\,c\,\alpha'} - \frac{1}{6r^3c^3\alpha'}.$$

Substituting:

$$I_1 + I_2 = 2c\,\theta\,c\,\alpha' \int\limits_{\alpha'=0}^{\alpha'=\Phi} \left\{\lambda_0\left[\frac{\pi}{4} - \frac{1}{2r\,c\,\alpha'} - \frac{1}{12r^3c^3\alpha'} +\right.\right.$$

$$+ \left(\frac{1}{2} - \frac{1}{4r^2c^2\alpha'}\right)\left(\frac{1}{r\,c\,\alpha'}\right)\Bigg]$$

$$+ \lambda_1 c\,\alpha'\left[\left(\frac{1}{3} - \frac{1}{6r^2c^2\alpha'} - \frac{1}{24r^4c^4\alpha'}\right)\left(2 + \frac{1}{r^2c^2\alpha'}\right)\right]$$

$$+ \lambda_2 c^2\alpha'\left[\left(\frac{3\pi}{16} - \frac{3}{8r\,c\,\alpha'} - \frac{1}{16r^3c^3\alpha'}\right) +\right.$$

$$+ \left(\frac{1}{8} - \frac{1}{16r^2c^2\alpha'}\right) + \left(\frac{3}{r\,c\,\alpha'} + \frac{2}{r^3c^3\alpha'}\right)\Bigg] +$$

$$+ \lambda_3 c^3\alpha'\left[\left(\frac{1}{15} - \frac{1}{30r^2c^2\alpha'}\right)\left(8 + \frac{4}{r^2c^2\alpha'}\right)\right] +$$

$$+ \lambda_4 c^4\alpha'\left[\left(\frac{5\pi}{32} - \frac{5}{16r\,c\,\alpha'}\right) + \left(\frac{1}{48}\right)\left(\frac{15}{r\,c\,\alpha'}\right)\right] +$$

$$\left.+ \lambda_5 c^5\alpha'\left[\left(\frac{1}{35}\right)(16)\right]\right\} d\alpha',$$

$$I_1 + I_2 = 2\,c\,\theta\,c\,\alpha' \int\limits_{\alpha'=0}^{\alpha'=\Phi} \left\{\lambda_0\left[\frac{\pi}{4} - \frac{1}{3r^3 c^3 \alpha'}\right] + \lambda_1\left[\frac{2}{3}\,c\,\alpha' - \frac{1}{4r^4 c^3 \alpha'}\right] + \right.$$

$$\left. + \lambda_2\left[\frac{3\pi}{16}\,c^2\,\alpha'\right] + \lambda_3\left[\frac{8}{15}\,c^3\,\alpha'\right] + \lambda_4\left[\frac{5\pi}{32}\,c^4\right] + \lambda_5\left[\frac{16}{35}\,c^5\,\alpha'\right]\right\} d\alpha'.$$

First integration:

$$I_1 + I_2 = 2c\,\theta\,c\,\alpha' \int\limits_{\alpha'=0}^{\alpha'=\Phi} \left\{\frac{1}{24r^4 c^3 \alpha'} + \frac{\pi}{4}\lambda_0 + \frac{2}{3}\lambda_1\,c\,\alpha' + \frac{3\pi}{16}\lambda_2\,c^2\,\alpha' + \right.$$

$$\left. + \frac{8}{15}\lambda_3\,c^3\,\alpha' + \frac{5\pi}{32}\lambda_4\,c^4\,\alpha' + \frac{16}{35}\lambda_5\,c^5\,\alpha'\right\} d\alpha'.$$

Integrating again, we have

$$I_1 + I_2 = 2c\,\theta\left\{\left(\frac{1}{12r^4}\right)\left(\frac{s\,\alpha'}{c\,\alpha'}\right) + \frac{\pi}{4}\lambda_0[s\,\alpha'] + \frac{2}{3}\lambda_1\left[\frac{\alpha'}{2} + \frac{s\,\alpha'}{2}\,c\,\alpha'\right] + \right.$$

$$+ \frac{3\pi\,\lambda_1}{16}\left[\frac{s\,\alpha'}{3}\right][2 + c^2\,\alpha'] + \frac{8}{15}\lambda_3\left[\frac{3\alpha'}{8} + \frac{s\,\alpha'}{8}(3\,c\,\alpha' + 2c^3\,\alpha')\right] +$$

$$+ \frac{5\pi}{32}\lambda_4\left[\frac{s\,\alpha'}{15}(8 + 4c^2\,\alpha' + 3c^4\,\alpha')\right] +$$

$$\left. + \frac{16}{35}\lambda_5\left[\frac{5\alpha'}{16} + \frac{s\alpha'}{48}(15c\,\alpha' + 10c^3\,\alpha' + 8c^5\,\alpha')\right]\right\}_{\alpha'=0}^{\alpha'=\Phi},$$

where
$$\cos\Phi = \frac{1}{r},$$

$$\sin\Phi = 1 - \frac{1}{2r^2} - \frac{1}{8r^4},$$

$$\Phi = \frac{\pi}{2} - \frac{1}{r} - \frac{1}{6r^3}.$$

Subtituting, we find

$$I_1 + I_2 = 2c\,\theta\left\{\frac{1}{12r^4}[r] + \frac{\pi}{4}\lambda_0\left[1 - \frac{1}{2r^2} - \frac{1}{8r^4}\right] + \right.$$

$$+ \frac{2}{3}\lambda_1\left[\left(\frac{\pi}{4} - \frac{1}{2r} - \frac{1}{12r^3}\right) + \left(\frac{1}{2} - \frac{1}{4r^2} - \frac{1}{16r^4}\right)\left(\frac{1}{r}\right)\right] +$$

$$+ \frac{3\pi\,\lambda_2}{16}\left[\frac{1}{3} - \frac{1}{6r^2} - \frac{1}{24r^4}\right]\left[2 + \frac{1}{r^2}\right] + \frac{8\lambda_3}{15}\left[\left(\frac{3\pi}{16} - \frac{3}{8r}\right) + \right.$$

$$\left.\left. + \left(\frac{1}{8} - \frac{1}{16r^2}\right)\left(\frac{3}{r}\right)\right] + \frac{5\pi}{32}\lambda_4\left[\frac{8}{15}\right] + \frac{16}{35}\lambda_5\left[\left(\frac{5\pi}{32}\right)\right]\right\},$$

$$I_1 + I_2 = 2c\,\theta\left\{\frac{1}{12r^3} + \lambda_0\left[\frac{\pi}{4} - \frac{\pi}{8r^2}\right] + \lambda_1\left[\frac{\pi}{6} - \frac{2}{9r^3}\right] + \lambda_2\left[\frac{\pi}{8}\right] + \right.$$

$$\left. + \lambda_3\left[\frac{\pi}{10}\right] + \lambda_4\left[\frac{\pi}{12}\right] + \lambda_5\left[\frac{\pi}{14}\right]\right\}.$$

Second integration yields:

$$I_1 + I_2 = c\,\theta\left\{\frac{\pi}{3} + \frac{\pi}{4r} + (0)\frac{1}{r^2} + \left(\frac{\pi}{6} - \frac{5}{18}\right)\frac{1}{r^3} + (0)\frac{1}{r^4}\right\}.$$

When the terminator is in the forward half of the satellite horizon, it is necessary to subtract a portion of the $(I_1 + I_2)$ integral. This "corrector," I_3', is given by

$$I_3' = \int\limits_{\alpha' = \frac{\pi}{2} - \theta}^{\alpha' = \Phi} (\cos\theta \cos\alpha' - \sin\theta \sin\alpha') \int\limits_{\delta = 0}^{\delta = \omega} \left(\sum_{i=0}^{5} \lambda_i \cos^i\alpha' \cos^{i+2}\delta\right) d\delta\, d\alpha'.$$

The limits and integral of the δ-integration are identical to the previous case so that

$$I_3' = \int\limits_{\alpha' = \pi/2 - \theta}^{\alpha'} (\cos\theta \cos\alpha' - \sin\theta \sin\alpha')\left\{\frac{1}{12 r^4 c^3 \alpha'} + \frac{\pi}{4}\lambda_0 + \frac{2}{3}\lambda_1 c\,\alpha' + \right.$$
$$\left. + \frac{3\pi}{16}\lambda_2 c^2 \alpha' + \frac{8\lambda_3}{15} c^3 \alpha' + \frac{5\pi}{32}\lambda_4 c^4 \alpha' + \frac{16}{35}\lambda_5 c^5 \alpha'\right\} d\alpha'.$$

The integral is

$$I_3' = \cos\theta \int\limits_{\alpha' = \pi/2 - \theta}^{\alpha' = \Phi} \left\{\frac{1}{12 r^4 c^2 \alpha'} + \frac{\pi}{4}\lambda_0 c\,\alpha' + \frac{2}{3}\lambda_1 c^2 \alpha' + \frac{3\pi}{16}\lambda_2 c^3 \alpha' + \right.$$
$$\left. + \frac{8\lambda_3}{15} c^4 \alpha' + \frac{5\pi}{32}\lambda_4 c^5 \alpha' + \frac{16}{35}\lambda_5 c^6 \alpha'\right\} d\alpha' -$$
$$- \sin\theta \int\limits_{\alpha' = \frac{\pi}{2} - \theta}^{\alpha' = \Phi} \left\{\frac{t\,\alpha'}{12 r^4 c^2 \alpha'} + \left(\frac{\pi}{4}\lambda_0 + \frac{2}{3}\lambda_1 c\,\alpha' + \frac{3\pi}{16}\lambda_2 c^2 \alpha' + \frac{8\lambda_3}{15} c^3 \alpha' + \right.\right.$$
$$\left.\left. + \frac{5\pi}{32}\lambda_4 c^4 \alpha' + \frac{16}{35}\lambda_5 c^5 \alpha'\right)\sin\alpha'\right\} d\alpha',$$

$$I_3' = \cos\theta\left\{\frac{1}{12 r^4}\frac{s\,\alpha'}{c\,\alpha'} + \frac{\pi}{4}\lambda_0 s\,\alpha' + \frac{1}{3}\lambda_1(\alpha' + s\,\alpha' c'\alpha') + \right.$$
$$+ \frac{\pi}{16}\lambda_2 (s\,\alpha')\,(2 + c^2 \alpha') + \frac{\lambda_3}{15}(3\alpha' + s\,\alpha'[3c\,\alpha' + 2c^3\alpha']) +$$
$$+ \frac{\pi}{96}\lambda_4 (8 + 4c^2\alpha' + 3c^4\alpha')\,s\,\alpha' + \frac{1}{105}\lambda_5 (15\alpha' + s\,\alpha'[15c\,\alpha' +$$
$$\left. + 10c^3\alpha' + 8c^5\alpha'])\right\}_{\alpha' = \pi/2 - \theta}^{\alpha' = \Phi} -$$
$$- \sin\theta\left\{\frac{t^2\alpha'}{24 r^4} - \left(\frac{\pi}{4}\lambda_0 c\,\alpha' + \frac{\lambda_1}{3} c^2 \alpha' + \frac{\pi}{16}\lambda_2 c^3 \alpha' + \frac{2\lambda_3}{15} c^4 \alpha' + \right.\right.$$
$$\left.\left. + \frac{\pi}{32}\lambda_4 c^5 \alpha' + \frac{8}{105}\lambda_5 c^6 \alpha'\right)\right\}_{\alpha' = \pi/2 - \theta}^{\alpha' = \Phi}.$$

Substituting limits gives us

$$I'_3 = c\,\theta \left\{ \frac{1}{12r^4}\left[r - \frac{c\,\theta}{s\,\theta}\right] + \frac{\pi}{4}\lambda_0\left[1 - \frac{1}{2r^2} - c\,\theta\right] + \right.$$

$$+ \frac{1}{3}\lambda_1\left(\left[\frac{\pi}{2} - \frac{1}{r} - \frac{1}{6r^3} - \frac{\pi}{2} + \theta\right] + \right.$$

$$\left. + \left[\frac{1}{r} - \frac{1}{2r^3} - c\,\theta\,s\,\theta\right]\right) + \frac{\pi}{16}\lambda_2\left(\left[1 - \frac{1}{2r^2}\right]\left[2 + \frac{1}{r^2}\right] - c\,\theta\,[2 + s^2\,\theta]\right) +$$

$$+ \frac{\lambda^3}{15}\left(\left[\frac{3\pi}{2} - \frac{3}{r} - \frac{3\pi}{2} + 3\theta\right] + \left[1 - \frac{1}{2r^2}\right]\left[\frac{3}{r}\right] - c\,\theta\,[3s\,\theta + 2s^3\,\theta]\right) +$$

$$+ \frac{\pi}{96}\lambda_4([1]\,[8] - c\,\theta[8 + 4s^2\,\theta + 3s^4\,\theta]) +$$

$$\left. + \frac{\lambda^5}{105}\left(\left[\frac{15\pi}{2} - \frac{15\pi}{2} + 15\theta\right] - c\,\theta[15s\,\theta + 10s^3\,\theta + 8s^5\,\theta]\right)\right\} +$$

$$+ s\,\theta\left\{-\frac{1}{24r^4}\left([r^2 - 1] - \frac{c^2\,\theta}{s^2\,\theta}\right) + \frac{\pi}{4}\lambda_0\left(\frac{1}{r} - s\,\theta\right) + \frac{\lambda_1}{3}\left(\frac{1}{r^2} - s^2\,\theta\right) + \right.$$

$$\left. + \frac{\pi}{16}\lambda_2\left(\frac{1}{r^3} - s^3\,\theta\right) - \frac{2\lambda_3}{15}s^4\,\theta - \frac{\pi}{32}\lambda_4\,s^5\,\theta - \frac{8\lambda_5}{105}s^6\,\theta\right\}.$$

Eliminating high order $\sin\theta$ terms we have:

$$I_3^{\backslash} = \frac{1}{12r^4}\left(r\,c\,\theta - \frac{c^2\,\theta}{s\,\theta}\right) + \frac{\pi}{4}\lambda_0\left[c\,\theta - \frac{c\,\theta}{2r^2} - c^2\,\theta\right] +$$

$$+ \frac{1}{3}\lambda_1\left(-\frac{2c\,\theta}{3r^3} + \theta\,c\,\theta - s\,\theta\,c^2\,\theta\right) + \frac{\pi}{16}\lambda_2(2c\,\theta - 3c^2\,\theta + c^4\,\theta) +$$

$$+ \frac{\lambda_3}{15}\left(3\theta\,c\,\theta - \frac{3c\,\theta}{2r^3} - 5s\,\theta\,c^2\,\theta + 2s\,\theta\,c^4\,\theta\right) +$$

$$+ \frac{\pi}{96}\lambda_4(8c\,\theta - 15c^2\,\theta + 10c^4\,\theta - 3c^6\,\theta) +$$

$$+ \frac{\lambda_5}{105}(15\theta\,c\,\theta - 33s\,\theta\,c^2\,\theta + 26s\,\theta^4\,\theta - 8s\,\theta\,c^6\,\theta) -$$

$$- \frac{1}{24r^4}\left(r^2\,s\,\theta - s\,\theta - \frac{c^2\,\theta}{s\,\theta}\right) + \frac{\pi}{4}\lambda_0\left(\frac{s\,\theta}{r} - 1 + c^2\,\theta\right) +$$

$$+ \frac{\lambda^1}{3}\left(\frac{s\,\theta}{r^2} - s\,\theta + s\,\theta\,c^2\,\theta\right) - \frac{2\lambda_3}{15}(s\,\theta - 2s\,\theta\,c^2 + s\,\theta\,c^4\,\theta) +$$

$$+ \frac{\pi}{16}\lambda_2\left(\frac{s\,\theta}{r^3} - 1 + 2c^2\,\theta - c^4\,\theta\right) - \frac{\pi}{32}\lambda_4(1 - 3c^2\,\theta + 3c^4\,\theta - c^6\,\theta) -$$

$$- \frac{8}{105}\lambda_5(s\,\theta - 3s\,\theta\,c^2\,\theta + 3s\,\theta\,c^4\,\theta - s\,\theta\,c^6\,\theta).$$

Collecting terms, we find

$$I_3' = \frac{1}{24}\left(-\frac{s\,\theta}{r^2} + \frac{2c\,\theta}{r^3} + \frac{1}{r^4}\left[s\,\theta - \frac{c^2\,\theta}{s\,\theta}\right]\right) +$$

$$+ \frac{\pi}{4}\lambda_0\left([c\,\theta - 1] + \frac{s\,\theta}{r} - \frac{c\,\theta}{2r^2}\right) +$$

$$+ \frac{\lambda_1}{3}\left([\theta\,c\,\theta - s\,\theta] + \frac{s\,\theta}{r^2} - \frac{2}{3}\frac{c\,\theta}{r^3}\right) + \frac{\pi\,\lambda_2}{16}\left([2c\,\theta - 1 - c^2\,\theta] + \frac{s\,\theta}{r^3}\right) +$$

$$+ \frac{\lambda_3}{15}([3\theta\,c\,\theta - s\,\theta\,c^2\,\theta - 2s\,\theta]) + \frac{\pi\,\lambda_4}{96}(-3 + 8c\,\theta - 6c^2\,\theta + c^4\,\theta) +$$

$$+ \frac{\lambda_5}{105}(15\theta\,c\,\theta - 9s\,\theta\,c^2\,\theta + 2s\,\theta\,c^4\,\theta - 8s\,\theta).$$

Substituting for λ_i gives us

$$I_3' = \frac{1}{24}\left(-\frac{s\,\theta}{r^2} + \frac{2c\,\theta}{r^3} + \frac{1}{r^4}\left[s\,\theta - \frac{c^2\,\theta}{s\,\theta}\right]\right) +$$

$$+ \left(\frac{3\pi}{8r^3} - \frac{\pi}{4r}\right)\left(c\,\theta - 1 + \frac{s\,\theta}{r} - \frac{c\,\theta}{2r^2}\right) +$$

$$+ \left(\frac{25}{8r^4} - \frac{3}{2r^2} + \frac{1}{3}\right)\left(\theta\,c\,\theta - s\,\theta + \frac{s\,\theta}{r^2} - \frac{2}{3}\frac{c\,\theta}{r^3}\right) +$$

$$+ \left(\frac{3\pi}{16r} - \frac{15\pi}{16r^3}\right)\left(2c\,\theta - 1 - c^2\,\theta + \frac{s\,\theta}{r^3}\right) +$$

$$+ \left(\frac{1}{2r^2} - \frac{35}{12r^4}\right)(3\theta\,c\,\theta - s\,\theta\,c^2\,\theta - 2s\,\theta) +$$

$$+ \frac{35\pi}{192r^3}(-3 + 8c\,\theta - 6c^2\,\theta + c^4\,\theta) +$$

$$+ \frac{3}{8r^4}(15\theta\,c\,\theta - 9s\,\theta\,c^2\,\theta + 2s\,\theta\,c^4\,\theta - 8s\,\theta).$$

Collecting terms yields

$$I_3' = \left(\frac{\theta\,c\,\theta}{3} - \frac{s\,\theta}{3}\right) + \frac{1}{r}\left(\frac{\pi}{16} + \frac{\pi\,c\,\theta}{8} - \frac{3\pi\,c^2\,\theta}{16}\right) +$$

$$+ \frac{1}{r^2}s\,\theta\left(-\frac{\pi}{4} + \frac{19}{24} - \frac{c^2\,\theta}{2}\right) +$$

$$+ \frac{1}{r^3}\left(\frac{\pi}{64} + \left[\frac{\pi}{12} - \frac{5}{36}\right]c\theta - \frac{5\pi}{32}c^2\,\theta + \frac{35\pi}{192}c^4\,\theta\right) +$$

$$+ \frac{1}{r^4}s\,\theta\left(-\frac{7}{4} + \frac{9\pi}{16} - \frac{11}{24}c^2\theta + \frac{3}{4}c^4\theta - \frac{c^2\,\theta}{24s^2\,\theta}\right).$$

When the terminator is in the rearward half of the satellite horizon, it is necessary to compute a new double integral such that

$$I_3 = \int_{\alpha' = \theta - \frac{\pi}{2}}^{\alpha' = \Phi} (\cos\theta\cos\alpha' + \sin\theta\sin\alpha') \int_{\delta = 0}^{\delta = \omega} \left(\sum_{i=0}^{5} \cos^i\alpha'\cos^{i+2}\delta\right) d\delta\, d\alpha'.$$

The limits and form of the δ-integration remain unchanged so that we can use the previously accomplished result:

First Integration:

$$I_3 = c\theta \int\limits_{\alpha'=\theta-\frac{\pi}{2}}^{\alpha'=\Phi} (\cos\theta \cos\alpha' + \sin\theta \sin\alpha') \left\{ \frac{1}{24 r^4 c^3 \alpha'} + \frac{\pi}{4} \lambda_0 + \frac{2}{3} \lambda_1 c\,\alpha' + \right.$$

$$\left. + \frac{3\pi}{16} \lambda_2 c^2 \alpha' + \frac{8}{15} \lambda_3 c^3 \alpha' + \frac{5\pi}{32} \lambda_4 c^4 \alpha' + \frac{16}{35} \lambda_5 c^5 \alpha' \right\} d\alpha'.$$

Expanding and collecting $(c\,\alpha')$ terms

$$I_3 = c\theta \int\limits_{\alpha'=\theta-\frac{\pi}{2}}^{\alpha'=\Phi} \left\{ \frac{1}{12 r^4 c^2 \alpha'} + \frac{\pi}{4} \lambda_0 c\,\alpha' + \frac{2}{3} \lambda_1 c^2 \alpha' + \frac{3\pi}{16} \lambda_2 c^3 \alpha' + \right.$$

$$\left. + \frac{8}{15} \lambda_3 c^4 \alpha' + \frac{5\pi}{32} \lambda_4 c^5 \alpha' + \frac{16}{35} \lambda_5 c^6 \alpha' \right\} d\alpha' +$$

$$+ s\theta \int\limits_{\alpha'=\theta-\frac{\pi}{2}}^{\alpha'=\Phi} \left(\left\{ \frac{t\,\alpha'}{12 r^4 c^2 \alpha'} \right\} - \left\{ \frac{\pi}{4} \lambda_0 + \frac{2}{3} \lambda_1 c\,\alpha' + \frac{3\pi}{16} \lambda_2 c^2 \alpha' + \right. \right.$$

$$\left. + \frac{8}{15} \lambda_3 c^3 \alpha' + \frac{5\pi}{32} \lambda_4 c^4 \alpha' + \frac{16}{35} \lambda_5 c^5 \alpha' \right\} s\,\alpha'\, d\alpha'.$$

Integrating, we have

$$I_3 = c\,\theta \left\{ \frac{t\,\alpha'}{12 r^4} + \frac{\pi}{4} \lambda_0\, s\,\alpha' + \frac{1}{3} \lambda_1 (\alpha' + s\,\alpha'\, c\,\alpha') + \right.$$

$$+ \frac{\pi}{16} \lambda_2\, s\,\alpha' (2 + c^2 \alpha') +$$

$$+ \frac{1}{15} \lambda_3 (3\alpha' + s\,\alpha' [3 c\,\alpha' + 2 c^3 \alpha']) +$$

$$+ \frac{\pi}{96} \lambda_4\, s\,\alpha' (8 + 4 c^2 \alpha' + 3 c^4 \alpha') +$$

$$\left. + \frac{1}{105} \lambda_5 (15\alpha' + s\,\alpha' [15 c\,\alpha' + 10 c^3 \alpha' + 8 c^5 \alpha']) \right\}_{\alpha'=\theta-\frac{\pi}{2}}^{\alpha'=\Phi}$$

$$+ s\,\theta \left\{ \frac{t^2 \alpha'}{24 r^4} - \left[\frac{\pi}{4} \lambda_0\, c\,\alpha' + \frac{1}{3} \lambda_1 c^2 \alpha' + \frac{\pi}{16} \lambda_2 c^3 \alpha' + \right. \right.$$

$$+ \frac{2}{15} \lambda_3 c^4 \alpha' +$$

$$\left. + \frac{3\pi}{96} \lambda_4 c^5 \alpha' + \frac{8}{105} \lambda_5 c^6 \alpha' \right\}_{\alpha'=\theta-\frac{\pi}{2}}^{\alpha'=\Phi}.$$

Substituting limits

$$I_3 = c\,\theta \left\{ \frac{1}{12 r^4} \left(r + \frac{c\,\theta}{s\,\theta} \right) + \frac{\pi}{4} \lambda_0 \left(1 - \frac{1}{2 r^2} + c\,\theta \right) + \right.$$

$$+ \frac{1}{3} \lambda_1 \left(\left[\frac{\pi}{2} - \frac{1}{r} - \frac{1}{6 r^3} - \theta + \frac{\pi}{2} + \frac{1}{r} - \frac{1}{2 r^3} + c\,\theta\, s\,\theta \right] \right) +$$

$$+ \frac{\pi}{16} \lambda_2 \left[\left(1 - \frac{1}{2 r^2} \right) \left(2 + \frac{1}{r^2} \right) + c\,\theta (2 + s^2\,\theta) \right] +$$

$$+ \frac{1}{15} \lambda_3 \left[\left(\frac{3\pi}{2} - \frac{3}{r} - 3\theta + \frac{3\pi}{2} \right) + \frac{3}{r} (1) + c\,\theta (3 s\,\theta + 2 s^3\,\theta) \right] +$$

$$+ \frac{\pi}{96} \lambda_4 [(1)\,(8) + c\,\theta (8 + 4 s^2\,\theta + 3 s^4\,\theta)] +$$

$$+ \frac{1}{105} \lambda_5 \left[\left(15 \frac{\pi}{2} - 15\theta + \frac{15\pi}{2} \right) + c\,\theta (15 s\,\theta + 10 s^3\,\theta + 8 s^5\,\theta) \right] \Bigg\} +$$

$$+ s\,\theta \left\{ \frac{r^2 - 1 - c^2\,\theta / s^2\,\theta}{24 r^4} + \frac{\pi}{4} \lambda_0 \left[s\,\theta - \frac{1}{r} \right] + \frac{1}{3} \lambda_1 \left[s^2\,\theta - \frac{1}{r^2} \right] + \right.$$

$$\left. + \frac{\pi}{16} \lambda_2 \left[s^3\,\theta - \frac{1}{r^3} \right] + \frac{2}{15} \lambda_3\, s^4\,\theta + \frac{3\pi}{96} \lambda_4\, s^5\,\theta + \frac{8}{105} \lambda_5\, s^6\,\theta \right\}.$$

Collecting $\left(\frac{1}{r}\right)$ terms and replacing multi-order $s\,\theta$ with $c\,\theta$

$$I_3 = \left(\frac{c\,\theta}{12 r^3} + \frac{c^2\,\theta / s\,\theta}{12 r^4} \right) + \frac{\pi}{4} \lambda_0 \left(c\,\theta - \frac{c\,\theta}{2 r^2} + c^2\,\theta \right) +$$

$$+ \frac{1}{3} \lambda_1 \left(\pi\, c\,\theta - \frac{2 c\,\theta}{3 r^3} - \theta\, c\,\theta + s\,\theta\, c^2\,\theta \right) + \frac{\pi\,\lambda_2}{16} (2 c\,\theta + 3 c^2\,\theta - c^4\,\theta) +$$

$$+ \frac{1}{15} \lambda_3 (3\pi\, c\,\theta - 3\theta\, c\,\theta + 5 s\,\theta\, c^2\,\theta - 2 s\,\theta\, c^4\,\theta) +$$

$$+ \frac{\pi}{96} \lambda_4 (8 c\,\theta + 15 c^2\,\theta + 10 c^4\,\theta + 3 c^6\,\theta) +$$

$$+ \frac{1}{105} \lambda_5 (15\pi\, c\,\theta - 15\theta\, c\,\theta + 33 s\,\theta\, c^2\,\theta - 26 s\,\theta\, c^4\,\theta + 8 s\,\theta\, c^6\,\theta) +$$

$$+ \left(\frac{s\,\theta}{24 r^2} - \frac{s\,\theta}{24 r^2} - \frac{c^2\,\theta / s\,\theta}{24 r^2} \right) + \frac{\pi}{4} \lambda_0 \left[1 - c^2\,\theta - \frac{s\,\theta}{r} \right] +$$

$$+ \frac{1}{3} \lambda_1 \left[s\,\theta - s\,\theta\, c^2\,\theta - \frac{s\,\theta}{r^2} \right] + \frac{\pi}{16} \lambda_2 \left[1 - 2 c^2\,\theta + c^4\,\theta - \frac{s\,\theta}{r^3} + \right.$$

$$+ \frac{1}{15} \lambda_3 [2 s\,\theta - 4 s\,\theta\, c^2\,\theta + 2 s\,\theta\, c^4\,\theta] +$$

$$+ \frac{\pi}{96} \lambda_4 (3 - 9 c^2\,\theta + 9 c^4\,\theta - 3 c^6\,\theta) +$$

$$+ \frac{1}{105} \lambda_5 (8 s\,\theta - 24 s\,\theta\, c^2\,\theta + 24 s\,\theta\, c^4\,\theta - 8 s\,\theta\, c^6\,\theta),$$

$$I_3 = \left(\frac{s\,\theta}{24r^2} + \frac{c\,\theta}{12r^3} - \frac{s\,\theta}{24r^4} + \frac{c^2\,\theta/s\,\theta}{24r^4}\right) + \frac{\pi}{4}\lambda_0\left(1 + c\,\theta - \frac{s\,\theta}{r} - \frac{c\,\theta}{2r^2}\right) +$$
$$+ \frac{\lambda_1}{3}\left([\pi - \theta]\,c\,\theta + s\,\theta - \frac{s\,\theta}{r^2} - \frac{2c\,\theta}{3r^3}\right) +$$
$$+ \frac{\pi\,\lambda_2}{16}\left(1 + 2c\,\theta + c^2\,\theta - \frac{s\,\theta}{r^3}\right) +$$
$$+ \frac{\lambda_5}{105}([15\pi - 15\theta]\,c\,\theta + 8s\,\theta + 9s\,\theta\,c^2\,\theta - 2s\,\theta\,c^4\,\theta).$$

Substituting for the λ_i

$$I_3 = \left(\frac{s\,\theta}{24r^2} + \frac{c\,\theta}{12r^3} + \frac{s\,\theta}{24r^4} + \frac{c^2\,\theta/s\,\theta}{24r^4}\right) +$$
$$+ \left(\frac{3\pi}{8r^3} - \frac{\pi}{4r}\right)\left([1 + c] - \frac{s\,\theta}{r} - \frac{c\,\theta}{2r^2}\right) +$$
$$+ \left(\frac{25}{8r^4} - \frac{3}{2r^2} + \frac{1}{3}\right)\left([\pi\,c\,\theta - \theta\,c\,\theta + s\,\theta] - \frac{s\,\theta}{r^2} - \frac{2c\,\theta}{3r^3}\right) +$$
$$+ \left(\frac{3\pi}{16r} - \frac{15\pi}{16r^3}\right)\left([1 + 2c\,\theta + c^2\,\theta] - \frac{s\,\theta}{r^3}\right) +$$
$$+ \left(\frac{1}{2r^2} - \frac{35}{12r^4}\right)(3\pi\,c\,\theta - 3\theta\,c\,\theta + 2s\,\theta + s\,\theta\,c^2\,\theta) +$$
$$+ \frac{35\pi}{192r^3}(3 + 8c\,\theta + 6c^2\,\theta - c^4\,\theta) +$$
$$+ \frac{3}{8r^4}(15\pi\,c\,\theta - 15\theta\,c\,\theta + 8s\,\theta + 9s\,\theta\,c^2\,\theta - 2s\,\theta\,c^4\,\theta).$$

Collecting we finally find

$$I_3 = \left(\frac{1}{3}[\pi - \theta]\,c\,\theta + \frac{s\,\theta}{3}\right) + \frac{1}{r}\left(-\frac{\pi}{16} + \frac{\pi\,c\,\theta}{8} + \frac{3\pi\,c^2\,\theta}{16}\right) +$$
$$+ \frac{1}{r^2}\,s\,\theta\left(\frac{\pi}{4} - \frac{19}{24} + \frac{c^2\,\theta}{2}\right) +$$
$$+ \frac{1}{r^3}\left(-\frac{\pi}{64} + \left[\frac{\pi}{12} - \frac{5}{36}\right]c\,\theta + \frac{5\pi}{32}\,c^2\,\theta - \frac{35\pi}{192}\,c^4\,\theta\right) +$$
$$+ \frac{1}{r^4}\left(\frac{7}{4} - \frac{9\pi}{16} + \frac{11}{24}\,c^2\,\theta - \frac{3}{4}\,c^4\,\theta + \frac{c^2\,\theta}{24s^2\,\theta}\right)s\,\theta.$$

In summary:

$$I_1 + I_2 = c\left[(\pi/3) + (\pi/4)\frac{1}{r} + (0)\frac{1}{r^2} + \left(\frac{\pi}{6} - \frac{5}{18}\right)\frac{1}{r^3} + (0)\frac{1}{r^4} + \cdot\cdot\right],$$

$$I_3' = \frac{1}{3}[\theta\,c - s] + \frac{1}{r}[\pi/16]\,[1 + 2c - 3c^2] +$$
$$+ \frac{1}{r^2}[s/4]\,[-\pi + 19/6 - 2c^2] +$$
$$+ \frac{1}{r^3}[\pi/64 + (c/36)\,(3\pi - 5) + (5\pi/192)\,(-6c^2 + 7c^4)] +$$
$$+ \frac{1}{r^4}[s/4][-7 + 9\pi/4 - (11/6)\,c^2 + 3c^4 - (1/6)\,c^2/s^2] + \cdots,$$

$$I_3 = \frac{1}{3}[(\pi - \theta)c + s] + \frac{1}{r}[\pi/16][-1 + 2c + 3c^2] +$$
$$+ \frac{1}{r^2}[s/4][\pi - 19/6 + 2c^2] +$$
$$+ \frac{1}{r^3}[-\pi/64 + (c/36)(3\pi - 5) + (5\pi/192)(6c^2 - 7c^4)] +$$
$$+ \frac{1}{r^4}[s/4][7 - 9\pi/4 + (11/6)c^2 - 3c^4 + (1/6)c^2/s^2] + \cdots$$

where $c \triangleq \cos\theta$ and $s \triangleq \sin\theta$.

But $I_3 \equiv I_1 + I_2 - I_3'$ so that I_3' can be eliminated from the equation at the top of this page. In terms of step functions the complete relationship for diffuse reflection then becomes:

$$f_{\substack{\text{diff.}\\ \text{total}}} = \frac{2k_1}{r^2}\left(\frac{dn}{dS\,dt}\right)_S \varepsilon'_{\text{diff.}} \{U(\theta_T - \theta)(I_1 + I_2) +$$
$$+ U(\theta - \theta_T)\,U(\theta_T' - \theta)\,I_3\}.$$

For very large r, the $1/r$ terms are negligible, $U(\theta_T - \theta) = 0$, and $U(\theta - \theta_T) = 1$

$$\therefore \left(\frac{dn}{dt\,dS'}\right)_\infty = k_1\left(\frac{dn}{dS\,dt}\right)_S \varepsilon'_{\text{diff.}} \frac{1}{r^2}\left\{\frac{2}{3}(\pi - \theta)\cos\theta + \frac{2}{3}\sin\theta\right\},$$

which agrees with LAMBERT's diffuse phase law provided in VAN ALLEN (1958) (i.e., the phase law for the visibility of a planet or satellite at a great distance).

k_1 has been introduced as a proportionality constant. It can be evaluated by noting that

$$\int_S \left(\frac{dn}{dS\,dt}\right)_S \varepsilon' \cos\alpha\cos\delta\,dS = \int \left(\frac{dn}{dt\,dS'}\right)_\infty dS';$$

i.e., the conservation of photons, where the integral is taken over a sphere at a great distance from the planet ($\varepsilon'_{\text{diff.}}$ cancels out).

Therefore, using $dS = \cos\delta\,d\alpha\,d\delta$,

$$\pi = k_1 \frac{2}{3}\frac{1}{r^2}\int_0^\pi \{(\pi - \theta)\cos\theta + \sin\theta\}\frac{2\pi r^2 \sin\theta\,d\theta}{dS'}$$

[the area element of the spherical surface S' is $(2\pi r\sin\theta)(r\,d\theta)$]

$$\therefore \frac{\pi}{k_1} = \frac{4\pi}{3}\int_0^\pi (\pi\sin\theta\cos\theta + \sin^2\theta - \theta\cos\theta\sin\theta)\,d\theta$$

$$\frac{1}{k_1} = \frac{4}{3}\left\{\left[\frac{\pi\sin^2\theta}{2}\right]_0^\pi + \left[\frac{\theta}{2} - \frac{\sin 2\theta}{4}\right]_0^\pi\right.$$
$$\left. - \frac{\theta}{4}\Big|_0^\pi + \left[+\frac{\theta}{2}\cos^2\theta\right]_0^\pi + \left[\frac{\sin 2\theta}{8}\right]_0^\pi\right\}$$
$$= \frac{4}{3}\left\{0 + \frac{\pi}{2} - \frac{\pi}{4} + \frac{\pi}{2}\right\} = \frac{4}{3}\left(\frac{3\pi}{4}\right) = \pi.$$

In similar fashion we find $k_2 = \frac{1}{2}\pi(1 - c_1/3)$.

A.2 Total Delayed IR Flux. The total delayed IR flux requires the introduction of a function to describe the equator to pole variation in emission. The function $1 - c_1 \sin^2\varphi$ was choosen both for computational simplicity and because, the emission near the poles offen drops off very sharply. The following analytical integrations were accomplished by J. Y. MIYAMOTO, B. D. WARD, and K. W. BEHNKE. The coordinate system was choosen inorder to parallel the diffuse reflection analysis (see Fig. 8).

$$\frac{dn}{dt\,dS'} = k_2(1 - \varepsilon'_{\text{diff.}} - \varepsilon_{\text{spec.}})\left(\frac{dn}{dt\,dS}\right)_S \int_S \frac{(1 - c_1 \sin^2\varphi)\cos\gamma_{1,2}}{\varrho^2}\, dS_{1,2}, \quad \text{(A-4)}$$

where

$$\cos\gamma_{1,2} = \frac{r}{\varrho}\left(\cos\beta_{1,2} - \frac{1}{r}\right) = \frac{r}{\varrho}\left(\cos\delta\cos\alpha'_{1,2} - \frac{1}{r}\right),$$

$$dS_{1,2} = \cos\delta\, d\alpha'_{1,2}\, d\delta.$$

From an examination of Fig. 8 the law of cosines gives us:

$$\cos(90^\circ - \varphi) = \cos(90^\circ - \delta)\cos\delta_{\text{eq}} + \sin(90^\circ - \delta)\sin\delta_{\text{eq}}\cos(90^\circ - \alpha),$$

so that

$$\sin\varphi = \sin\delta\cos\delta_{\text{eq}} + \cos\delta\sin\delta_{\text{eq}}\sin\alpha.$$

By taking advantage of the east-west symmetry we will carry out our integration over S_2 for convenience. Thus:

$$\alpha = \frac{\pi}{2} + \alpha'_2 \quad \text{and} \quad \sin\alpha = \sin\left(\frac{\pi}{2} + \alpha'_2\right) = \cos\alpha'_2.$$

Hereafter, for simplicity, we will drop the subscript to α'.

The integral under consideration is:

$$\tilde{I} = 2 \int_{\alpha'_1 = 0}^{\alpha'_2 = \cos^{-1}\left(\frac{1}{r}\right)} \int_{\delta_1 = -\cos^{-1}\left(\frac{1}{r\cos\alpha'}\right)}^{\delta_2 = +\cos^{-1}\left(\frac{1}{r\cos\alpha'}\right)} (1 - c_1\sin^2\varphi)\frac{r}{\varrho^3}\left(\cos\delta\cos\alpha' - \frac{1}{r}\right) \times$$
$$\times \cos\delta\, d\delta\, d\alpha.$$

Using the previously developed expansions, we can write this integral as:

$$I = 2\int\!\!\int (1 - \sin^2\varphi)\frac{1}{r^2}\left\{\left(-\frac{1}{r} + \frac{3}{2}\frac{1}{r^3} - \frac{15}{8}\frac{1}{r^5}\right)\cos\delta + \right.$$
$$+ \left(1 - \frac{9}{2}\frac{1}{r^2} + \frac{75}{8}\frac{1}{r^4}\right)\cos\alpha'\cos^2\delta + \left(\frac{3}{r} - \frac{15}{r^3} + \frac{105}{4r^5}\right)\cos^2\alpha'\cos^3\delta +$$
$$+ \left(\frac{15}{2r^2} - \frac{175}{4r^4}\right)\cos^3\alpha'\cos^4\delta + \left(\frac{35}{2r^3} - \frac{315}{8r^5}\right)\cos^4\alpha'\cos^5\delta +$$
$$\left. + \left(\frac{315}{8r^4}\right)\cos^5\alpha'\cos^6\delta\right\} d\delta\, d\alpha'.$$

Since

$$\sin^2\varphi = \cos^2\delta_{eq}\sin^2\delta + 2\sin\delta_{eq}\cos\delta_{eq}\cos\alpha'\sin\delta\cos\delta + \\ + \sin^2\delta_{eq}\cos^2\alpha'\cos^2\delta,$$

we find

$$I = I_1 - I_2,$$

where

$$I_1 \triangleq 2\iint \frac{\cos\gamma\cos\delta}{\varrho^2}\,d\delta\,d\alpha' = 2\iint \frac{1}{r^2}\{\hat{\lambda}_1\cos\delta + \hat{\lambda}_2\cos\alpha'\cos^2\delta + \\ + \hat{\lambda}_3\cos^2\alpha'\cos^3\delta + \hat{\lambda}_4\cos^3\alpha'\cos^4\delta + \hat{\lambda}_5\cos^4\alpha'\cos^5\delta + \\ + \hat{\lambda}_6\cos^5\alpha\cos^6\delta\}\,d\delta\,d\alpha,$$

$$I_2 \triangleq 2\iint \frac{c_1\sin^2\varphi\cos\gamma\cos\delta}{\varrho^2}\,d\delta\,d\alpha' = \iint \frac{c_1}{r^2}\Big\{\hat{A}^2[\hat{\lambda}_1\sin^2\delta\cos\delta + \\ + \hat{\lambda}_2\cos\alpha'\sin^2\delta\cos^2\delta + \hat{\lambda}_3\cos^2\alpha'\sin^2\delta\cos^3\delta + \hat{\lambda}_4\cos^3\alpha'\sin^2\delta\cos^4\delta + \\ + \hat{\lambda}_5\cos^4\alpha'\sin^2\delta\cos^5\delta + \hat{\lambda}_6\cos^5\alpha'\sin^2\delta\cos^6\delta] + \\ + 2\hat{A}\hat{B}[\hat{\lambda}_1\cos\alpha'\sin\delta\cos^2\delta + \hat{\lambda}_2\cos^2\alpha'\sin\delta\cos^3\delta + \\ + \hat{\lambda}_3\cos^3\alpha'\sin\delta\cos^4\delta + \hat{\lambda}_4\hat{B}\cos^4\alpha'\sin\delta\cos^5\delta + \hat{\lambda}_5\cos^5\alpha'\sin\delta\cos^6\delta + \\ + \hat{\lambda}_6\cos^6\alpha'\sin\delta\cos^7\delta] + B^2[\hat{\lambda}_1\cos^2\alpha'\cos^3\delta + \hat{\lambda}_2\cos^3\alpha'\cos^4\delta + \\ + \hat{\lambda}_3\cos^4\alpha'\cos^5\delta + \hat{\lambda}_4\cos^5\alpha'\cos^6\delta + \hat{\lambda}_5\cos^6\alpha'\cos^7\delta + \hat{\lambda}_6\cos^7\alpha' + \\ + \cos^8\delta]\,d\delta\,d\alpha',$$

$$\hat{A} \triangleq \cos\delta_{eq} \qquad \hat{\lambda}_1 \triangleq -\frac{1}{r} + \frac{3}{2}\frac{1}{r^3} - \frac{15}{8}\frac{1}{r^5}$$

$$\hat{\lambda}_3 \triangleq 3\frac{1}{r} - 15\frac{1}{r^3} + \frac{105}{4}\frac{1}{r^5} \qquad \hat{\lambda}_5 \triangleq \frac{32}{2}\frac{1}{r^3} - \frac{315}{8}\frac{1}{r^5},$$

$$\hat{B} \triangleq \sin\delta_{eq} \qquad \hat{\lambda}_2 \triangleq 1 - \frac{9}{2}\frac{1}{r^2} + \frac{75}{8}\frac{1}{r^4}$$

$$\hat{\lambda}_4 \triangleq \frac{15}{2}\frac{1}{r^2} - \frac{175}{4}\frac{1}{r^4} \qquad \hat{\lambda}_6 \triangleq \frac{315}{8}\frac{1}{r^4}.$$

Now, we will first evaluate I_2. We have

$$P_2 = c_1\frac{\cos\gamma\cos\delta}{\varrho^2}\sin^2\varphi = \frac{c_1}{r_2}\{\hat{A}^2[(\hat{\lambda}_1 + \hat{\lambda}_3\cos^2\alpha' + \hat{\lambda}_5\cos^4\alpha')\sin^2\delta\cos\delta + \\ + (-\hat{\lambda}_3\cos^2\alpha' - 2\hat{\lambda}_5\cos^4\alpha')\sin^4\delta\cos\delta + (\hat{\lambda}_5\cos^4\alpha')\sin^6\delta\cos\delta + \\ + (\hat{\lambda}_2\cos\alpha')\cos^2\delta + (-\hat{\lambda}_2\cos\alpha' + \hat{\lambda}_4\cos^3\alpha')\cos^4\delta + (-\hat{\lambda}_4\cos^3\alpha' + \\ + \hat{\lambda}_6\cos^5\alpha')\cos^6\delta + (-\hat{\lambda}_3\cos^5\alpha')\cos^8\delta] + 2\hat{A}\hat{B}[(\hat{\lambda}_1\cos\alpha')\cos^2\delta\sin\delta + \\ + (\hat{\lambda}_2\cos^2\alpha')\cos^3\delta\sin\delta + (\hat{\lambda}_3\cos^3\alpha')\cos^4\delta\sin\delta + (\hat{\lambda}_4\cos^4\alpha')\cos^5\delta\sin\delta + \\ + (\hat{\lambda}_5\cos^5\alpha')\cos^6\delta\sin\delta + (\hat{\lambda}_6\cos^6\alpha')\cos^7\delta\sin\delta] + \hat{B}^2[(\hat{\lambda}_1\cos^2\alpha')\cos^3\delta + \\ + (\hat{\lambda}_2\cos^3\alpha')\cos^4\delta + (\hat{\lambda}_3\cos^4\alpha')\cos^5\delta + (\hat{\lambda}_4\cos^5\alpha')\cos^6\delta + (\hat{\lambda}_5\cos^6\alpha')\times \\ \times\cos^7\delta + (\hat{\lambda}_6\cos^7\alpha')\cos^8\delta]\}$$

and integrating with respect to δ,

$$\int P_2\,d\delta = \int_{\delta_2}^{\delta_2} c_1 \frac{\cos\gamma\cos\delta}{\varrho^2}\sin^2\varphi\,d\delta = \frac{c_1}{r^2}\Big\{\hat{A}^2\Big[\frac{1}{3}(\hat{\lambda}_1 + \hat{\lambda}_3\cos^2\alpha' +$$

$$+\hat{\lambda}_5\cos^4\alpha')\,[\sin^3\delta\,|_{\delta_1}^{\delta_2} + \frac{1}{5}(-\hat{\lambda}_3\cos^2\alpha' - 2\hat{\lambda}_5\cos^4\alpha')\Big[\sin^5\delta\,|_{\delta_1}^{\delta_2} +$$

$$+\frac{1}{7}(\hat{\lambda}_5\cos^4\alpha')\,[\sin^7\delta\,|_{\delta_1}^{\delta_2} + (\hat{\lambda}_2\cos\alpha')\Big[\frac{\delta}{2} + \frac{1}{2}\sin\delta\cos\delta\,|_{\delta_1}^{\delta_2} +$$

$$+(-\hat{\lambda}_2\cos\alpha' + \hat{\lambda}_4\cos^3\alpha')\Big[\frac{3}{8}\delta + \frac{5}{8}\sin\delta\cos\delta - \frac{1}{4}\cos\delta\sin^3\delta\,|_{\delta_1}^{\delta_2} +$$

$$+(-\hat{\lambda}_4\cos^3\alpha' + \hat{\lambda}_6\cos^5\alpha')\Big[\frac{5}{16}\delta + \frac{11}{16}\sin\delta\cos\delta - \frac{13}{24}\cos\delta\sin^3\delta +$$

$$+\frac{1}{6}\cos\delta\sin^5\delta\,|_{\delta_1}^{\delta_2} + (-\hat{\lambda}_6\cos^5\alpha')\Big[\frac{35}{128}\delta + \frac{93}{128}\sin\delta\cos\delta -$$

$$-\frac{163}{192}\sin^3\delta\cos\delta + \frac{25}{48}\sin^5\delta\cos\delta - \frac{1}{8}\sin^7\delta\cos\delta\,|_{\delta_1}^{\delta_2}\Big] +$$

$$+2\hat{A}\hat{B}\Big[-\frac{1}{3}(\hat{\lambda}_1\cos\alpha')\cos^3\delta\,|_{\delta_1}^{\delta_2} - \frac{1}{4}(\hat{\lambda}_2\cos^2\alpha')\cos^4\delta\,|_{\delta_1}^{\delta_2} -$$

$$-\frac{1}{5}(\hat{\lambda}_3\cos^3\alpha'))\cos^5\delta\,|_{\delta_1}^{\delta_2} - \frac{1}{6}(\hat{\lambda}_4\cos^4\alpha')\cos^6\delta\,|_{\delta_1}^{\delta_2} -$$

$$-\frac{1}{7}(\hat{\lambda}_5\cos^5\alpha')\cos^7\delta\,|_{\delta_1}^{\delta_2} - \frac{1}{8}(\hat{\lambda}_6\cos^6\alpha')\cos^8\delta\,|_{\delta_1}^{\delta_2}\Big] +$$

$$+\hat{B}^2(\hat{\lambda}_1\cos^2\alpha')\Big[\sin\delta - \frac{1}{3}\sin^3\delta\,|_{\delta_1}^{\delta_2} + (\hat{\lambda}_2\cos^3\alpha')\Big[\frac{3}{8}\delta +$$

$$+\frac{5}{8}\sin\delta\cos\delta - \frac{1}{4}\cos\delta\sin^3\delta\,|_{\delta_1}^{\delta_2} + (\hat{\lambda}_3\cos^4\alpha')\Big[\sin\delta - \frac{2}{3}\sin^3\delta +$$

$$+\frac{1}{5}\sin^5\delta\,|_{\delta_1}^{\delta_2} + (\hat{\lambda}_4\cos^5\alpha')\Big[\frac{5}{16}\delta + \frac{11}{16}\sin\delta\cos\delta - \frac{13}{24}\cos\delta\sin^3\delta +$$

$$+\frac{1}{6}\cos\delta\sin^5\delta\,|_{\delta_1}^{\delta_2} + (\hat{\lambda}_5\cos^6\alpha')\Big[\sin\delta - \sin^3\delta + \frac{3}{5}\sin^5\delta -$$

$$-\frac{1}{7}\sin^7\delta\,|_{\delta_1}^{\delta_2} + (\hat{\lambda}_6\cos^7\alpha')\Big[\frac{35}{128}\delta + \frac{93}{128}\sin\delta\cos\delta - \frac{163}{192}\sin^3\delta\cos\delta +$$

$$+\frac{25}{48}\sin^5\delta\cos\delta - \frac{1}{8}\sin^7\delta\cos\delta\,|_{\delta_1}^{\delta_2}\Big]\Big]\Big\},$$

where, noting that $\delta_1 = -\cos^{-1}\left(\frac{1}{r\cos\alpha'}\right)$ and $\delta_2 = +\cos^{-1}\left(\frac{1}{r\cos\alpha'}\right)$, we have

$$\delta\,|_{\delta_1}^{\delta_2} = \pi - \frac{1}{r\cos\alpha'} - \frac{1}{3r^3\cos^3\alpha'} + 0\left(\frac{1}{r^5}\right),$$

$$\sin\delta\,|_{\delta_1}^{\delta_2} = 2 - \frac{1}{r^2}\,\frac{1}{\cos^2\alpha'} - \frac{1}{4r^4\cos^4\alpha'}\cdots,$$

$$\sin^3\delta\,|_{\delta_1}^{\delta_2} = 2 - \frac{3}{r^2\cos^2\alpha'} + \frac{3}{4r^4\cos^4\alpha'}\cdots,$$

$$\sin^5\delta\,\big|_{\delta_1}^{\delta_2} = 2 - \frac{5}{r^2\cos^2\alpha'} + \frac{15}{4r^4\cos^4\alpha'}\cdots,$$

$$\sin^7\delta\,\big|_{\delta_1}^{\delta_2} = 2 - \frac{7}{r^2\cos^2\alpha'} + \frac{35}{4r^4\cos^4\alpha'}\cdots,$$

$$\sin\delta\cos\delta\,\big|_{\delta_1}^{\delta_2} = \frac{2}{r\cos\alpha'} - \frac{1}{r^3\cos^3\alpha'} - \frac{1}{4r^5\cos^5\alpha'}\cdots,$$

$$\sin^3\delta\cos\delta\,\big|_{\delta_1}^{\delta_2} = \frac{2}{r\cos\alpha'} - \frac{3}{r^3\cos^3\alpha'} + \frac{3}{4r^5\cos^5\alpha'}\cdots,$$

$$\sin^5\delta\cos\delta\,\big|_{\delta_1}^{\delta_2} = \frac{2}{r\cos\alpha'} - \frac{5}{r^3\cos^3\alpha'} + \frac{15}{4r^5\cos^5\alpha'}\cdots,$$

$$\sin^7\delta\cos\delta\,\big|_{\delta_1}^{\delta_2} = \frac{2}{r\cos\alpha'} - \frac{7}{r^3\cos^3\alpha'} + \frac{35}{4r^5\cos^5\alpha'}\cdots,$$

$$\cos^n\delta\,\big|_{\delta_1}^{\delta_2} = 0 \qquad n = 0, 1, 2, \ldots.$$

Substituting these into the equation for $\int P_2\,d\delta$ yields

$$\begin{aligned}
P_2' = \int P_2\,d\delta = \frac{c_1}{r^2}\Big\{\hat{A}^2\Big[\frac{1}{3}(\hat\lambda_1 + \hat\lambda_3\cos^2\alpha' + \hat\lambda_5\cos^4\alpha')\Big(2 - \frac{3}{r^2\cos^2\alpha'}\Big) + \\
+ \frac{1}{5}(-\hat\lambda_3\cos^2\alpha' - 2\hat\lambda_5\cos^4\alpha')\Big(2 - \frac{5}{r^2\cos^2\alpha'}\Big) + \frac{1}{7}(\hat\lambda_5\cos^4\alpha')(2) + \\
+ (\hat\lambda_2\cos\alpha')\Big(\frac{\pi}{2} - \frac{2}{3r^3\cos^3\alpha'} + (-\hat\lambda_2\cos\alpha' + \hat\lambda_4\cos^3\alpha')\Big(\frac{3\pi}{8}\Big) + \\
+ (-\hat\lambda_4\cos^3\alpha' + \hat\lambda_6\cos^5\alpha')\Big(\frac{5\pi}{16}\Big) + (-\hat\lambda_6\cos^5\alpha')\Big(\frac{35}{128}\pi\Big)\Big] + \\
+ \hat{B}^2\Big[(\hat\lambda_1\cos^2\alpha')\Big(\frac{4}{3}\Big) + (\hat\lambda_2\cos^2\alpha')\Big(\frac{3\pi}{8}\Big) + (\hat\lambda_3\cos^4\alpha')\Big(\frac{16}{15}\Big) + \\
+ (\hat\lambda_4\cos^5\alpha')\Big(\frac{5\pi}{16}\Big) + (\hat\lambda_5\cos^6\alpha')\Big(\frac{32}{35}\Big) + (\hat\lambda_6\cos^7\alpha')\Big(\frac{35}{128}\pi\Big)\Big]\Big\}.
\end{aligned}$$

Expanding and collecting in terms of powers of $\cos\alpha'$, we obtain

$$\begin{aligned}
P_2' = \frac{c_1}{r_2}\Big\{\hat{A}^2\Big[\Big(-\frac{1}{r^2}\hat\lambda_1 - \frac{2}{3r^3}\hat\lambda_2\Big)\Big(\frac{1}{\cos^2\alpha'}\Big) + \Big(\frac{2}{3}\hat\lambda_1\Big) + \Big(\frac{\pi}{8}\hat\lambda_2\Big)(\cos\alpha') + \\
+ \Big(\frac{1}{r^2}\hat\lambda_5 + \frac{4}{15}\hat\lambda_3\Big)(\cos^2\alpha') + \Big(\frac{\pi}{16}\hat\lambda_4\Big)(\cos^3\alpha') + \frac{16}{105}\hat\lambda_5\Big)(\cos^4\alpha') + \\
+ \Big(\frac{5\pi}{128}\hat\lambda_6\Big)(\cos^5\alpha')\Big] + \hat{B}^2\Big[\Big(\frac{4}{3}\hat\lambda_1\Big)(\cos^2\alpha') + \Big(\frac{3\pi}{8}\hat\lambda_2\Big)(\cos^3\alpha') + \\
+ \Big(\frac{16}{15}\hat\lambda_3\Big)(\cos^4\alpha') + \Big(\frac{5\pi}{16}\hat\lambda_4\Big)(\cos^5\alpha') + \frac{32}{35}\hat\lambda_5\Big)(\cos^6\alpha') + \\
+ \Big(\frac{35\pi}{128}\hat\lambda_6\Big)(\cos^7\alpha')\Big]\Big\}.
\end{aligned}$$

Now, integrating with respect to α', we find

$$I_2 = 2\int_0^{\alpha_2' = \cos^{-1}\frac{1}{r}} P_2' d\alpha' = \frac{2c_1}{r^2}\Big\{\hat{A}^2\Big[\Big(-\frac{1}{r_2}\hat{\lambda}_1 - \frac{2}{3r^3}\hat{\lambda}_2\Big)\Big]\Big[\tan\alpha'\Big|_0^{\alpha_2'} +$$

$$+\Big(\frac{2}{3}\hat{\lambda}_1\Big)\Big[\alpha'\Big|_0^{\alpha_2'} + \Big(\frac{\pi}{8}\hat{\lambda}_2\Big)[\sin\alpha'] + \Big(\frac{1}{r^2}\hat{\lambda}_5 + \frac{4}{15}\hat{\lambda}_3\Big)\times$$

$$\times\Big[\frac{\alpha'}{2} + \frac{1}{2}\sin\alpha'\cos\alpha'\Big|_0^{\alpha_2'} + \Big(\frac{\pi}{16}\hat{\lambda}_4\Big)\Big[\sin\alpha' - \frac{1}{3}\sin^3\alpha'\Big|_0^{\alpha_2'} +$$

$$+\Big(\frac{16}{105}\hat{\lambda}_5\Big)\Big[\frac{3}{8}\alpha' + \frac{5}{8}\sin\alpha'\cos\alpha' - \frac{1}{4}\cos\alpha'\sin^3\alpha'\Big|_0^{\alpha_2'} +$$

$$+\Big(\frac{5\pi}{128}\hat{\lambda}_6\Big)\Big[\sin\alpha' - \frac{2}{3}\sin^3\alpha' + \frac{1}{5}\sin^5\alpha'\Big|_0^{\alpha_2'}\Big] + \hat{B}^2\Big[\Big(\frac{4}{3}\hat{\lambda}_1\Big)\Big[\frac{\alpha'}{2} +$$

$$+\frac{1}{2}\sin\alpha'\cos\alpha'\Big|_0^{\alpha_2'} + \Big(\frac{3\pi}{8}\hat{\lambda}_2\Big)\Big[\sin\alpha' - \frac{1}{3}\sin^3\alpha'\Big|_0^{\alpha_2'} + \frac{16}{15}\hat{\lambda}_3\Big)\Big[\frac{3}{8}\alpha' +$$

$$+\frac{5}{8}\sin\alpha'\cos\alpha' - \frac{1}{4}\cos\alpha'\sin^3\alpha'\Big|_0^{\alpha_2'} + \Big(\frac{5\pi}{16}\hat{\lambda}_4\Big)\Big[\sin\alpha' - \frac{2}{3}\sin^3\alpha' +$$

$$+\frac{1}{5}\sin^5\alpha'\Big|_0^{\alpha_2'} + \Big(\frac{32}{35}\hat{\lambda}_5\Big)\Big[\frac{5}{16}\alpha' + \frac{11}{16}\sin\alpha'\cos\alpha' - \frac{13}{24}\cos\alpha'\sin^3\alpha' +$$

$$+\frac{1}{6}\cos\alpha'\sin^5\alpha'\Big|_0^{\alpha_2'} + \frac{35\pi}{128}\hat{\lambda}_6\Big)\Big[\sin\alpha' - \sin^3\alpha' + \frac{3}{5}\sin^5\alpha' - \frac{1}{7}\sin^7\alpha'\Big|_0^{\alpha_2'}\Big]\Big\}.$$

Using,

$$\alpha'\Big|_0^{\cos^{-1}\frac{1}{r}} = \frac{\pi}{2} - \frac{1}{r} - \frac{1}{6r^3}\cdots, \qquad \sin\alpha'\cos\alpha'\Big| = \frac{1}{r} - \frac{1}{2r^3} - \frac{3}{8r^5}\cdots,$$

$$\sin\alpha'\Big| = 1 - \frac{1}{2r^2} - \frac{1}{8r^4}\cdots, \qquad \sin^3\alpha'\cos\alpha'\Big| = \frac{1}{r} - \frac{3}{2r^3} + \frac{3}{8r^5}\cdots,$$

$$\sin^3\alpha'\Big| = 1 - \frac{3}{2r^2} + \frac{3}{8r^4}\cdots, \qquad \sin^5\alpha'\cos\alpha'\Big| = \frac{1}{r} - \frac{5}{2r^3} + \frac{15}{8r^5}\cdots,$$

$$\sin^5\alpha'\Big| = 1 - \frac{5}{2r^2} + \frac{15}{8r^4}\cdots, \qquad \tan\alpha'\Big| = r - \frac{1}{2}\frac{1}{r} - \frac{1}{8r^3}\cdots,$$

$$\sin^7\alpha'\Big| = 1 - \frac{7}{2r^2} + \frac{35}{8r^4}\cdots$$

and substituting in the λ' s and collecting in powers of $\left(\frac{1}{r}\right)$, we find

$$I_2 = \frac{2c_1}{r^2}\Big\{\hat{A}^2\Big[\frac{\pi}{8} + \frac{1}{r}\Big(-\frac{2}{15}\pi\Big) + \frac{1}{r^2}\Big(1 - \frac{5}{16}\pi\Big) + \frac{1}{r^3}(0) +$$

$$+\frac{1}{r^4}\Big(\frac{24}{135} + \frac{167}{384}\pi\Big)\Big] + \hat{B}^2\Big[\frac{\pi}{4} + \frac{1}{r}\Big(\frac{4}{15}\pi\Big) + \frac{1}{r^2}\Big(\frac{\pi}{8}\Big)\frac{1}{r^3}(0) +$$

$$+\frac{1}{r^4}\Big(\frac{4}{9} - \frac{23}{192}\pi\Big)\Big]\Big\}.$$

For I_2, we have.

$$P_1 = \frac{1}{r^2}\{\lambda_1 \cos\delta + \lambda_2 \cos\alpha' \cos^2\delta + \lambda_3 \cos^2\alpha' \cos^3\delta + \lambda_4 \cos^3\alpha' \cos^4\delta + \\ + \lambda_5 \cos^4\alpha' \cos^5\delta + \lambda_3 \cos^5\alpha' \cos^6\delta.\}$$

Integrating with respect to δ,

$$P_1' = \int P_1\, d\delta = \frac{1}{r^2}\Big\{(\lambda_1)\,[\sin\delta] + (\lambda_2 \cos\alpha')\left[\frac{\delta}{2} + \frac{1}{2}\sin\delta\cos\delta\right] + \\ + (\lambda_3 \cos^2\alpha')\left[\sin\delta - \frac{1}{3}\sin^3\delta\right] + \\ + (\lambda_4 \cos^3\alpha')\left[\frac{3}{8}\delta + \frac{5}{8}\sin\delta\cos\delta - \frac{1}{4}\cos\delta\sin^3\delta\right] + \\ + (\lambda_5 \cos^4\alpha')\left[\sin\delta - \frac{2}{3}\sin^3\delta + \frac{1}{5}\sin^5\delta\right] + \\ + (\lambda_6 \cos^5\alpha')\Big[\frac{5}{16}\delta + \frac{11}{16}\sin\delta\cos\delta - \frac{13}{24}\cos\delta\sin^3\delta + \\ + \frac{1}{6}\cos\delta\sin^5\delta\Big]\Big\}.$$

Evaluating the limits and collecting terms in powers of $\cos\alpha'$,

$$P_1' = \frac{1}{r^2}\Big\{(\lambda_1)\left(2 - \frac{1}{r^2\cos^2\alpha'}\right) + (\lambda_2 \cos\alpha')\Big(\frac{\pi}{2} - \frac{1}{r\cos\alpha'} - \frac{1}{6r^3\cos^3\alpha'} + \\ + \frac{1}{r\cos\alpha'} - \frac{1}{2r^3\cos^3\alpha'}\Big) + (\lambda_3\cos^2\alpha')\Big(2 - \frac{1}{r^2\cos^2\alpha'} - \frac{2}{3} + \\ + \frac{1}{r^2\cos^2\alpha'}\Big) + (\lambda_4\cos^3\alpha')\left(\frac{3}{8}\pi - \frac{3}{4r\cos\alpha'} + \frac{5}{4r\cos\alpha'} - \frac{2}{4r\cos\alpha'}\right) + \\ + (\lambda_5\cos^4\alpha')\Big[2 - \frac{4}{3} + \frac{2}{5}\Big) + (\lambda_6\cos^5\alpha')\left(\frac{5}{16}\pi\right)\Big\}$$

$$= \frac{1}{r^2}\Big\{(2\lambda_1) - \left(\frac{1}{r^2}\lambda_1\right)\left(\frac{1}{\cos^2\alpha'}\right) + \left(\frac{\pi}{2}\lambda_2\right)\cos\alpha' - \\ - \left(\frac{2}{3}\frac{1}{r^3}\lambda_2\right)\frac{1}{\cos^2\alpha'} + \left(\frac{4}{3}\lambda_3\right)\cos^2\alpha' + \left(\frac{3\pi}{8}\lambda_4\right)\cos^3\alpha' + \\ + \left(\frac{16}{15}\lambda_5\right)\cos^4\alpha' + \left(\frac{5\pi}{16}\lambda_6\right)\cos^5\alpha'\Big\}$$

$$= \frac{1}{r^2}\Big\{\left(-\frac{1}{r^2}\lambda_1 - \frac{2}{3}\frac{1}{r^3}\lambda_2\right)\left(\frac{1}{\cos^2\alpha'}\right) + (2\lambda_1) + \left(\frac{\pi}{2}\lambda_2\right)(\cos\alpha') + \\ + \left(\frac{4}{3}\lambda_3\right)\cos^2\alpha' + \left(\frac{3\pi}{8}\lambda_4\right)\cos^3\alpha' + \\ + \left(\frac{16}{15}\lambda_5\right)\cos^4\alpha' + \left(\frac{5\pi}{16}\lambda_6\right)\cos^5\alpha'\Big\}.$$

Integrating with respect to α' and substituting in the values of λ' s,

$$I_1 = 2\int P_1' \, d\alpha' = \frac{2}{r^2}\left\{\left(\frac{1}{3}\frac{1}{r^3} + \frac{3}{2}\frac{1}{r^5}\right)\left(r - \frac{1}{2}\frac{1}{r}\right) + \right.$$

$$+ \left(-\frac{2}{r} + 3\frac{1}{r^3}\right)\left(\frac{\pi}{2} - \frac{1}{r} - \frac{1}{6r^3}\right) +$$

$$+ \left(\frac{\pi}{2} - \frac{9\pi}{4}\frac{1}{r^2} + \frac{75\pi}{16}\frac{1}{r^3}\right) \times$$

$$\times \left(1 - \frac{1}{2r^2} - \frac{1}{8r^4}\right) + \left(4\frac{1}{r} - 20\frac{1}{r^3}\right) \times$$

$$\times \left(\frac{\pi}{4} - \frac{1}{2r} - \frac{1}{12r^3} + \frac{1}{2r} - \frac{1}{4r^3}\right) +$$

$$+ \left(\frac{45\pi}{16}\frac{1}{r^4} - \frac{3\cdot 175}{4\cdot 8}\pi\frac{1}{r^4}\right)\left(1 - \frac{1}{2r^2} - \frac{1}{3} + \frac{1}{2r^2}\right) +$$

$$+ \left(\frac{8\cdot 35}{15}\frac{1}{r^3}\right)\left(\frac{3\pi}{16} - \frac{3}{8}\frac{1}{r} + \frac{5}{8}\frac{1}{r} + - \frac{1}{4}\frac{1}{r}\right) +$$

$$\left. + \left(\frac{5\cdot 15\cdot 21}{8\cdot 16}\pi\frac{1}{r^4}\right)\left(1 - \frac{2}{3} + \frac{1}{5}\right)\right\}.$$

Collecting in terms of powers of $\left(\frac{1}{r}\right)$, we have

$$I_1 = \frac{2}{r^2} \times$$

$$\times \left\{\frac{\pi}{2} + \frac{1}{r}(0) + \frac{1}{r^2}\left(\frac{7}{3} - \frac{5}{8}\pi\right) + \frac{1}{r^3}(0) + \frac{1}{r^4}\left(-\frac{8}{3} + \frac{11}{8}\pi\right)\right\}.$$

Then $I = I_1 - I_2$, where

$$I_1 = \frac{2}{r^2} \times$$

$$\times \underbrace{\left\{\frac{\pi}{2} + \frac{1}{r}(0) + \frac{1}{r^2}\left(\frac{7}{3} - \frac{5}{8}\pi\right) + \frac{1}{r^3}(0) + \frac{1}{r^4}\left(-\frac{8}{3} + \frac{11}{8}\pi\right)\right\}}_{C_1},$$

$$I_2 = \frac{2c_1}{r^2} \times$$

$$\times \underbrace{\left\{\cos^2\delta_{eq}\left[\frac{\pi}{8} + \frac{1}{r}\left(-\frac{2}{15}\pi\right) + \frac{1}{r^2}\left(1 - \frac{5}{16}\pi\right) + \frac{1}{r^3}(0) + \frac{1}{r^4}\left(\frac{8}{45} - \frac{167}{384}\pi\right)\right]\right.}_{C_2}$$

$$\left. + \sin^2\delta_{eq}\underbrace{\left[\frac{\pi}{4} + \frac{1}{r}\left(\frac{4}{15}\pi\right) + \frac{1}{r^2}\left(\frac{\pi}{8}\right) + \frac{1}{r^3}(0) + \frac{1}{r^4}\left(\frac{4}{9} - \frac{23}{192}\pi\right)\right]}_{C_3}\right\},$$

or

$$I = \iint \frac{(1 - c_1 \sin^2\varphi) \cos\gamma \cos\delta}{\varrho^2}\, d\alpha'\, d\delta = \frac{2}{r^2} \left\{ \left[\frac{\pi}{2} + \frac{1}{r}(0) + \right.\right.$$
$$+ \frac{1}{r^2}\left(\frac{7}{3} - \frac{5}{8}\pi\right) + \frac{1}{r^3}(0) + \frac{1}{r^4}\left(-\frac{8}{3} + \frac{11}{8}\pi\right)\Big] -$$
$$- c_1 \cos^2\delta_{eq}\left[\left(\frac{\pi}{8} + \frac{1}{r}\left(-\frac{2}{15}\pi\right) + \frac{1}{r^2}\left(1 - \frac{5}{16}\pi\right) + \frac{1}{r^3}(0) + \right.\right.$$
$$\left.\left. + \frac{1}{r^4}\left(\frac{8}{45} + \frac{167}{384}\pi\right)\right] - c_1 \sin^2\delta_{eq}\left[\left(\frac{\pi}{4}\right) + \frac{1}{r}\left(\frac{4}{15}\pi\right) + \right.$$
$$\left.\left. + \frac{1}{r^2}\left(\frac{\pi}{8}\right) + \frac{1}{r^3}(0) + \frac{1}{r^4}\left(\frac{4}{9} - \frac{23}{192}\pi\right)\right]\right\}$$
$$= \frac{2}{r^2}\left\{\left[\frac{\pi}{2} - c_1\left(\frac{\pi}{8}\cos^2\delta_{eq} + \frac{\pi}{4}\sin^2\delta_{eq}\right)\right] + \right.$$
$$+ \frac{1}{r}\left[0 - c_1\left(-\frac{2}{15}\pi\cos^2\delta_{eq} + \frac{4}{15}\pi\sin^2\delta_{eq}\right)\right] +$$
$$+ \frac{1}{r^2}\left[\left(\frac{7}{3} - \frac{5}{8}\pi\right) - c_1\left(\left(1 - \frac{5}{16}\pi\right)\cos^2\delta_{eq} + \frac{\pi}{8}\sin^2\delta_{eq}\right)\right] +$$
$$+ \frac{1}{r^3}[0 - c_1(0)] + \frac{1}{r^4}\left[\left(-\frac{8}{3} + \frac{11}{8}\pi\right) - \right.$$
$$\left.\left. - c_1\left(\left(\frac{8}{45} + \frac{167}{384}\pi\right)\cos^2\delta_{eq} + \left(\frac{4}{9} - \frac{23}{192}\pi\right)\sin^2\delta_{eq}\right)\right]\right\}$$
$$= \frac{2}{r^2}\left\{\left[\frac{\pi}{2} - c_1\frac{\pi}{8}(1 + \sin^2\delta_{eq})\right] + \frac{1}{r}\left[c_1\frac{2\pi}{15}(1 - 3\sin^2\delta_{eq})\right] + \right.$$
$$+ \frac{1}{r^2}\left[\left(\frac{7}{3} - \frac{5}{8}\pi\right) - c_1\left(\left\{1 - \frac{5\pi}{16}\right\} - \left\{1 - \frac{7\pi}{16}\right\}\sin^2\delta_{eq}\right)\right] +$$
$$+ \frac{1}{r^3}[0] + \frac{1}{r^4}\left[\left(-\frac{8}{3} + \frac{11}{8}\pi\right) - c_1\left(\left\{\frac{8}{45} + \frac{167}{384}\pi\right\} + \right.\right.$$
$$\left.\left.\left. + \left\{\frac{36}{135} - \frac{213}{384}\pi\right\}\sin^2\delta_{eq}\right)\right]\right\}.$$

The derivation of the delayed IR radial component was accomplished by R. SMITH and W. K. BEHNKE. It is similar in approach and will not be presented.

A.3 Transverse and Radial Component. Consider next the radial and transverse components of the reflected visible and prompt IR photon flux. Let ξ be defined as the angle between $\boldsymbol{\varrho}$ and $\boldsymbol{r}$ (see Fig. 7). Thus the radial component of photon flux through the incremental spacecraft are a dS' will be

$$\frac{dn}{dt\,dS'} = \varepsilon_{\text{diff}}\, k_1 \left(\frac{dn}{dS\,dt}\right)_S \frac{\cos\alpha_{1,2}\cos\delta\cos\gamma\cos\xi\, dS_{1,2}}{\varrho^2}. \quad \text{(A-5)}$$

If ζ is defined as the angle between the $X - Y$ plane of Fig. 6 (i.e., the plane formed by the Sun's direction, $\boldsymbol{U}$, and the satellite direction, $\boldsymbol{r}$) and the plane of $\boldsymbol{a}$, ρ, and $\boldsymbol{r}$ (shown in Fig. 7), then the transverse component of photon flux is

$$\frac{dn}{dt\,dS'} = \varepsilon'_{\text{diff}}\,k_1 \left(\frac{dn}{dS\,dt}\right)_S \frac{\cos\alpha_{1,2}\cos\delta\cos\gamma\sin\xi\cos\zeta\,dS_{1,2}}{\varrho^2}. \quad \text{(A-6)}$$

From Figs. 6 and 7,

$$\sin\xi = \frac{1}{r}\sin(\pi-\gamma) = \frac{1}{r}\sin\gamma \quad \text{or} \quad \sin\xi = \frac{1}{\varrho}\sin\beta,$$

$$\cos\xi = (r^2+\varrho^2-1)/2r\varrho, \qquad \text{and from spherical triangles,}$$

$$\cos\zeta = (\cos\delta - \cos\beta\cos\alpha')/\sin\beta\sin\alpha'.$$

The normal component of photon flux in this sun-body center-spacecraft system cancels out; accordingly, only the radial and transverse components are considered in the analysis.

First consider the formulation of $\cos\gamma\cos\xi$ from Eq. (A-5). Since $r = \varrho\cos\xi + \cos\beta$, we have

$$\cos\xi = \frac{1}{\varrho}(r-\cos\beta)$$

and we have already seen that

$$\cos\gamma = \frac{r}{\varrho}\left(\cos\beta - \frac{1}{r}\right).$$

Thus,

$$\cos\gamma\cos\xi = \frac{r}{\varrho^2}\left(r\cos\beta - 1 - \cos^2\beta + \frac{\cos\beta}{r}\right). \quad \text{(A-7)}$$

Second, consider $\cos\gamma\sin\xi\cos\zeta$ from Eq. (A-6).

From trigonometry

$$\sin\xi = \frac{1}{\varrho}\sin\beta$$

and, as we have seen

$$\cos\zeta = (\cos\delta - \cos\beta\cos\alpha')/\sin\beta\sin\alpha'. \quad \text{(A-8)}$$

Thus

$$\cos\gamma\sin\xi\cos\zeta = \frac{1}{\varrho^2}\left(\cos\beta - \frac{1}{r}\right)(\cos\delta - \cos\beta\cos\alpha')/\sin\alpha'.$$

Finally, we introduce LEGENDRE polynomials and analytically integrate term by term.

The LEGENDRE polynomial expansion for the factor r^4/ϱ^4 is given by:

$$\begin{aligned}\frac{r^4}{\varrho^4} = 1 &+ \frac{1}{r}(4\cos\alpha'\cos\delta) + \frac{2}{r^2}(6\cos^2\alpha'\cos^2\delta - 1) + \\ &+ \frac{4}{r^3}(8\cos^3\alpha'\cos^3\delta - 3\cos\alpha'\cos\delta) + \\ &+ \frac{1}{r^4}(80\cos^4\alpha'\cos^4\delta - 48\cos^2\alpha'\cos^2\delta + 3) + \\ + \frac{1}{r^5}&(192\cos^5\alpha'\cos^5\delta - 160\cos^3\alpha'\cos^3\delta + 24\cos\alpha'\cos\delta) + \cdots. \quad \text{(A-9)}\end{aligned}$$

In the case of the transverse component we shall not differentiate between S_1 and S_2 and integrate directly across S_1 and S_2 through use of α and δ (not $\alpha'_{1,2}$, δ' i.e., let $\alpha' = \alpha - \theta$). In this case according to an analysis of J. Y. MIYAMOTO, K. W. BEHNKE, E. F. STOOPS, and B. D. WARD, we have

$$\left(\frac{dn}{dt\,dS'}\right) = k_1\,\varepsilon'_{\text{diff.}}\left(\frac{dn}{dS\,dt}\right)_S\left[\frac{\cos\alpha\cos\delta}{\varrho^2}\,\frac{r}{\varrho}\cos(\alpha-\theta)\cos\delta - \frac{1}{r}\right]$$
$$\left(\frac{1}{\varrho}\sin\beta\right)\left[\frac{\cos\delta - \cos\beta\cos(\alpha-\theta)}{\sin\beta\sin(\alpha-\theta)}\right]dS,$$

where

$$dS = \cos\delta\,d\alpha\,d\delta,$$
$$\cos\beta = \cos\delta\cos(\alpha-\theta).$$

Thus, we rewrite the above as

$$\left(\frac{dn}{dt\,dS'}\right)_T = k_1\,\varepsilon'_{\text{diff.}}\left(\frac{dn}{dS\,dt}\right)\cos\alpha\cos^3\delta\left[\cos(\alpha-\theta)\cos\delta - \frac{1}{r}\right]\left(\frac{r}{\varrho^4}\right)$$
$$\sin(\alpha-\theta)\,d\delta\,d\alpha.$$

Using

$$\left(\frac{r}{\varrho^4}\right) = \frac{1}{r^3}\left\{1 + \frac{1}{r}(4p) + \frac{1}{r^2}(12p^2-2) + \frac{1}{r^3}(32p^3-12p) + \right.$$
$$\left. + \frac{1}{r^4}(80p^4 - 48p^2 + 3) + \cdots\right\},$$

where

$$p \triangleq \cos(\alpha-\theta)\cos\delta.$$

Substituting into the preceding equation (and collecting up to the r^6 term), we have

$$\left(\frac{dn}{dt\,dS'}\right)_T = 2k_1\left(\frac{dn}{dS\,dt}\right)_S\int_{\alpha=\alpha_1}^{\alpha=\alpha_2}\int_{\delta=0}^{\delta=\cos^{-1}\left(\frac{1}{r\cos(\alpha-\theta)}\right)}\left\{\cos\theta\left[\left(-\frac{1}{r^4}+\frac{2}{r^6}\right)\right.\right.$$
$$\sin(\alpha-\theta)\cos(\alpha-\theta)\cos^3\delta + \left(\frac{1}{r^3}-\frac{6}{r^5}\right)\sin(\alpha-\theta)\times$$
$$\times\cos^2(\alpha-\theta)\cos^4\delta + \left(\frac{4}{r^4}-\frac{24}{r^6}\right)\sin(\alpha-\theta)\cos^3(\alpha-\theta)\cos^5\delta +$$
$$+\left(\frac{12}{r^5}\right)\sin(\alpha-\theta)\cos^4(\alpha-\theta)\cos^6\delta +$$
$$\left.+\left(\frac{32}{r^6}\right)\sin(\alpha-\theta)\cos^5(\alpha-\theta)\cos^7\delta\right] -$$
$$-\sin\theta\left[\left(-\frac{1}{r^4}+\frac{2}{r^6}\right)\sin^2(\alpha-\theta)\cos^3\delta + \left(\frac{1}{r^3}-\frac{6}{r^5}\right)\right.$$
$$\sin^2(\alpha-\theta)\cos(\alpha-\theta)\cos^4\delta + \left(\frac{4}{r^4}-\frac{24}{r^6}\right)\sin^2(\alpha-\theta)$$
$$\cos^2(\alpha-\theta)\cos^5\delta + \left(\frac{12}{r^5}\right)\sin^2(\alpha-\theta)\cos^3(\alpha-\theta)\cos^6\delta +$$
$$\left.\left.+\left(\frac{32}{r^6}\right)\sin^2(\alpha-\theta)\cos^4(\alpha-\theta)\cos^7\delta\right]\right\}d\delta\,d\alpha.$$

Now,

$$\int \cos^3\delta\, d\delta = \sin\delta - \frac{1}{3}\sin^3\delta,$$

$$\int \cos^4\delta\, d\delta = \frac{3}{8}\delta + \frac{5}{8}\sin\delta\cos\delta - \frac{1}{4}\cos\delta\sin^3\delta,$$

$$\int \cos^5\delta\, d\delta = \sin\delta - \frac{2}{3}\sin^3\delta + \frac{1}{5}\sin^5\delta,$$

$$\int \cos^6\delta\, d\delta = \frac{5}{15}\delta + \frac{11}{16}\sin\delta\cos\delta - \frac{13}{24}\cos\delta\sin^3\delta + \frac{1}{6}\cos\delta\sin^5\delta,$$

$$\int \cos^7\delta\, d\delta = \sin\delta - \sin^2\delta + \frac{3}{5}\sin^5\delta - \frac{1}{7}\sin^7\delta.$$

Since

$$\text{with}\quad l = \cos^{-1}\left(\frac{1}{r\cos(\alpha-\theta)}\right),$$

$$\delta\Big|_0^l = \cos^{-1}\left(\frac{1}{r\cos(\alpha-\theta)}\right),$$

$$\sin\delta\Big|_0^l = \frac{[r^2\cos^2(\alpha-\theta)-1]^{1/2}}{r\cos(\alpha-\theta)},$$

$$\sin^3\delta\Big|_0^l = \frac{[r^2\cos^2(\alpha-\theta)-1]^{3/2}}{r^3\cos^3(\alpha-\theta)},$$

$$\sin^5\delta\Big|_0^l = \frac{[r^2\cos^2(\alpha-\theta)-1]^{5/2}}{r^5\cos^5(\alpha-\theta)},$$

$$\sin^7\delta\Big|_0^l = \frac{[r^2\cos^2(\alpha-\theta)-1]^{7/2}}{r^7\cos^7(\alpha-\theta)},$$

$$\sin\delta\cos\delta\Big|_0^l = \frac{[r^2\cos^2(\alpha-\theta)-1]^{1/2}}{r^2\cos^2(\alpha-\theta)},$$

$$\sin^3\delta\cos\delta\Big|_0^l = \frac{[r^2\cos^2(\alpha-\theta)-1]^{3/2}}{r^4\cos^4(\alpha-\theta)},$$

$$\sin^5\delta\cos\delta\Big|_0^l = \frac{[r^2\cos^2(\alpha-\theta)-1]^{5/2}}{r^6\cos^6(\alpha-\theta)}.$$

We have

$$\hat{I}_3 \triangleq \int \cos^3\delta\, d\delta = \frac{[r^2\cos^2(\alpha-\theta)-1]^{1/2}}{r\cos(\alpha-\theta)} - \frac{1}{3}\,\frac{[r^2\cos^2(\alpha-\theta)-1]^{3/2}}{r^3\cos^3(\alpha-\theta)},$$

$$\hat{I}_4 \triangleq \int \cos^4\delta\, d\delta = \frac{3}{8}\left[\cos^{-1}\left(\frac{1}{r\cos(\alpha-\theta)}\right)\right] + \frac{5}{8}\,\frac{[r^2\cos^2(\alpha-\theta)-1]^{1/2}}{r^2\cos^2(\alpha-\theta)} + {}$$
$$- \frac{1}{4}\,\frac{[r^2\cos^2(\alpha-\theta)-1]^{3/2}}{r^4\cos^4(\alpha-\theta)},$$

$$\hat{I}_5 \triangleq \int \cos^5\delta\, d\delta = \frac{[r^2\cos^2(\alpha-\theta)-1]^{1/2}}{r\cos(\alpha-\theta)} - \frac{2}{3}\,\frac{[r^2\cos^2(\alpha-\theta)-1]^{3/2}}{r^3\cos^3(\alpha-\theta)} + {}$$
$$+ \frac{1}{5}\,\frac{[r^2\cos^2(\alpha-\theta)-1]^{5/2}}{r^5\cos^5(\alpha-\theta)},$$

$$\hat{I}_6 \triangleq \int \cos^6 \delta \, d\delta = \frac{5}{16}\left[\cos^{-1}\left(\frac{1}{r\cos(\alpha-\theta)}\right)\right] +$$

$$+ \frac{11}{16}\frac{[r^2\cos^2(\alpha-\theta)-1]^{1/2}}{r^2\cos^2(\alpha-\theta)} - \frac{13}{24}\frac{[r^2\cos^2(\alpha-\theta)-1]^{3/2}}{r^4\cos^4(\alpha-\theta)} +$$

$$+ \frac{1}{6}\frac{[r^2\cos^2(\alpha-\theta)-1]^{5/2}}{r^6\cos^6(\alpha-\theta)},$$

$$\hat{I}_7 \triangleq \int \cos^7 \delta \, d\delta = \frac{[r^2\cos^2(\alpha-\theta)-1]^{1/2}}{r\cos(\alpha-\theta)} - \frac{[r^2\cos^2(\alpha-\theta)-1]^{3/2}}{r^3\cos^3(\alpha-\theta)} +$$

$$+ \frac{3}{5}\frac{[r^2\cos^2(\alpha-\theta)-1]^{5/2}}{r^5\cos^5(\alpha-\theta)} - \frac{1}{7}\frac{[r^2\cos^2(\alpha-\theta)-1]^{7/2}}{r^7\cos^7(\alpha-\theta)}.$$

By the binomial expansion,

$$[r^2\cos^2(\alpha-\theta)-1]^{n/2} = r^n\cos^n(\alpha-\theta)\left[1 - \frac{n}{2r^2}\left(\frac{1}{\cos(\alpha-\theta)}\right)^2 + \right.$$

$$\left. + \frac{n}{4}\left(\frac{n}{2}-1\right)\frac{1}{r^4}\left(\frac{1}{\cos(\alpha-\theta)}\right)^4 + 0\left(\frac{1}{r^6}\right)\right].$$

Also,

$$\cos^{-1}\left(\frac{1}{r\cos(\alpha-\theta)}\right) = \frac{\pi}{2} - \left\{\frac{1}{r\cos(\alpha-\theta)} + \frac{1}{6r^3}\left[\frac{1}{\cos(\alpha-\theta)}\right]^3 + 0\left(\frac{1}{r^5}\right)\right\}.$$

Substituting the above, we can write (note the cancellations)

$$\int_{\alpha_1}^{\alpha_2} \sin(\alpha-\theta)\cos(\alpha-\theta)\,\hat{I}_3\,d\alpha = \int_{\alpha_1}^{\alpha_2} \sin(\alpha-\theta)\cos(\alpha-\theta),$$

$$\left\{\left[1 - \frac{1}{2r^2}\left(\frac{1}{\cos(\alpha-\theta)}\right)^2\right] - \frac{1}{3}\left[1 - \frac{3}{2r^2}\left(\frac{1}{\cos\alpha-\theta)}\right)^2\right]\right\} d\alpha$$

$$= \frac{2}{3}\int_{\alpha_1}^{\alpha_2} \sin(\alpha-\theta)\cos(\alpha-\theta)\,d\alpha = -\frac{1}{3}\cos^2(\alpha-\theta)\Big|_{\alpha_1}^{\alpha_2}$$

$$\int_{\alpha_1}^{\alpha_2} \sin(\alpha-\theta)\cos^2(\alpha-\theta)\,\hat{I}_4\,d\alpha = \int_{\alpha_1}^{\alpha_2} \sin(\alpha-\theta)\cos^2(\alpha-\theta) \times$$

$$\times\left\{\frac{3}{8}\left[\frac{\pi}{2} - \frac{1}{r\cos(\alpha-\theta)} - \frac{1}{6r^3}\left(\frac{1}{\cos(\alpha-\theta)}\right)^3\right] + \right.$$

$$+ \frac{5}{8}\left[\frac{1}{r\cos(\alpha-\theta)} - \frac{1}{2r^3}\left(\frac{1}{\cos(\alpha-\theta)}\right)^3\right] -$$

$$\left. - \frac{1}{4}\left[\frac{1}{r\cos(\alpha-\theta)} - \frac{3}{2r^3}\left(\frac{1}{\cos(\alpha-\theta)}\right)^3\right]\right\} d\alpha$$

$$= \frac{3\pi}{16}\int_{\alpha_1}^{\alpha_2} \sin(\alpha-\theta)\cos^2(\alpha-\theta)\,d\alpha = -\frac{\pi}{16}\cos^3(\alpha-\theta)\Big|_{\alpha_1}^{\alpha_2}.$$

$$\int_{\alpha_1}^{\alpha_2} \sin(\alpha-\theta)\cos^3(\alpha-\theta)\,\hat{I}_5\,d\alpha = \int_{\alpha_1}^{\alpha_2} \sin(\alpha-\theta)\cos^3(\alpha-\theta)\,\times$$

$$\times\left\{1-\frac{1}{2r^2\cos^2(\alpha-\theta)}-\frac{2}{3}\left[1-\frac{3}{2r^2\cos^2(\alpha-\theta)}\right],\right.$$

$$\left.+\frac{1}{5}\left[1-\frac{5}{2r^2\cos^2(\alpha-\theta)}\right]\right\}d\alpha = \frac{8}{15}\int_{\alpha_1}^{\alpha_2}\sin(\alpha-\theta)\cos^3(\alpha-\theta)\,d\alpha$$

$$= -\frac{2}{15}\cos^4(\alpha-\theta)\Big|_{\alpha_1}^{\alpha_2},$$

$$\int_{\alpha_1}^{\alpha_2}\sin(\alpha-\theta)\cos^4(\alpha-\theta)\,\hat{I}_6\,d\delta = \int_{\alpha_1}^{\alpha_2}\sin(\alpha-\theta)\cos^4(\alpha-\theta)\,\times$$

$$\times\left\{\frac{5}{16}\left[\frac{\pi}{2}-\frac{1}{r\cos(\alpha-\theta)}\right]+\frac{11}{16}\left[\frac{1}{r\cos(\alpha-\theta)}\right]-\frac{13}{24}\left[\frac{1}{r\cos(\alpha-\theta)}\right]+\right.$$

$$\left.+\frac{1}{6}\left[\frac{1}{r\cos(\alpha-\theta)}\right]\right\}d\alpha = \frac{5\pi}{32}\int_{\alpha_1}^{\alpha_2}\sin(\alpha-\theta)\cos^4(\alpha-\theta)\,d\alpha$$

$$= -\frac{\pi}{32}\cos^5(\alpha-\theta)\Big|_{\alpha_1}^{\alpha_2},$$

$$\int_{\alpha_1}^{\alpha_2}\sin(\alpha-\theta)\cos^5(\alpha-\theta)\,\hat{I}_7\,d\alpha = \int_{\alpha_1}^{\alpha_2}\sin(\alpha-\theta)\cos^5(\alpha-\theta),$$

$$\left\{1-1+\frac{3}{5}[1]-\frac{1}{7}[1]\right\}d\alpha = \frac{16}{35}\int_{\alpha_1}^{\alpha_2}\sin(\alpha-\theta),$$

$$\cos^5(\alpha-\theta)\,d\alpha = -\frac{8}{105}\cos^6(\alpha-\theta)\Big|_{\alpha_1}^{\alpha_2},$$

$$\int_{\alpha_1}^{\alpha_2}\sin^2(\alpha-\theta)\,\hat{I}_3\,d\alpha = \frac{2}{3}\int_{\alpha_1}^{\alpha_2}\sin^2(\alpha-\theta)\,d\alpha = \frac{1}{3}\left[\alpha-\frac{1}{2}\sin 2(\alpha-\theta)\right]_{\alpha_1}^{\alpha_2},$$

$$\int_{\alpha_1}^{\alpha_2}\sin^2(\alpha-\theta)\cos(\alpha-\theta)\,\hat{I}_4\,d\alpha = \frac{3\pi}{16}\int_{\alpha_1}^{\alpha_2}\sin^2(\alpha-\theta)\cos(\alpha-\theta)\,d\alpha$$

$$= \frac{\pi}{16}\sin^3(\alpha-\theta)\Big|_{\alpha_1}^{\alpha_2},$$

$$\int_{\alpha_1}^{\alpha_2}\sin^2(\alpha-\theta)\cos^2(\alpha-\theta)\,\hat{I}_5\,d\alpha = \frac{8}{15}\int_{\alpha_1}^{\alpha_2}\sin^2(\alpha-\theta)\cos^2(\alpha-\theta)\,d\alpha$$

$$= \frac{1}{15}\left[\alpha-\frac{1}{4}\sin 4(\alpha-\theta)\right]_{\alpha_1}^{\alpha_2},$$

$$\int_{\alpha_1}^{\alpha_2} \sin^2(\alpha-\theta)\cos^3(\alpha-\theta)\,\hat{I}_6\,d\alpha = \frac{5\pi}{32}\int_{\alpha_1}^{\alpha_2}\sin^2(\alpha-\theta)\cos^3(\alpha-\theta)\,d\alpha$$

$$= \frac{\pi}{32}\left[\sin^3(\alpha-\theta)\cos^2(\alpha-\theta)+\frac{2}{3}\sin^3(\alpha-\theta)\right]_{\alpha_1}^{\alpha_2},$$

$$\int_{\alpha_1}^{\alpha_2} \sin^2(\alpha-\theta)\cos^4(\alpha-\theta)\,\hat{I}_7\,d\alpha = \frac{16}{35}\int_{\alpha_1}^{\alpha_2}\sin^2(\alpha-\theta)\cos^4(\alpha-\theta)\,d\alpha$$

$$= \frac{16}{35}\left[\frac{\alpha}{16}+\frac{1}{6}\sin^3(\alpha-\theta)\cos^3(\alpha-\theta)-\frac{1}{64}\sin 4(\alpha-\theta)\right]_{\alpha_1}^{\alpha_2}.$$

Substituting and collecting terms up to the r^6 term,

$$\left(\frac{dn}{dt\,dS'}\right)_T = 2k_1\left(\frac{dn}{dS\,dt}\right)_S\Bigg\{\cos\theta\left[\frac{1}{r^3}\left(-\frac{\pi}{16}\cos^3(\alpha-\theta)\right)+\right.$$

$$+\frac{1}{r^4}\left(\frac{1}{3}\cos^2(\alpha-\theta)-\frac{8}{15}\cos^4(\alpha-\theta)\right)+$$

$$+\frac{1}{r^5}\frac{3\pi}{8}\left(\cos^3(\alpha-\theta)-\cos^5(\alpha-\theta)\right)+$$

$$\left.+\frac{1}{r^6}\left(-\frac{2}{3}\cos^2(\alpha-\theta)+\frac{16}{5}\cos^4(\alpha-\theta)-\frac{256}{105}\cos^6(\alpha-\theta)\right)\right]_{\alpha_1}^{\alpha_2}-$$

$$-\sin\theta\left[\frac{1}{r^3}\left(\frac{\pi}{16}\sin^3(\alpha-\theta)\right)+\frac{1}{r^4}\left(-\frac{1}{15}(\alpha-\theta)+\frac{1}{6}\sin 2(\alpha-\theta)-\right.\right.$$

$$\left.-\frac{1}{15}\sin 4(\alpha-\theta)\right)+\frac{1}{r^5}\left(-\frac{\pi}{8}\sin^3(\alpha-\theta)+\right.$$

$$\left.+\frac{3\pi}{8}\sin^3(\alpha-\theta)\cos^2(\alpha-\theta)\right)+\frac{1}{r^6}\left(-\frac{2}{105}(\alpha-\theta)-\right.$$

$$\left.\left.-\frac{1}{3}\sin 2(\alpha-\theta)+\frac{6}{15}\sin 4(\alpha-\theta)+\frac{256}{105}\sin^3(\alpha-\theta)\cos^3(\alpha-\theta)\right)\right]_{\alpha_1}^{\alpha_2}\Bigg\}.$$

For no terminator, i.e., $-\frac{\pi}{2} \leqq \alpha_1 = \theta - \Phi$, $\alpha_2 = \theta + \Phi \leqq \frac{\pi}{2}$ $\left(\text{where } \Phi = \cos^{-1}\left(\frac{1}{r}\right)\right)$

$$\cos^n(\alpha-\theta)\Big|_{\alpha_1=\theta+\Phi}^{\alpha_2=\theta+\Phi} = 0 \quad \text{for} \quad n = 1, 2, 3, \ldots,$$

$$\sin^3(\alpha-\theta)\Big|_{\theta-\Phi}^{\theta+\Phi} = 2\sin^3\Phi = 2\left(1-\frac{1}{r^2}\right)^{3/2} = 2-\frac{3}{r^2}+0\left(\frac{1}{r^4}\right),$$

$$\sin 2(\alpha-\theta)\Big|_{\theta-\Phi}^{\theta+\Phi} = 2\sin 2\Phi = 4\sin\Phi\cos\Phi = \frac{4}{r}\left(1-\frac{1}{r^2}\right)^{1/2}$$

$$= \frac{4}{r}+0\left(\frac{1}{r^3}\right),$$

$$\sin 4(\alpha-\theta)\Big|_{\theta-\Phi}^{\theta+\Phi} = 2\sin 4\Phi = 8\sin\Phi\cos\Phi - 16\sin^3\Phi\cos\Phi$$

$$= \frac{8}{r}\left[1-0\left(\frac{1}{r^2}\right)\right] = -\frac{8}{r}+0\left(\frac{1}{r^3}\right),$$

$$\alpha\Big|_{\theta-\Phi}^{\theta+\Phi} = 2\Phi = 2\cos^{-1}\left(\frac{1}{r}\right) = 2\left[\frac{\pi}{2} - \frac{1}{r} + 0\left(\frac{1}{r^3}\right)\right] = \pi - \frac{2}{r} + 0\left(\frac{1}{r^3}\right),$$

$$\sin^3(\alpha-\theta)\cos^2(\alpha-\theta)\Big|_{\theta-\Phi}^{\theta+\Phi} = 2\sin^3\Phi\cos^2\Phi = 2(\sin^3\Phi - \sin^5\Phi)$$

$$= 2\left[1 + 0\left(\frac{1}{r^2}\right)\right],$$

$$-2\left[1 + 0\left(\frac{1}{r^2}\right)\right] = 0 + 0\left(\frac{1}{r^2}\right),$$

$$\sin^3(\alpha-\theta)\cos^3(\alpha-\theta)\Big|_{\theta-\Phi}^{\theta+\Phi} = 2\sin^3\Phi\cos^3\Phi = \frac{2}{r}(\sin^3\Phi - \sin^5\Phi)\cos\Phi$$

$$= 0 + 0\left(\frac{1}{r^3}\right).$$

Substituting the above yields:

$$\left(\frac{dn}{dt\,dS'}\right)_{T'} = -2k_1\left(\frac{dn}{dS\,dt}\right)_S \sin\theta\left[\frac{1}{r^3}\left(\frac{\pi}{8}\right) + \frac{1}{r^4}\left(-\frac{\pi}{15}\right) + \right.$$

$$\left. + \frac{1}{r^5}\left(\frac{4}{3} - \frac{7\pi}{16}\right) + \frac{1}{r^6}\left(-\frac{2\pi}{105}\right) + 0\left(\frac{1}{r^7}\right)\right].$$

When, terminator is involved, i.e., $\alpha_2 = \theta + \Phi > \frac{\pi}{2}$,

$$\cos^n(\alpha-\theta)\Big|_{\theta-\Phi}^{\pi/2} = \cos^n\left(\frac{\pi}{2} - \theta\right) - \cos^n\Phi = \sin^n\theta - \frac{1}{r^n},$$

$$n = 2, 3, 4, 5, 6,$$

$$\sin^3(\alpha-\theta)\Big|_{\theta-\Phi}^{\pi/2} = \sin^3\left(\frac{\pi}{2} - \theta\right) + \sin^3\Phi = \cos^3\theta + 1 -$$

$$-\frac{3}{2}\left(\frac{1}{r^2}\right) + 0\left(\frac{1}{r^4}\right),$$

$$\sin 2(\alpha-\theta)\Big|_{\theta-\Phi}^{\pi/2} = \sin 2\left(\frac{\pi}{2} - \theta\right) + \sin 2\Phi = \sin 2\theta + \sin 2\Phi =$$

$$2\sin\theta\cos\theta + \frac{2}{r} + 0\left(\frac{1}{r^3}\right),$$

$$\sin 4(\alpha-\theta)\Big|_{\theta-\Phi}^{\pi/2} = \sin 4\left(\frac{\pi}{2} - \theta\right) + \sin 4\Phi = -\sin 4\theta + \sin 4\Phi =$$

$$-4\sin\theta\cos\theta + 8\sin^3\theta\cos\theta - \frac{4}{r} + 0\left(\frac{1}{r^3}\right),$$

$$\sin^3(\alpha-\theta)\cos^2(\alpha-\theta)\Big|_{\theta-\Phi}^{\pi/2} = \sin^3\left(\frac{\pi}{2} - \theta\right)\cos^2\left(\frac{\pi}{2} - \theta\right) +$$

$$+ \sin^3\Phi\cos^2\Phi = \cos^3\theta\sin^2\theta + 0\left(\frac{1}{r^2}\right),$$

$$\sin^3(\alpha-\theta)\cos^3(\alpha-\theta)\Big|_{\theta-\Phi}^{\pi/2} = \sin^3\left(\frac{\pi}{2} - \theta\right)\cos^3\left(\frac{\pi}{2} - \theta\right) +$$

$$+ \sin^3\Phi\cos^3\Phi = \cos^3\theta\sin^3\theta + 0\left(\frac{1}{r^3}\right),$$

$$\alpha\Big|_{\theta-\Phi}^{\pi/2} = \frac{\pi}{2} - \theta + \Phi = \pi - \theta - \frac{1}{r} + 0\left(\frac{1}{r^3}\right).$$

Substituting the above

$$\left(\frac{dn}{dt\,dS'}\right)_T = 2k_1\left(\frac{dn}{dS\,dt}\right)_S\left\{\cos\theta\left[\frac{1}{r^3}\left(-\frac{\pi}{16}\sin^3\theta\right)+\right.\right.$$

$$+\frac{1}{r^4}\left(\frac{1}{3}\sin^2\theta-\frac{8}{15}\sin^4\theta\right)+\frac{1}{r^5}\frac{3\pi}{8}(\sin^3\theta-\sin^5\theta)+$$

$$\left.+\frac{1}{r^6}\left(\frac{\pi}{16}-\frac{1}{3}-\frac{2}{3}\sin^2\theta+\frac{16}{5}\sin^4\theta-\frac{256}{105}\sin^6\theta\right)\right]-$$

$$-\sin\theta\left[\frac{1}{r^3}\frac{\pi}{16}(\cos^3\theta+1)+\frac{1}{r^4}\left(-\frac{\pi}{15}+\frac{1}{15}\theta+\right.\right.$$

$$\left.+\frac{3}{5}\sin\theta\cos\theta-\frac{8}{15}\sin^3\theta\cos\theta\right)+\frac{1}{r^5}\left(\frac{2}{3}-\frac{7\pi}{32}-\frac{\pi}{8}\cos^3\theta+\right.$$

$$\left.+\frac{3\pi}{8}\cos^3\theta\sin^2\theta\right)+\frac{1}{r^6}\left(-\frac{2\pi}{105}+\frac{2}{105}\theta-\frac{142}{105}\sin\theta\cos\theta+\right.$$

$$\left.\left.\left.+\frac{48}{35}\sin^3\theta\cos\theta+\frac{256}{105}\cos^3\theta\sin^3\theta\right)\right]\right\},$$

or rearranging,

$$\left(\frac{dn}{dt\,dS'}\right)_T = 2k_1\left(\frac{dn}{dS\,dt}\right)_S\left\{\frac{1}{r^3}\left(-\frac{\pi}{16}\sin\theta\right)(1+\cos\theta)+\right.$$

$$+\frac{1}{r^4}\left(\frac{1}{15}\sin\theta\right)[(\pi-\theta)-\sin 2\theta]+\frac{1}{r^5}(\sin\theta)\left(\frac{7\pi}{32}-\frac{2}{3}+\frac{\pi}{8}\cos^3\theta\right)+$$

$$\left.+\frac{1}{r^6}\left[\frac{2}{105}(\pi-\theta)\sin\theta+\left(\frac{\pi}{16}-\frac{9}{35}\right)\cos\theta+\frac{8}{15}\cos^3\theta-\frac{64}{105}\cos^5\theta\right]\right\}.$$

Denoting

$$I_4 \triangleq \left\{\frac{1}{r}\left(\frac{\pi}{8}\right)+\frac{1}{r^2}\left(-\frac{\pi}{15}\right)+\frac{1}{r^3}\left(-\frac{7\pi}{16}+\frac{4}{3}\right)+\right.$$

$$\left.+\frac{1}{r^4}\left(-\frac{2\pi}{105}\right)+\cdots\right\}\frac{1}{\sin\theta},$$

$$I_5 \triangleq \frac{1}{r}\left(+\frac{\pi}{16}\right)\left(\frac{1+\cos\theta}{\sin\theta}\right)+\frac{1}{r^2}\left(\frac{1}{15}\right)\left(\frac{-(\pi-\theta)+2\sin 2\theta}{\sin\theta}\right)$$

$$+\frac{1}{r^3}\left(-\frac{7\pi}{32}+\frac{2}{3}-\frac{\pi}{8}\cos^3\theta\right)\frac{1}{\sin\theta}+$$

$$+\frac{1}{r^4}\left(-2\,\frac{[\pi-\theta]}{105}\sin\theta+\left[\frac{9}{35}-\frac{\pi}{16}\right]\cos\theta-\frac{8}{15}\cos^3\theta+\frac{64}{105}\cos^5\theta\right)+\cdots.$$

we have

$$\left(\frac{1}{\sin^2\theta}\right)\left(\frac{dn}{dt\,dS'}\right)_T = \frac{2}{r^2}k_1\,\varepsilon'_{\text{diff.}}\left(\frac{dn}{dS\,dt}\right)_S$$

$$\{U(\theta_T-\theta)\,I_4+U(\theta'_T-\theta)\,U(\theta-\theta_T)\,I_5\},$$

where

$$\theta_T = \frac{\pi}{2}-\cos^{-1}\left(\frac{1}{r}\right) \qquad \theta'_T = \frac{\pi}{2}+\cos^{-1}\left(\frac{1}{r}\right)$$

[note the sign change due to the convention established by Eq. (7)].

According to the analysis of Col. R. R. LOCHRY and verified by B. SMITH[1], the radial component of the photon flux is given by

$$(f_r)_{\text{diff.}} = \frac{2k_1}{r^2}\,\varepsilon'_{\text{diff}}\left(\frac{dn}{dS\,dt}\right)_S U(\theta_T' - \theta)\,[I_6 - U(\theta - \theta_T)I_7],$$

where

$$I_6 \triangleq c\left[(\pi/3) + \frac{1}{r}(\pi/4) + \frac{1}{r^2}(-\pi/15) + \frac{1}{r^3}(\pi/12 - 5/18) + \right.$$
$$\left. + \frac{1}{r^4}(-\pi/105)\right] + \cdots,$$

$$I_7 \triangleq \frac{1}{3}[\theta c - s] + \frac{1}{r}(\pi/16)(1 + 2c - 3c^2) + \frac{1}{r^2}[(107/120 - \pi/4)s -$$
$$- (\theta/15)c - (8/15)s c^2] + \frac{1}{r^3}[c(\pi/24 - 5/36) - (\pi/8)c^2 + (5\pi/24)c^4] +$$
$$+ \frac{1}{r^4}[(11\pi/16 - 361/168)s - \theta c/105 - (8/21)s c^2 + (32/35)s c^4 -$$
$$- (1/24)c^2/s] + \cdots.$$

It should be noted that the calculation fo f_{total} for diffuse reflection essentially involved integrals of the symbolic form $\int\limits_{S_{1,2}} |df|$, where df represents the incremental flux at S' occasioned by the elements dS_1 or dS_2. Such an integration truly gives the total incident photon flux (from all directions due to the diffusely reflecting object) but is *not* equal to the absolute magnitude of the total flux vector $|f|$, i.e., not equal to

$$\Bigl|\int\limits_{S_{1,2}} df\Bigr|$$

Thus

$$f^2_{\substack{\text{mag.}\\ \text{diff.}}} \neq (f_r)^2_{\text{diff.}} + (f_t)^2_{\text{diff.}}.$$

A.4 Normal Component of Delayed IR Flux. The normal component of delayed IR flux is obtained by an analytical integration due to R. G.

[1] The development of the above equations is identical to that used for the total flux component with the exception that the values of $\lambda_i, i = 0, 1, \ldots, 5$, are replaced by μ_i.

$$\mu_0 = \frac{2}{r^3} - \frac{1}{r}, \qquad \mu_3 = \frac{8}{r^2} - \frac{56}{r^4},$$

$$\mu_1 = \frac{13}{r^4} - \frac{5}{r^2} + 1, \quad \mu_4 = \frac{20}{r^3},$$

$$\mu_2 = \frac{3}{r} - \frac{18}{r^3}, \qquad \mu_5 = \frac{48}{r^4}.$$

TOTTEN. The coordinate system and approach are the same as they were for the integration of the total IR flux. Note that because the spacecraft is on a meridian in both equatorial and $I'\,J'$ coordinates of Fig. 8 the orthogonal component will vanish and the only non-radial component will be the normal one, which is tangent to a meridian (see Fig. 9). From Figs. 6 and 7 we observe from the law of sines that

$$\sin\zeta = \frac{\sin\delta}{\sin\theta},$$

so that $\cos\gamma \sin\xi \sin\zeta = \frac{r}{\varrho^2}\left(\sin\delta\cos\delta\cos\alpha' - \frac{1}{r}\sin\delta\right)$.

Thus

$$(f_n)_{\mathrm{IR}} = (1 - \varepsilon'_{\mathrm{diff.}} - \varepsilon_{\mathrm{spec}})\left(\frac{dn}{dS\,dt}\right)_S k_2 \int\limits_S \frac{(1 - c_1\sin^2\varphi)\cos\gamma\sin\xi\sin\zeta}{\varrho^2}\,dS$$

becomes (note that the isotropic term, 1, does not contribute to the normal flux and that again $\sin^2\varphi = \sin^2\delta\cos^2\delta_{\mathrm{eq}} + \cos^2\delta\sin^2\delta_{\mathrm{eq}}\cos^2\alpha' + 2\sin\delta\cos\delta\sin\delta_{\mathrm{eq}}\cos\delta_{\mathrm{eq}}\cos\alpha'$)

$$\int\limits_S \frac{r}{\varrho^4}\Big[(-c_1\sin^2\delta\cos^2\delta_{\mathrm{eq}} - c_1\cos^2\delta\sin^2\delta_{\mathrm{eq}}\cos^2\alpha' -$$

$$- 2c_1\sin\delta_{\mathrm{eq}}\cos\delta_{\mathrm{eq}}\sin\delta\cos\delta\cos\alpha')\left(\sin\delta\cos\delta\cos\alpha' - \frac{1}{r}\sin\delta\right)dS$$

$$= \int\limits_S \frac{r}{\varrho^4}\Big[-c_1\sin^3\delta\cos\delta\cos\alpha'\cos^2\delta_{\mathrm{eq}} - c_1\cos^3\delta\sin\delta\cos^3\alpha'\sin^2\delta_{\mathrm{eq}}$$

$$- 2c_1\sin^2\delta\cos^2\delta\cos^2\alpha'\sin\delta_{\mathrm{eq}}\cos\delta_{\mathrm{eq}} - \frac{1}{r}(\delta - c_1\sin^3\delta\cos^2\delta_{\mathrm{eq}} -$$

$$- c_1\cos^2\delta\sin\delta\cos^2\alpha'\sin^2\delta_{\mathrm{eq}} - 2c_1\sin^2\delta\cos\delta\cos\alpha'\sin\delta_{\mathrm{eq}}\cos\delta_{\mathrm{eq}})\Big]\,dS$$

$$= \int\limits_S \frac{r}{\varrho^4}\Big[-c_1\sin^3\delta\cos^2\delta\cos\alpha'\cos^2\delta_{\mathrm{eq}} - c_1\cos^4\delta\sin\delta\cos^3\alpha'\sin^2\delta_{\mathrm{eq}} -$$

$$- 2c_1\cos^3\delta\sin^2\delta\cos^2\alpha'\sin\delta_{\mathrm{eq}}\cos\delta_{\mathrm{eq}} - \frac{1}{r} -$$

$$- (c_1\sin^3\delta\cos\delta\cos^2\delta_{\mathrm{eq}} - c_1\cos^3\delta\sin\delta\cos^2\alpha'\sin^2\delta_{\mathrm{eq}} -$$

$$- 2c_1\cos^2\delta\sin^2\delta\cos\alpha'\sin\delta_{\mathrm{eq}}\cos\delta_{\mathrm{eq}})\Big]\,d\delta\,d\alpha'.$$

From Eq. (A-9) we introduce the relationship for $\frac{r}{\varrho^4}$ and by eliminating the odd functions, which cancel for δ integrated from the same plus to the same minus value (i.e., $\int_0^{\Phi} s^n c^m + \int_{-\Phi}^{0} s^n c^m = 0$ where n is odd), we find

$$-\int_S \Big[2c_1 c^3\delta\, s^2\delta\, c^2\alpha'\, s\,\delta_{\text{eq}}\, c\,\delta_{\text{eq}}\left(\frac{r}{\varrho^4}\right) +$$
$$+\frac{1}{r}(-2c_1 c^2\delta\, s^2\delta\, c\,\alpha'\, s\,\delta_{\text{eq}}\, c\,\delta_{\text{eq}})\left(\frac{r}{\varrho^4}\right)\Big]d\delta\, d\alpha'$$
$$=\frac{-c_1 s\,\delta_{\text{eq}}\, c\,\delta_{\text{eq}}}{r^2}\int\Big[\frac{1}{r}(2c^3\delta\, s^2\delta\, c^2\alpha') +$$
$$+\frac{1}{r^2}(8c^4\delta\, s^2\delta\, c^3\alpha' - 2c^2\delta\, s^2\delta\, c\,\alpha') +$$
$$+\frac{1}{r^3}(24c^5\delta\, s^2\delta\, c^4\alpha' - 4c^3\delta\, s^2\delta\, c^2\alpha' - 8c^3\delta\, s^2\delta\, c^2\alpha') +$$
$$+\frac{1}{r^4}(64c^6\delta\, s^2\delta\, c^5\alpha' - 24c^4\delta\, s^2\delta\, c^3\alpha' - 24c^4\delta\, s^2\delta\, c^3\alpha' +$$
$$+4c^2\delta\, s^2\delta\, c\,\alpha') + \frac{1}{r^5}(160c^7\delta\, s^2\delta\, c\,\alpha' - 96c^5\delta\, s^2\delta\, c^4\alpha' +$$
$$+6c^3\delta\, s^2\delta\, c^2\alpha' - 64c^5\delta\, s^2\delta\, c^4\alpha' + 24c^3\delta\, s^2\delta\, c^2) +$$
$$+\frac{1}{r^6}(384c^8\delta s^2 c^7\alpha' - 320c^6\delta s^2\delta c^5\alpha' + 48c^4\delta s^2\delta c^3\alpha' - 160c^6\delta s^2\delta c^5\alpha' +$$
$$+96c^4\delta s^2\delta c^3\alpha' - 6c^2\delta s^2\delta c\,\alpha') + \cdots\Big]d\alpha'\, d\delta,$$
$$=\frac{-c_1 s\,\delta_{\text{eq}}\, c\,\delta_{\text{eq}}}{r^2}\int_S\Big[\frac{1}{r}(2c^3\delta s^2\delta c^2\alpha') + \frac{1}{r^2}(8c^4\delta\, s^2\delta\, c^3\alpha' - 2c^2\delta\, s^2\delta\, c\alpha') +$$
$$+\frac{1}{r^3}(24c^5\delta s^2\delta\, c^4\alpha' - 12c^3\delta\, s^2\delta\, c^2\alpha') + \frac{1}{r^4}(64c^6\delta\, s^2\delta\, c^5\alpha' -$$
$$-48c^4\delta\, s^2\delta\, c^3\alpha' + 4c^2\delta\, s^2\delta\, c\,\alpha') +$$
$$+\frac{1}{r^5}(160c^7\delta\, s^2\delta\, c^6\alpha' - 160c^5\delta\, s^2\delta\, c^4\alpha' + 30c^3\delta\, s^2\delta\, c^2\alpha') +$$
$$+\frac{1}{r^6}(384c^8\delta s^2\delta c^7\alpha' - 480c^6\delta\, s^2\delta\, c^5\alpha' + 144c^4\delta\, s^2\delta\, c^3\alpha' -$$
$$-6c^2\delta\, s^2\delta c\,\alpha') + \cdots\Big]d\alpha'\, d\delta$$
$$=-c_1\frac{s\,\delta_{\text{eq}}\, c\,\delta_{\text{eq}}}{r^2}\int_S\Big[\left(-\frac{2}{r^2}+\frac{4}{r^4}-\frac{6}{r^6}\right)c^2\delta\, s^2\delta\, c\,\alpha' +$$
$$+\left(\frac{2}{r}-\frac{12}{r^3}+\frac{30}{r^5}\right)c^3\delta\, s^2\delta\, c^2\alpha' + \left(\frac{8}{r^2}-\frac{48}{r^4}+\frac{144}{r^6}\right)c^4\delta\, s^2\delta\, c^3\alpha' +$$
$$+\left(\frac{24}{r^3}-\frac{160}{r^5}\right)c^5\delta\, s^2\delta\, c^4\alpha' + \left(\frac{64}{r^4}-\frac{480}{r^6}\right)c^6\delta\, s^2\delta\, c^5\alpha' +$$
$$+\left(\frac{160}{r^5}\right)c^7\delta\, s^2\delta\, c^6\alpha' + \left(\frac{384}{r^6}\right)c^8\delta s^2\delta c^7\alpha'\Big]d\alpha'\, d\delta,$$
$$=c_1\frac{s\,\delta_{\text{eq}}\, c\,\delta_{\text{eq}}}{r^2}\int_S(\bar{\lambda}_1 c^2\delta\, s^2\delta\, c\,\alpha' + \bar{\lambda}\, c^3\delta\, s^2\delta\, c^2\alpha' + \bar{\lambda}_3 c^4\delta\, s^2\delta\, c^3\alpha' +$$
$$+\bar{\lambda}_4 c^5\delta\, s^2\delta\, c^4\alpha' + \bar{\lambda}_5 c^6\delta\, s^2\delta\, c^5\alpha' + \bar{\lambda}_6 c^7\delta\, s^2\delta\, c^6\alpha',$$

where

$$\bar{\lambda}_1 \triangleq -\frac{2}{r^2} + \frac{4}{r^4} - \frac{6}{r^6} + \cdots, \qquad \bar{\lambda}_4 \triangleq \frac{24}{r^3} - \frac{160}{r^5} + \cdots,$$

$$\bar{\lambda}_2 \triangleq \frac{2}{r} - \frac{12}{r^3} + \frac{30}{r^5} + \cdots, \qquad \bar{\lambda}_5 \triangleq \frac{64}{r^4} - \frac{480}{r^6} + \cdots,$$

$$\bar{\lambda}_3 \triangleq \frac{8}{r^2} - \frac{48}{r^4} + \frac{144}{r^6} + \cdots, \qquad \bar{\lambda}_6 \triangleq \frac{160}{r^5} + \cdots, \qquad \bar{\lambda}_7 \triangleq \frac{384}{r^6} + \cdots$$

Placing limits on the integrals; we have

$$\frac{-2c_1\, s\,\delta_{\mathrm{eq}}\, c\,\delta_{\mathrm{eq}}}{r^2}\left\{\int\limits_{\alpha'=\cos^{-1}\left(\frac{1}{r}\right)}^{0}\ \int\limits_{\delta=0}^{\cos^{-1}\left(\frac{1}{r\cos\alpha'}\right)} [\,]\, d\delta\, d\alpha' + \right.$$

$$\left. + \int\limits_{\alpha'=0}^{\cos^{-1}\left(\frac{1}{r}\right)}\ \int\limits_{\delta=-\cos^{-1}\left(\frac{1}{r\cos\alpha'}\right)}^{0} [\,]\, d\delta\, d\alpha'\right\}.$$

Integrating first with respect to δ, we find

$$I_N = -\frac{2c_1\, s\,\delta_{\mathrm{eq}}\, c\,\delta_{\mathrm{eq}}}{r^2}\left\{\int\limits_{\alpha'=\cos^{-1}\left(\frac{1}{r}\right)}^{0}\left[\bar{\lambda}_1\, c\,\alpha\left(-\frac{s\,\delta\, c^3\,\delta}{4} + \frac{\delta}{8} + \frac{s\,\delta\, c\,\delta}{8}\right) + \right.\right.$$

$$+ \bar{\lambda}_2\, c^2\,\alpha'\left(-\frac{s\,\delta\, c^4\,\delta}{5} + \frac{s\,\delta}{8} - \frac{s^3\,\delta}{15}\right) +$$

$$+ \bar{\lambda}_3\, c^3\,\alpha'\left(-\frac{s\,\delta\, c^5\,\delta}{6} + \frac{\delta}{16} + \frac{5s\,\delta\, c\,\delta}{48} - \frac{s^3\,\delta\, c\,\delta}{24}\right) +$$

$$+ \bar{\lambda}_4\, c^4\,\alpha'\left(\frac{s\,\delta c^6\,\delta}{7} + \frac{c^4\,\delta\, s\,\delta}{35} + \frac{4s\,\delta}{35} - \frac{4s^3\,\delta}{105}\right) +$$

$$+ \bar{\lambda}_5\, c^5\,\alpha'\left(\frac{s\,\delta c^7\,\delta}{8} + \frac{3\delta}{64} + \frac{5s\,\delta c\,\delta}{64} - \frac{s^3\,\delta c\,\delta}{24}\right) +$$

$$\left. + \bar{\lambda}_6\, c^6\,\alpha'\left(\frac{s\,\delta c^8\,\delta}{9} + \frac{c^4\,\delta s\,\delta}{45} + \frac{4s\,\delta}{45} - \frac{4s^3\,\delta}{135}\right) + \cdots\right]_0^{\cos^{-1}\left(\frac{1}{r\cos\alpha'}\right)} d\alpha' +$$

$$\left. + \int\limits_{\alpha'=0}^{\cos^{-1}\left(\frac{1}{r}\right)} [\,]_{-\cos^{-1}\left(\frac{1}{r\cos\alpha}\right)}^{0}\, d\alpha'\right\}.$$

Rearranging terms yields:

$$\frac{-2c_1\, s\,\delta_{\rm eq}\, c\,\delta_{\rm eq}}{r^2}\left\{\int\limits_{\alpha'=\cos^{-1}\left(\frac{1}{r}\right)}^{0}\left[\left(\frac{\bar\lambda_1\, c\,\alpha'}{8}+\frac{\bar\lambda_3\, c^3\,\alpha'}{16}+\frac{3\,\bar\lambda_5\, c^5\,\alpha'}{64}\right)\delta+\right.\right.$$

$$+\left(\frac{\bar\lambda_2\, c^2\,\alpha'}{5}+\frac{4\,\bar\lambda_4\, c^4\,\alpha'}{35}+\frac{4\,\bar\lambda_6\, c^6\,\alpha'}{45}\right)s\,\delta+$$

$$+\left(\frac{\bar\lambda_1\, c\,\alpha'}{8}+\frac{5\,\bar\lambda_3\, c^3\,\alpha'}{48}+\frac{5\,\bar\lambda_5\, c^5\,\alpha'}{64}\right)s\,\delta\, c\,\delta+$$

$$+\left(-\frac{\bar\lambda^2\, c_2\,\alpha'}{15}-\frac{4\,\bar\lambda_4\, c^4\,\alpha'}{105}-\frac{4\,\bar\lambda_6\, c^6\,\alpha'}{135}\right)s^3\,\delta+\left(-\frac{\bar\lambda_1\, c\,\alpha'}{5}\right)s\,\delta\, c^3\,\delta+$$

$$+\left(-\frac{\bar\lambda^2\, c_2\,\alpha'}{5}+\frac{\bar\lambda_4\, c^4\,\alpha'}{35}+\frac{\bar\lambda_6\, c^6\,\alpha'}{45}\right)s\,\delta\, c^4\,\delta+$$

$$+\left(-\frac{\bar\lambda_3\, c^3\,\alpha'}{6}\right)s\,\delta\, c^5\,\delta+$$

$$+\left(-\frac{\bar\lambda_3\, c^3\,\alpha'}{24}-\frac{\bar\lambda_5\, c^5\,\alpha'}{24}\right)s^3\,\delta\, c\,\delta+\cdots\Big]_0^{\cos^{-1}\left(\frac{1}{r\,c\,\alpha'}\right)}d\alpha'+$$

$$\left.+\int\limits_{\alpha'=0}^{\cos^{-1}\left(\frac{1}{r}\right)}\Big[\;\Big]_{-\cos\left(\frac{1}{r\,c\,\alpha'}\right)}^{0}d\alpha'\right\}.$$

Substituting and neglecting terms higher than $\frac{1}{r^5}$, and recognizing that $\sin(-\delta)=-\sin\delta$ while $\cos(-\delta)=\cos\delta$; gives us

$$\frac{-2c_1\, s\,\delta_{\rm eq}\, c\,\delta_{\rm eq}}{r^2}\left\{\int\limits_{\alpha'=\cos^{-1}\left(\frac{1}{r}\right)}^{0}\left[2\left(\frac{\bar\lambda_1\,\pi}{16}\,c\,\alpha'-\frac{\bar\lambda_1}{8r}-\frac{\bar\lambda_1}{48\,r^3\,c^2\,\alpha'}+\right.\right.\right.$$

$$\left.+\frac{\bar\lambda_3\,\pi}{32}\,c\,\alpha'-\frac{\bar\lambda_3\, c^2\,\alpha'}{16\,r}-\frac{\bar\lambda_3}{96\,r^3}+\frac{3\,\bar\lambda_5\,\pi}{128}\,c^5\,\alpha'-\frac{3\,\bar\lambda_5\, c^4\,\alpha'}{64\,r}\right)+$$

$$+2\left(\frac{\bar\lambda_2\, c^2\,\alpha'}{5}-\frac{\bar\lambda_2}{40\,r^4\,c^2\,\alpha'}+\frac{4\,\bar\lambda_4\, c^4\,\alpha'}{35}\right)+$$

$$\left.\left.+2\left(-\frac{\bar\lambda_2\, c^2\,\alpha'}{15}-\frac{\bar\lambda_2}{40\,r^4\,c^2\,\alpha'}-\frac{4\,\bar\lambda_4\, c^4\,\alpha'}{106}\right]d\alpha'\right\}\right.$$

$$=-\frac{4c_1\, s\,\delta_{\rm eq}\, c\,\delta_{\rm eq}}{r^2}\left\{\int\limits_{\alpha'=\cos^{-1}\left(\frac{1}{r}\right)}^{0}\left[\left(-\frac{\bar\lambda_1}{8r}-\frac{\bar\lambda_3}{96\,r^3}\right)+\frac{1}{c^2\,\alpha'}\left(-\frac{\bar\lambda_1}{48\,r^3}-\frac{\bar\lambda_2}{20\,r^4}\right)+\right.\right.$$

$$+c\,\alpha'\left(\frac{\bar\lambda_1\,\pi}{16}+\frac{\bar\lambda_3\,\pi}{32}\right)+c^2\,\alpha'\left(-\frac{\bar\lambda_3}{16r}+\frac{\bar\lambda_2}{5}-\frac{\bar\lambda_2}{15}\right)+$$

$$\left.\left.+c^4\,\alpha'\left(-\frac{3\,\bar\lambda_5}{64r}+\frac{4\,\bar\lambda_4}{35}-\frac{4\,\bar\lambda_4}{105}\right)+c^5\,\alpha'\left(\frac{3\,\bar\lambda_5\,\pi}{128}\right)+\cdots\right]d\alpha'\right\}.$$

Integrating with respect to α', yields:

$$\frac{-4c_1\, s\,\delta_{\text{eq}}\, c\,\delta_{\text{eq}}}{r^2}\Bigg\{\Bigg[\left(-\frac{\bar\lambda_1}{8r}-\frac{\bar\lambda_3}{96r^3}\right)\alpha' + \tan\alpha'\left(-\frac{\bar\lambda_1}{48r^3}-\frac{\bar\lambda_2}{20r^4}\right)+$$

$$+\, s\,\alpha'\left(\frac{\bar\lambda_1\pi}{16}+\frac{\bar\lambda_3\pi}{32}\right)+\left(\frac{\alpha'}{2}+\frac{s\,\alpha'\,c\,\alpha'}{2}\right)\left(-\frac{\bar\lambda_3}{16r}+\frac{2\bar\lambda_2}{15}\right)+$$

$$+\left(\frac{3}{8}\alpha'+\frac{5}{8}s\,\alpha'\,c\,\alpha'-\frac{s^3\alpha'\,c\,\alpha'}{8}\right)\left(-\frac{3\bar\lambda_5}{64r}+\frac{8\bar\lambda_4}{105}\right)+$$

$$+\left(\frac{c^4\alpha'\,s\,\alpha'}{5}+\frac{4}{5}s\,\alpha'-\frac{4}{15}s^3\alpha'\right)\left(\frac{3\bar\lambda_5\pi}{128}\right)+\cdots\Bigg]^{0}_{\alpha'=\cos^{-1}\left(\frac{1}{r}\right)}\Bigg\}.$$

Collecting terms we find:

$$\frac{-4c_1\, s\,\delta_{\text{eq}}\, c\,\delta_{\text{eq}}}{r^2}\Bigg\{\Bigg[\alpha'\left(-\frac{\bar\lambda_1}{8r}-\frac{\bar\lambda_3}{96r^3}-\frac{\bar\lambda_3}{32r}+\frac{\bar\lambda_2}{15}-\frac{9\bar\lambda_5}{512}+\frac{\bar\lambda_4}{35}\right)+$$

$$+\, s\,\alpha'\left(\frac{\bar\lambda_1\pi}{16}+\frac{\bar\lambda_3\pi}{32}+\frac{3\bar\lambda_5\pi}{160}\right)+s\,\alpha'\,c\,\alpha'\left(-\frac{\bar\lambda_3}{32r}+\frac{\bar\lambda_2}{15}+\frac{15\bar\lambda_5}{512r}+\frac{\bar\lambda_4}{21}\right)+$$

$$+\tan\alpha'\left(-\frac{\bar\lambda_1}{48r^3}-\frac{\bar\lambda_2}{20r^4}\right)+s^3\alpha'\,c\,\alpha'\left(-\frac{3\bar\lambda_5}{512r}-\frac{\bar\lambda_4}{105}\right)+$$

$$+\, c^4\alpha' s\,\alpha'\left(\frac{3\bar\lambda_5\pi}{640}\right)+s^3\alpha'\left(-\frac{\bar\lambda_5\pi}{160}\right)+\cdots\Bigg]^{0}_{\alpha'=\cos^{-1}\left(\frac{1}{r}\right)}\Bigg\},$$

where

$$\alpha' = \cos^{-1}\left(\frac{1}{r}\right)=\frac{\pi}{2}-\frac{1}{r}-\frac{1}{6r^3}\cdots,$$

$$c\,\alpha'=\frac{1}{r}$$

$$s\,\alpha' = 1-\frac{1}{2r^2}-\frac{1}{8r^4}\cdots,$$

$$s^3\alpha' = 1-\frac{3}{2r^2}+\frac{3}{8r^4}\cdots.$$

Substituting (neglecting terms higher than $\frac{1}{r^4}$), we have

$$\frac{-4c_1\, s\,\delta_{\text{eq}}\, c\,\delta_{\text{eq}}}{r^2}\Bigg\{\Bigg[-\frac{\pi\bar\lambda_1}{16r}+\frac{\bar\lambda_1}{8r^2}+\frac{\bar\lambda_1}{48r^4}-\frac{\pi\bar\lambda_3}{192r^3}+\frac{\bar\lambda_3}{96r^4}-\frac{\pi\bar\lambda_3}{64r}+$$

$$+\frac{\bar\lambda_3}{192r^4}+\frac{\pi\bar\lambda_2}{30}-\frac{\bar\lambda_2}{90r^3}-\frac{9\pi\bar\lambda_5}{1024}+\frac{\pi\bar\lambda_4}{70}-\frac{\bar\lambda_4}{35r}+\frac{\bar\lambda_1\pi}{16}+\frac{\bar\lambda_3\pi}{32}+$$

$$+\frac{3\bar\lambda_5\pi}{160}-\frac{\bar\lambda_1\pi}{32r^2}-\frac{\bar\lambda_3\pi}{64r^2}-\frac{3\bar\lambda_5\pi}{320r^2}+\frac{15\bar\lambda_5}{512r^2}+\frac{\bar\lambda_4}{21r}+\frac{\bar\lambda_3}{64r^4}-\frac{15\bar\lambda_5}{30r^3}-$$

$$-\frac{\bar\lambda_4}{42r^3}-\frac{\bar\lambda_1}{48r^2}-\frac{\bar\lambda_2}{20r^3}+\frac{\bar\lambda_1}{96r^4}-\frac{3\bar\lambda_5}{512r^2}-\frac{\bar\lambda_4}{105r}+\frac{9\bar\lambda_5}{1024r^4}-\frac{3\bar\lambda_4}{210r^3}-$$

$$-\frac{\bar\lambda_5\pi}{160}+\frac{3\bar\lambda_5\pi}{320r^2}+\cdots\Bigg]\Bigg\}$$

$$=\frac{4c_1\, s\,\delta_{\text{eq}}\, c\,\delta_{\text{eq}}}{r^2}\left[\bar\lambda_1\left(\frac{\pi}{16}-\frac{\pi}{16r}+\frac{1}{8r^2}-\frac{1}{48r^2}+\frac{1}{48r^4}+\frac{1}{96r^4}\right)+\right.$$

$$\left.+\bar\lambda_2\left(\frac{\pi}{30}-\frac{1}{90r^3}-\frac{1}{30r^3}-\frac{1}{20r^3}\right)+\bar\lambda_3\left(\frac{\pi}{32}-\frac{\pi}{64r}-\frac{\pi}{64r^2}\right)+\cdots\right],$$

$$=\frac{4c_1\, s\,\delta_{\text{eq}}\, c\,\delta_{\text{eq}}}{r^2}\left[\left(\frac{\pi}{15}\right)\frac{1}{r}+\left(\frac{\pi}{8}\right)\frac{1}{r^2}+\left(-\frac{\pi}{3}\right)\frac{1}{r^3}\left(-\frac{71}{360}+\frac{9\pi}{8}\right)\frac{1}{r^4}+\right.$$

$$\left.+\left(\frac{9}{16}+\frac{7\pi}{8}\right)\frac{1}{r^5}+\cdots\right].$$

Appendix B

Analytical Determination of Specularly Reflected Photon Flux

From Fig. 7 we see that

$$\boldsymbol{a} = (\boldsymbol{U}_{\odot} + \boldsymbol{L})/|\boldsymbol{U}_{\odot} + \boldsymbol{L}|,$$

where $\boldsymbol{L}$ is a unit vector in the direction of ϱ. Note that $\boldsymbol{a}$ defines the position on the planet where sunlight is specularly reflected to the spacecraft. This equation is determinant for $\Theta < \pi$. Since specular reflection occurs for $0 \leqq \Theta \leqq \Theta_{penumbra}$ and since $\Theta_{penumbra} < \pi$ the equation is always valid. Also from Fig. 10 we find that

$$\boldsymbol{r} = \boldsymbol{\varrho} + \boldsymbol{a} = \varrho \boldsymbol{L} + \boldsymbol{a}$$

or

$$\boldsymbol{a} = \boldsymbol{r} - \varrho \boldsymbol{L}.$$

Thus elimination of $\boldsymbol{a}$ yields:

$$(\boldsymbol{r} - \varrho \boldsymbol{L})\,|\boldsymbol{U}_{\odot} + \boldsymbol{L}| = \boldsymbol{U}_{\odot} + \boldsymbol{L},$$

where

$$|\boldsymbol{U}_{\odot} + \boldsymbol{L}| = \sqrt{2}\sqrt{1 + \boldsymbol{L} \cdot \boldsymbol{U}_{\odot}}. \tag{B-1}$$

Since the angle of incidence is equal to the angle of reflection,

$$\boldsymbol{U}_{\odot} \cdot \boldsymbol{a} = \boldsymbol{L} \cdot \boldsymbol{a}$$

accordingly,

$$\boldsymbol{U}_{\odot} \cdot (\boldsymbol{r} - \varrho \boldsymbol{L}) = \boldsymbol{L} \cdot (\boldsymbol{r} - \varrho \boldsymbol{L}) = \boldsymbol{L} \cdot \boldsymbol{r} - \varrho$$

or

$$\boldsymbol{L} \cdot \boldsymbol{r} - \varrho = \boldsymbol{U}_{\odot} \cdot \boldsymbol{r} - \varrho \boldsymbol{L} \cdot \boldsymbol{U}_{\odot}. \tag{B-2}$$

A dot product of Eq. (B-1) with $\boldsymbol{L}$ yields

$$\boldsymbol{L} \cdot \boldsymbol{r} - \varrho = \frac{\boldsymbol{U}_{\odot} \cdot \boldsymbol{L} + 1}{\sqrt{2}\sqrt{1 + (\boldsymbol{L} \cdot \boldsymbol{U}_{\odot})}}. \tag{B-3}$$

Equating Eqs. (B-2) and (B-3) yields

$$\boldsymbol{U}_{\odot} \cdot \boldsymbol{r} - \varrho(\boldsymbol{U}_{\odot} \cdot \boldsymbol{L}) = \frac{(\boldsymbol{U}_{\odot} \cdot \boldsymbol{L}) + 1}{\sqrt{2}\sqrt{1 + (\boldsymbol{U}_{\odot} \cdot \boldsymbol{L})}}. \tag{B-4}$$

Eq. (B-4) represents one equation in the unknowns ϱ and $\boldsymbol{U}_{\odot} \cdot \boldsymbol{L}$ since $\boldsymbol{U}_{\odot} \cdot \boldsymbol{r} = r \cos\theta$ is known.

In general, during an orbit, θ will pass through zero at some point, thus any indeterminacies arising in this situation must be avoided. For this reason the solution of Eq. (B-4) for ϱ must be avoided. A rearrangement of Eq. (B-4) yields the cubic

$$(2r^2\cos^2\theta - 1) + (\boldsymbol{U}_\odot \cdot \boldsymbol{L})(4\varrho\, r\cos\theta + 2r^2\cos^2\theta - 2) + \\ + (\boldsymbol{U}_\odot \cdot \boldsymbol{L})^2 (2\varrho^2 + 3\varrho\, r\cos\theta - 1) + \\ + (\boldsymbol{U}_\odot \cdot \boldsymbol{L})^3 (2\varrho^2) = 0. \tag{B-5}$$

Our procedure is to guess ϱ and solve Eq. (B-5) for $\boldsymbol{U}_\odot \cdot \boldsymbol{L}$. Since $1 \geqq \boldsymbol{U}_\odot \cdot \boldsymbol{L} \geqq -1$ two of the three roots can be discarded. Having the first estimates:

$$\varrho_{(1)} \quad \text{and} \quad (\boldsymbol{U}_\odot \cdot \boldsymbol{L})_{(1)}$$

one enters the component Eqs. (B-1), a rearrangement of which yields

$$\boldsymbol{r}\sqrt{2}\sqrt{1 + (\boldsymbol{U}_\odot \cdot \boldsymbol{L})_{(1)}} - \boldsymbol{U}_\odot = \boldsymbol{L}\left[1 + \varrho_{(1)}\sqrt{2}\sqrt{1 + (\boldsymbol{U}_\odot \cdot \boldsymbol{L})_{(1)}}\right]$$

and solves for the first estimate of $\boldsymbol{L}_{(1)}$, i.e.,

$$\boldsymbol{L}_{(1)} = \frac{\boldsymbol{r}\sqrt{2}\sqrt{1 + (\boldsymbol{U}_\odot \cdot \boldsymbol{L})_{(1)}} - \boldsymbol{U}_\odot}{1 + \varrho_{(1)}\sqrt{2}\sqrt{1 + (\boldsymbol{U}_\odot \cdot \boldsymbol{L})_{(1)}}}.$$

Having $\boldsymbol{L}_{(1)}$ one forms

$$(\boldsymbol{U}_\odot \cdot \boldsymbol{L}_{(1)}) - (\boldsymbol{U}_\odot \cdot \boldsymbol{L})_{(1)} = g(\varrho). \tag{B-6}$$

We wish to find the root of this equation. NEWTON-RAPHSON interpolation is indicated where the derivative is best obtained numerically.

An alternative approach, developed by JOHN ONDRASIK, involves the solution of a fourth order equation for ϱ.

One defines i as the angle of incidence (i.e., the angle between $\boldsymbol{a}$ and $\boldsymbol{U}_\odot$), β as the angle between $\boldsymbol{a}$ and $\boldsymbol{\varrho}$, and α as the angle between $\boldsymbol{a}$ and $\boldsymbol{r}$. From the law of sines

$$\begin{aligned} \sin\alpha &= \frac{\varrho}{r}\sin\beta, \\ &= \frac{\varrho}{r}\sin(\pi - \theta + \alpha) = -\frac{\varrho}{r}\sin(\alpha - \theta), \\ &= \frac{\varrho}{r}(\sin\theta\cos\alpha - \sin\alpha\cos\theta). \end{aligned}$$

Thus

$$\sin\alpha = \frac{\varrho}{r}\left\{\frac{\sin\theta}{\left(1 + \frac{\varrho}{r}\cos\theta\right)}\right\}\cos\alpha. \tag{B-7}$$

Also from the law of sines

$$\sin\alpha = \frac{\varrho}{a}\sin\gamma$$

(where γ is defined in Fig. 7; note that $\alpha = \theta - i$ and $\beta = \pi - i = \pi - \theta + \alpha$, so that $\gamma = -2\alpha$).

Using trigonometric identities, we find

$$\sin\alpha = \frac{\varrho}{a}(\sin\theta\cos 2\alpha - \sin 2\alpha\cos\theta),$$

$$= \frac{\varrho}{a}[\sin\theta(2\cos^2\alpha - 1) - 2\sin\alpha\cos\alpha\cos\theta],$$

so that

$$\sin\alpha = \frac{\varrho}{a}\left\{\frac{\sin\theta\,(2\cos^2\alpha - 1)}{1 + \frac{\varrho}{a}\,2\cos\alpha\cos\theta}\right\}. \tag{B-8}$$

Equating Eqs. (B-7) and (B-8) yields

$$2r\cos^2\alpha - a\cos\alpha - \varrho\cos\theta - r = 0,$$

therefore,

$$\cos\alpha = \frac{1}{4r}\{a \pm \sqrt{a^2 + 8r(\varrho\cos\theta + r)}\}. \tag{B-9}$$

From the law of cosines:

$$\cos\alpha = (a^2 + r^2 - \varrho^2)/2ar. \tag{B-10}$$

Equating Eqs. (B-9) and (B-10) yields

$$4r(a^2 + r^2 - \varrho^2) - 2a^2 r = \pm 2ar\sqrt{a^2 + 8r(\varrho\cos\theta + r)}.$$

Squaring this gives the final fourth order equation to be solved for ϱ:

$$\varrho^4 - (a^2 + 2r^2)\varrho^2 - 2a^2 r\varrho\cos\theta + r^2(r^2 - a^2) = 0.$$

Consider the checks:

for $\theta = 0$ (so that $\alpha = 0$) we have

$r = \varrho - a$, while for $\theta = \pi/2 + \alpha$,

$r^2 = \varrho^2 - a^2(\cos\alpha = a/r$ and $\sin\alpha = \varrho/r)$.

References

ACORD, J. D., and J. L. NICKLAS: Theoretical and Practical Aspects of Solar Pressure Attitude Control for Interplanetary Spacecraft, NAA Peprint No. 63–627, 1963.

ALDRICH, L. B., and W. H. HOOVER: Annals Astrophysical Observatory, Smithsonian Institution, 7, 1954.

ALLEN, C. W.: Astrophysical Quantities, University of London, London 1955.

BAKER, R. M. L., JR.: Drag Interactions of Meteorites with the Earth's Atmosphere, Dissertation, University of California at Los Angeles, May 1958.

BAKER, R. M. L., JR., and M. W. MAKEMSON: An Introduction to Astrodynamics, New York: Academic Press 1960.

BALLINGER, J. C., J. C. ELIZALDE and E. H. CHRISTENSEN: Thermal Environment of Interplanetary Space, SAE 1961 National Aeronautics Meeting, SAE 344 B, 1961.

BRYANT, R. W.: The Effect of Solar Radiation Pressure on the Motion of an Artificial Satellite, Astronomical J. **66**, No. 8, 430 (1961).

CAMAC, W. G., and K. K. EDWARDS: "Effect of Surface Thermal Radiation Characteristics on the Temperature Control Problem in Satellites," in Surface Effects on Space Craft Material (ed. by F. J. CLAUSS), New York/London: John Wiley 1960.

CLANCY, THOMAS F., and TH. P. MITCHELL: Effects of Radiation Forces on the Attitude of an Artificial Earth Satellite, AIAA J. **2**, No. 8, 517 (1964).

CLAUSS, F. J.: Surface Effect on Space Craft Material, New York/London: John Wiley 1960.

CUNNINGHAM, F. G.: Earth Reflected Solar Radiation Input to Spherical Satellites, ARS J., p. 1033, July 1962.

DANJON, A.: Albedo, "Color, and Polarization of the Earth", in The Earth As a Planet (ed. by G. P. KUIPER), Univ. of Chicago Press, pp. 726—738, 1954.

DENNISON, A. J., JR.: Illumination of a Space Vehicle Surface Due to Sunlight Reflected from Earth, ARS J. **32**, 635 (1962).

FORSTER, K., and R. M. L. BAKER, JR.: Linearized Drag Analysis and Orbit Determination, presented as Preprint No. 63—28 at the AIAA Astrodynamics Conference, New Haven, Connecticut, August 9—21, 1963.

FRYE, W. E., and E. V. B. STEARNS: Stabilization and Attitude Control of Satellite Vehicles, ARS J. **29**, No. 12, 927 (1959).

GARWIN, L. I.: Solar Sailing—A Practical Method of Propulsion within the Solar System, Jet Propulsion **28**, 188—190 (1958).

GEYLING, F. T.: Drag Displacements and Decay of Near Circular Satellite Orbits, Congr. Int. Astronaut. Fed., Stockholm, Sweden, 1960.

HAPKE, B.: J. Geophys. Res. **68**, No. 15 (1963).

HOWARD, J. N., J. I. F. KING and P. R. GAST: "Thermal Radiation," in Handbook of Geophysics, Revised Edition, New York: Macmillan 1961.

HRYCAK, P.: Effects of Secondary Radiation on an Orbiting Satellite, ARS J., p. 1294, August 1962.

IVES, N. E.: The Effect of Solar Radiation Pressure on the Attitude Control of an Artificial Earth Satellite, Great Britain, Aeronautical Res. Council, ARC-R and M-3332, April 1961, NASA N 64-10053 (London, HMSO).

JAFFE, L.: Project Echo Results, Astronautics **6**, 32 (1961).

JOHNSON, F. S.: The Solar Constant, J. Meteorology **11**, 431 (1954).

KARRENBERG, H. K., E. LEVIN and D. LEWIS: Variation of Satellite Position With Uncertainties in the Mean Atmospheric Density, ARS J., p. 576, April 1962.

KATZ, A. J.: Determination of Thermal Radiation Incident upon the Surfaces of an Earth Satellite in an Elliptical Orbit, presented at the IAS National Summer Meeting, Los Angeles, June 28—July 1, 1960, IAS Paper No. 60—58.

KOSKELA, PAUL E.: Orbital Effects of Solar Radiation Pressure on an Earth Satellite, J. of the Astronautical Sciences **9**, No. 3, (1962).

KOZAI, Y.: Effects of Solar Radiation Pressure on the Motion of an Artificial Satellite, Smithsonian Institution Astrophysical Observatory Special Report No. 56, January 30, 1961.

—: Effects of Solar Radiation Pressure on the Motion of an Artificial Satellite, Smithsonian Contributions to Astrophysics **6**, 109 (1963).

—: On the Effects of the Sun and the Moon upon the Motion of a Close Earth Satellite, Smithsonian Contributions to Astrophysics, **6**, 47 (1963).

KREITH, F.: Radiation Heat Transfer for Spacecraft and Solar Power Plant Design, International Textbook Comp. 1962.

LAWDEN, D. F.: Optimal Escape from a Circular Orbit, Astronaut. Acta **4**, 218—233 (1958).

LEVIN, E.: Reflected Radiation Received by an Earth Satellite, ARS J. 1328, September 1962.

Lockheed California Company: Orbit Analysis of Communication Satellites, Work done under Contract AF 30(602)–3238, Proj. No. 4519, Task No. 451901 for Rome Air Development Center, June 1964.

Moon, P.: Proposed Standard Solar Radiation Curves for Engineering Use, J. Franklin Inst. **230**, 583 (1940).

Muhleman, D. O., D. B. Holdridge, R. L. Carpenter and K. C. Oslund: Observed Solar Pressure Perturbations of Echo I, Science **131**, 1487, Nov. 1960.

Musen, P., and R. Bryant: Perturbations in Perigee Height of Vanguard I, Science **131**, 935 (1960).

Musen, P.: The Influence of the Solar Radiation Pressure on the Motion of An Artificial Satellite, J. Geophys. Res. **65**, No. 5, 1391 (1960).

Newton, R. R.: Stabilizing a Spherical Satellite by Radiation Pressure, RAS J. **30**, No. 12, 1175 (1960).

Parkinson, R. W., H. M. Jones and I. I. Shapiro: Effects of Solar Radiation Pressure on Earth Satellite Orbits, Science **131**, 920 (1960).

Pierce, D. A.: A Rapid Method for Determining the Percentage of a Circular Orbit in the Shadow of the Earth, J. Astronaut. Sci. **10**, No. 3 (1963).

Pitkin, E. T.: Integration and Optimization of Sustained-Thrust Rocket Orbits, Dissertation, University of California at Los Angeles, May 1964.

Robertson, H. P.: Dynamical Effects of Radiation in the Solar System, Monthly Notices of the Royal Astronomical Society **97**, 423 (1937).

Sehnal, L.: "The Effect of the Re-Radiation of the Sunlight from the Earth on the Motion of Artificial Satellites", in The Use of Artificial Satellites For Geodesy, ed. by G. Veis, Amsterdam: North Holland Publishing Company, 1963.

Shapiro, I. I., and H. M. Jones: Loss of Mass in Echo Satellite, Science **133**, No. 3452, 579 (1961).

Shapiro, I.: "The Prediction of Satellite Orbits," in IUTAM Symposium, Dynamics of Satellites, Paris, May 28–31, 1962, Berlin/Göttingen/Heidelberg: Springer 1963.

Shapiro, I.: Effects of Sunlight Pressure on Air Density Determinations Involving Cylindrical Satellites, J. Geophys. Res. **68**, No. 19, 5349 (1963).

Shapiro, I., and H. M. Jones: Perturbations of the Orbits of the Echo Balloon, Science **131**, 1484 (1960).

Sohn, R. L.: Attitude Stabilization by Means of Solar Radiation Pressure, ARS J. **29**, No. 5, 371 (1959).

Stair, R., and R. G. Johnston: Preliminary Spectroradiometric Measurements of the Solar Constant, J. Res. Nat. Bur. Standards **57**, 205 (1956).

Swalley, F. E.: Thermal Radiation Incident on an Earth Satellite, National Aeronautics and Space Administration Technical Note TN D-1524, dated December 1962.

Ule, L. A.: Orientation of Spinning Satellites by Radiation Pressure, AIAA J. **1**, No. 7 (1963).

van Allen, J. A.: Scientific Uses of Earth Satellites, University of Michigan Press, 1958.

Wyatt, S. P.: "The Effect of Terrestrial Radiation Pressure on Satellite Orbits," in IUTAM Symposium, Dynamics of Satellites, Paris, May 28–31, 1962, Berlin/Göttingen/Heidelberg: Springer 1963.

Wyatt, S. P.: The Effect of Radiation Pressure on the Secular Acceleration of Satellites, Smithsonian Contributions to Astrophysics **6**, 113 (1963).

Zadunaisky, P. E., J. I. Shapiro and H. M. Jones: Experimental and Theoretical Results on the Orbit of Echo I, Smithsonian Institution Astrophysical Observatory Special Report No. 61, dated 10 March 1961.

Discussion

As Dr. BAKER pointed out that the albedo being known with an error of about 5 %, it is not worth carrying the results of the theory to a better precision. Mrs. MASSEVITCH suggests applying the theory to the Moon for example (the albedo of which is better known) and then to obtain from this a value of the albedo of the satellite.

Dr. KING-HELE would like to mention two difficulties common to both papers, by either Dr. SEHNAL or Dr. BAKER. The first is the deterioration of the satellite surface: for example white paint may tend to become brown so that the reflectivity of the satellite may decrease considerably in a manner which is difficult to specify. The second problem is the determination of the Earth shadow. If there happen to be clouds 10 km high at the point where the sun's rays graze the horizon, the shadow height will be increased by over 10 km and, if the satellite is moving almost parallel to the shadow cone the position of the eclipse point can be altered by 100 km or more.

Passive Spin Propulsion of Large Flexible Spherically Shaped Satellites by the Solar Radiation Field

By

John Mar and **Frank R. Vigneron**

Defence Research Telecommunications Establishment, Ottawa, Canada

Abstract. An analysis showing that a non-rigid spherically shaped orbiting body may undergo passive spin in the sun's electromagnetic and radiation pressure fields is proposed. The possibility of such an occurrence is illustrated by considering the spin behaviour of Echo II satellite.

Résumé. On montre qu'un corps sphérique non rigide peut être soumis à une accélération du mouvement de rotation du fait du champ électromagnétique et de la pression de radiation solaire. On illustre ceci par le comportement du satellite Echo II.

1. Introduction

Since the launching of Alouette I (Canada's ionospheric sounding satellite which first introduced the "STEM" [*1*] principle to extend crossed dipole antennas measuring 75 feet and 150 feet tip-to-tip in orbit), the behaviour and dynamics of non-rigid orbiting bodies has been the subject of study at the Defence Research Telecommunications Establishment (DRTE), Ottawa, Canada [*2*]. In studying the dynamic behaviour of Alouette I, the dependence of body motions on the thermally induced transient structural distortions of the antennas became evident [*3, 4*]. Etkin, [*5*], recently demonstrated that a principal factor in the spin decay of Alouette I and S 48 satellites is a torque caused by solar radiation pressure. It arises because of periodic solar-induced thermal bending of the long antennae, with a phase lag due to heat storage. The same concept is extended to the dynamic study of a large flexible spherically shaped satellite. An example case to be treated in this report is Echo II.

The analysis to follow, suggests that a large thin-walled spherical satellite will distort in proportion to the temperature distribution on its skin, and hence will assume an asymmetry in the form of a "bulge" in the direction of maximum solar heating. The body with this shape will be unstable in the pressure of the solar radiation field, as illustrated in Fig. 1a. The solar pressure will cause the body to rotate its bulge to achieve stability (Fig. 1b), and consequently the body will become asymmetric with respect to the incident radiation.

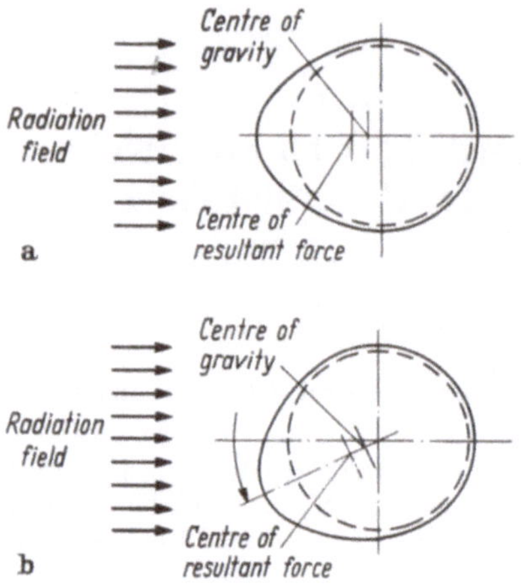

Fig. 1a and b. Configurations of the deformed sphere in radiation pressure field.

Since the effect of the "bulge" is to produce a shift in the centre of pressure from the centre of gravity by an amount $\boldsymbol{L}$, the resultant radiation force $\boldsymbol{F}$ on the body produces a propulsion torque $\boldsymbol{F} \times \boldsymbol{L}$. The torque will cause the body to accelerate in rotational velocity, with subsequent leveling out of the temperature distribution around the shell and restoration of body symmetry. The rotational velocity of the body will be stable when the retarding torque (due to eddy current damping or magnetic hysteresis damping) equals the propulsion torque.

The time of spacecraft exposure to radiation force is given by the time in sun for a given orbit. The acceleration and deceleration of spin rate of the Echo II is governed almost entirely by the percent sun variation, as will be shown in the following work.

2. Temperature Distribution on a Rotating Hollow Sphere

2.1 Derivation of the Temperature Equation

A reference surface element on the sphere is heated predominantly by direct solar radiation and internal radiation, and loses heat by radiation to the surrounding space. The skin cross section of Echo II is shown in Fig. 2. Heat conduction will have a small effect on the temperature distribution, and will be neglected. The skin construction of spherical passive communications satellites of this size is not conducive to good heat conduction.

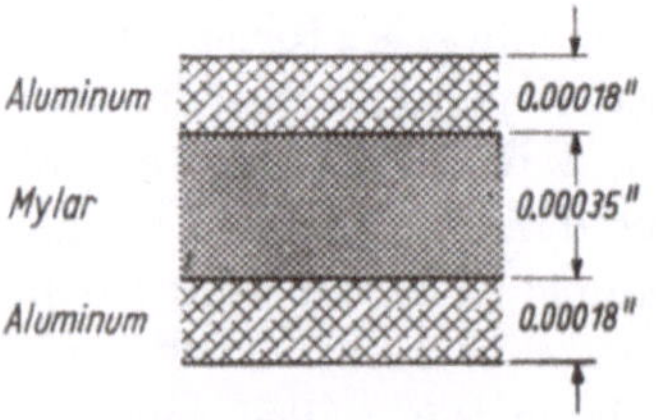

Fig. 2. Skin composite of Echo II satellite.

(Echo II is fabricated in the form of 82 thermally insulated segments (gores).)

Consider a reference element of area dA, on the outer skin of the satellite in the coordinate system shown in Fig. 3. The I, J, K, triad rotates with the sphere at velocity ω with respect to the fixed set $\boldsymbol{i}, \boldsymbol{j}, \boldsymbol{k}$. The components of the unit normal $\boldsymbol{n}$ of dA and the sun vector $\boldsymbol{S}$ with respect to the fixed set of axes are:

$$\boldsymbol{n} = \big(\cos\Phi \sin(\psi - \omega t)\big)\, i + \big(\cos\Phi \cos(\psi - \omega t)\big)\, j + (\sin\Phi)\, k, \tag{1}$$

$$\boldsymbol{S} = \boldsymbol{j}. \tag{2}$$

The angle β is then

$$\cos\beta = |\boldsymbol{n} \cdot \boldsymbol{S}| = \cos\Phi \cos(\psi - \omega t).$$

The heat rate into the elemental area, dA, is

$\alpha\, S \cos\Phi \cos^+(\psi - \omega t)\, dA$	direct solar radiation,
$\cos^+$	refers to a rectified cosine function, defined for the sun-lit region of the sphere,
$e\,\sigma\,\bar{T}^4\, dA$	internal radiation.

The heat rate out of dA is

$2e\,\sigma\,T^4\, dA$	radiation to space.

The difference between the heat rate in and the heat rate out equals the rate of heat stored, and hence

$$\varrho\, d c \frac{dT}{dt} = \alpha\, S \cos\Phi \cos^+(\psi - \omega t) + e\,\sigma\,\bar{T} - 2e\,\sigma\,T^4, \tag{3}$$

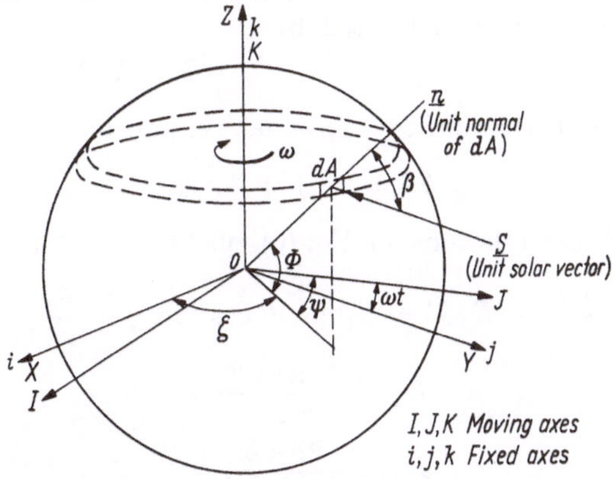

Fig. 3. Coordinate system of a reference surface element on the rotating sphere.

where ϱ is the density of the skin, d is the thickness, c is the specific heat, α is the solar absorptivity, e is the infrared emissivity, S is the solar constant, σ is Stefan-Boltzmann's constant, t is time, T is temperature and $\bar{T}$ is the mean temperature of the sphere.

2.2 Perturbation Solution of the Temperature Equation

When $\dot{\omega} = 0$, the "steady state" case, the temperature distribution will be constant and will lag the heat source. The explicit dependence of T on time can be eliminated by employing the relation between the coordinates [6].

$$2\pi\eta = \psi - \omega t \tag{4a}$$

and

$$t = t', \tag{4b}$$

where η is a non dimensional angle between the sun vector $\boldsymbol{S}$ and the fixed coordinate system, as illustrated in Fig. 4.
Hence $T = T(\eta, t', \Phi)$. and

$$\frac{dT}{dt} = \frac{\partial T}{\partial \eta}\frac{\partial \eta}{\partial t} + \frac{\partial T}{\partial \Phi}\frac{\partial \Phi}{\partial t} + \frac{\partial T}{\partial t'}\frac{\partial t'}{\partial t}.$$

Since $\frac{\partial \Phi}{\partial t} = 0$, and $\frac{\partial T}{\partial t'} = 0$ for the steady state case, and from (4a),

$$\frac{\partial \eta}{\partial t} = -\frac{\omega}{2\pi}$$

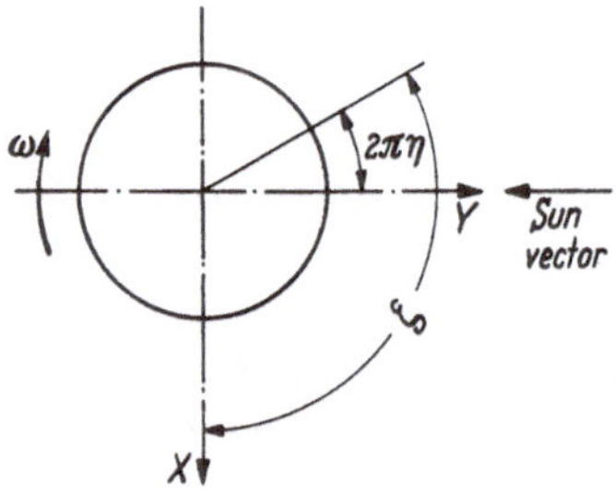

Fig. 4. Rotating ring element on a sphere.

$$\frac{dT}{dt} = -\frac{\omega}{2\pi}\frac{\partial T}{\partial \eta}. \tag{5}$$

Substituting Eq. (5) in (3),

$$\frac{-\omega \varrho d c}{2\pi}\frac{\partial T}{\partial \eta} = \alpha S \cos\Phi \cos^{+} 2\pi\eta + e\sigma \bar{T}^4 - 2e\sigma T^4 \tag{6}$$

The temperature may be approximated by

$$T = \bar{T}(1 + \hat{T}), \tag{7}$$

where $\hat{T}$ is a perturbation. By Taylors' expansion theorem,

$$T^4 = \bar{T}^4(1 + 4\hat{T}).$$

Substituting these relations in Eq. (6), yields

$$\frac{\partial T}{\partial \eta} - K_1 \hat{T} = K_2 \cos\Phi \cos^{+} 2\pi\eta + \frac{K_1}{8}, \tag{8}$$

where

$$K_1 = \frac{16\pi e\sigma \bar{T}^3}{\omega \varrho d c},$$

and

$$K_2 = \frac{-2\pi\alpha S}{\omega \varrho d c \bar{T}}.$$

The mean temperature $\bar{T}$ may be obtained by equating the total heat received by the sphere to the total heat reradiated, at steady state. From (3)

$$\int_A \alpha S \cos\Phi \cos^{+}\psi \, dA - \int_A 2e\sigma T^4 \, dA + \int_A e\sigma \bar{T}^4 \, dA = 0.$$

Substituting (7) in the above equation yields

$$\int_A \alpha\, S \cos\Phi \cos^+\psi\, dA - \int_A \sigma\, e\, \bar{T}^4\, dA - 8 e\, \sigma\, \bar{T}^4 \int_A \hat{T}\, dA = 0. \quad (9)$$

The integral $\int_A \hat{T}\, dA$ is evaluated by considering the equality

$$\bar{T} A = \int_A \hat{T}\, dA,$$

from which

$$\bar{T} A = \bar{T} \int_A dA + \int_A \hat{T}\, dA.$$

Hence

$$\int_A \hat{T}\, dA = 0. \quad (10)$$

Combining Eqs. (9) and (10), and performing the integration over the sphere yields:

$$\bar{T} = \left(\frac{\alpha\, S}{4 e\, \sigma}\right)^{1/4}. \quad (11)$$

Eq. (8) may be rewritten, upon defining the coordinate

$$\lambda = 2\pi\eta + \frac{\pi}{2},$$

as

$$\frac{\partial T}{\partial \lambda} - \frac{K_1}{2\pi} = \frac{K_2}{2\pi} \cos\Phi \sin^+\xi + \frac{K_1}{16\pi}. \quad (12)$$

The solution of (12), with the boundary condition of periodicity in the interval $0 \leqq T \leqq 2\pi$, (i.e., $T(0) = T(2\pi)$), is, in two intervals,

$$\hat{T} = \frac{2\pi K_2 \cos\Phi}{K_1^2 + 4\pi^2} \left(\frac{e^{\frac{K_1}{2\pi}\xi}}{1 - e^{K_1/2}}\right) - \frac{K_2 \cos\Phi \sin(\xi - \xi_0)}{\sqrt{K_1^2 + 4\pi^2}} - \frac{1}{8},$$

where

$$\tan\xi_0 = \frac{2\pi}{K_1}, \qquad 0 \leqq \xi \leqq \pi$$

for

$$\pi \leqq \xi \leqq 2\pi,$$

$$\hat{T} = \frac{2\pi K_2 \cos\Phi}{K_1^2 + 4\pi^2} \left(\frac{e^{\frac{K_1}{2\pi}(\xi - \pi)}}{1 - e^{K_1/2}}\right) - \frac{1}{8}. \quad (13)$$

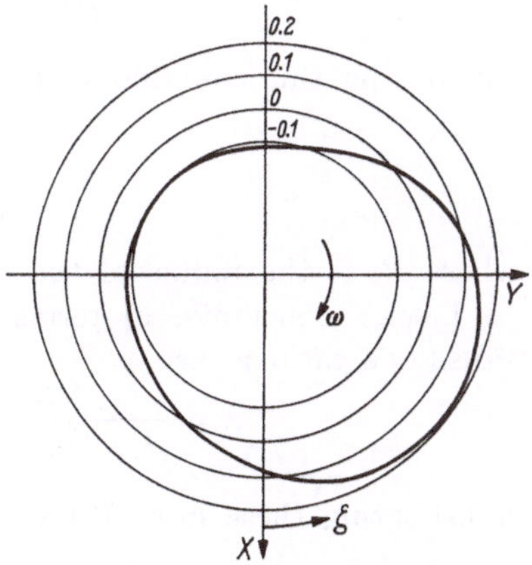

Fig. 5. Temperature perturbation $\frac{\hat{T}}{\bar{T}}$ v_s ξ.

The solution for the values $K_1 = 7.04$, $K_2 = -2.82$, and $\cos\Phi = 1$, is shown in Fig. 5.

3. Displacements of a Spherical Shell with Temperature

The displacements in an isotropic elastic solid with a quasi-static temperature distribution can be derived from a potential, Ψ, which is a solution of [7]:

$$\nabla^2 \Psi = \left(\frac{3\lambda + 2\mu}{\lambda + 2\mu}\right) \alpha_T T, \tag{14}$$

where λ and μ are LAMÉ's constants, α_T is the coefficient of linear expansion, and T is the temperature.

The displacements, $\bar{u}$, are then given by

$$\bar{u} = \nabla \Psi. \tag{15}$$

For a thin spherical shell, the radial stress, σ_r, must vanish at the inner and outer boundaries. The radial stress in terms of the displacements is

$$\sigma_r = 2\mu \frac{\partial u_r}{\partial r} + \lambda \nabla \cdot \bar{u} - (3\lambda + 2\mu) \alpha_T T. \tag{16}$$

Substituting (14) and (15) into (16), gives

$$\sigma_r = 2\mu \left(\frac{\partial u_r}{\partial r} - \frac{3\lambda + 2\mu}{\lambda + 2\mu} \alpha_T T\right). \tag{17}$$

If the radial stress is approximated to zero in the shell, Eq. (17) yields

$$\frac{\partial u_r}{\partial r} = \frac{3\lambda + 2\mu}{\lambda + 2\mu} \alpha_T T.$$

Integrating this equation with $u_r = 0$, at $T = 0$, yields

$$u_r = \frac{3\lambda + 2\mu}{\lambda + 2\mu} \alpha_T T R_0, \tag{18}$$

where R_0 is the radius of the sphere.

LAMÉ's constants, in terms of the modulus of elasticity E, and POISSON's ratio ν, are

$$\lambda = \frac{\nu E}{(1 + \nu)(1 - 2\nu)}, \qquad \mu = \frac{E}{2(1 + \nu)}.$$

Substituting these into (18) yields

$$u_r = \left(\frac{1 + \nu}{1 - \nu}\right) \alpha_T T R_0. \tag{19}$$

In the case of the spherical shell subjected to solar radiation, the temperature may be given approximately by equation (13).

4. Torque Resulting from Solar Radiation Pressure

A torque will result if the center of gravity of the deformed sphere and the resultant solar radiation force are not coincident, as shown in Fig. 6.

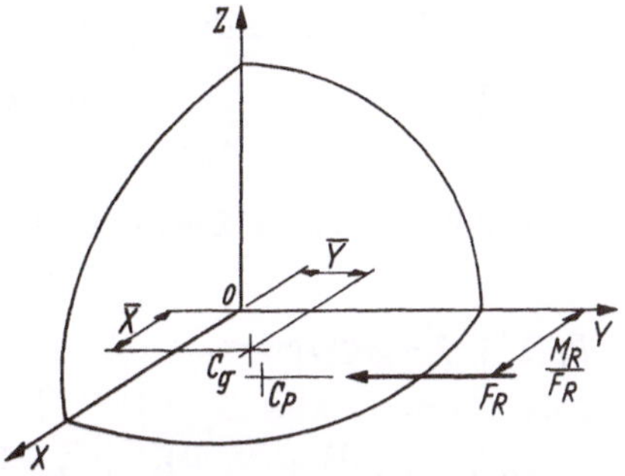

Fig. 6. Displacement of the centre of pressure from the centre of gravity.

4.1 Position of the Center of Gravity

The center of gravity of the deformed sphere will remain in the $x - y$ plane (Fig. 6) and may be calculated by

$$\bar{x} = \int_M x\, dm \Big/ \int_M dm\,, \qquad \bar{y} = \int_M y\, dm \Big/ \int_M dm\,. \tag{20}$$

It is evident that

$$dm = \frac{\varrho\, d\, r^2 \cos\Phi\, d\Phi\, d\xi}{\cos\gamma}\,, \tag{21}$$

where ϱ is the density, d is the thickness, r, Φ, ξ are spherical coordinates, and γ is the angle between the radius vector and the unit normal to the surface of the elemental area. The equation of the deformed sphere is, from (19)

$$r = R_0' + \varepsilon R_0' \cos\Phi \sin(\xi + \xi_0) + \lambda R_0' \cos\Phi\, e^{P\xi} \quad \text{for} \quad 0 \leqq \xi \leqq \pi\,, \tag{22a}$$

$$r = R_0' + \lambda R_0' \cos\Phi\, e^{P(\xi-\pi)} \quad \text{for} \quad \pi \leqq \xi \leqq 2\pi\,, \tag{22b}$$

where

$$R_0' = R_0\left(1 + \frac{7}{8}\,\frac{1+\nu}{1-\nu}\,\alpha_T \bar{T}\right),$$

$$\lambda = \frac{2\pi K_2 \bar{T}}{K_1^2 + 4\pi^2}\left(\frac{1}{1 - e^{K_1/2}}\right)\left(\frac{1+\nu}{1-\nu}\right)\alpha_T\,,$$

$$\varepsilon \equiv \frac{-K_2}{\sqrt{K_1^2 + 4\pi^2}}\left(\frac{1+\nu}{1-\nu}\right)\alpha_T\,.$$

For small displacements, the variations in $\bar{T}$ will have a negligible effect on the torque, as the dilatation does not alter the symmetry of the surface. Consequently for calculating the torque, the surface may be represented by,

$$r = R_0[1 + \varepsilon \cos\Phi \sin(\xi + \xi_0) + \lambda \cos\Phi\, e^{p\,\xi}] \quad \text{for } 0 \leqq \xi \leqq \pi, \tag{23a}$$

$$r = R_0(1 + \lambda \cos\Phi\, e^{p\,(\xi-\pi)}) \quad \text{for } \pi \leqq \xi \leqq 2\pi. \tag{23b}$$

Let

$$V = r - R_0[1 + \varepsilon \cos\Phi \sin(\xi + \xi_0) + \lambda \cos\Phi\, e^{p\,\xi}] \quad \text{for } 0 \leqq \xi \leqq \pi$$

and

$$V = r - R_0(1 + \lambda \cos\Phi\, e^{p\,(\xi-\pi)}) \quad \text{for } \pi \leqq \xi \leqq 2\pi.$$

Then

$$\frac{dv}{dn}\boldsymbol{n} = \nabla V = i\frac{\partial V}{\partial x} + j\frac{\partial V}{\partial y} + k\frac{\partial V}{\partial z}, \tag{24}$$

where $\boldsymbol{n}$ is the unit normal to the surface.

Also

$$\frac{\partial V}{\partial x} = \frac{\partial V}{\partial r}\frac{\partial r}{\partial x} + \frac{\partial V}{\partial \Phi}\frac{\partial \Phi}{\partial x} + \frac{\partial V}{\partial \xi}\frac{\partial \xi}{\partial x} \tag{25}$$

and similarily for $\frac{\partial V}{\partial y}$ and $\frac{\partial V}{\partial z}$.

From Fig. 3,

$$\begin{aligned} x &= r\cos\Phi\cos\xi, \\ y &= r\cos\Phi\sin\xi, \\ z &= r\sin\Phi, \\ r^2 &= x^2 + y^2 + z^2, \\ \sin\Phi &= \frac{z}{\sqrt{x^2+y^2+z^2}}, \\ \cos\xi &= \frac{x}{\sqrt{x^2+y^2}}. \end{aligned} \tag{26}$$

Combining Eqs. (23), (24), (25) and (26) gives the following equation for the unit normal

$$\begin{aligned} \frac{dV}{dn}\boldsymbol{n} = {} & i\left\{\cos\Phi\cos\xi - \frac{R_0}{r}\cos\xi\sin^2\Phi(\varepsilon\sin(\xi+\xi_0) + \lambda\, e^{p\,\xi})\right\} + \\ & + j\left\{\cos\Phi\sin\xi - \frac{R_0}{r}\sin\xi\sin^2\Phi(\varepsilon\sin(\xi+\xi_0) + \lambda\, e^{p\,\xi})\right\} + \\ & + k\left\{\sin\Phi + \frac{R_0}{r}\sin\Phi\cos\Phi(\varepsilon\sin(\xi+\xi_0) + \lambda\, e^{p\,\xi})\right\}. \end{aligned} \tag{27}$$

The modulus $\frac{dV}{dn}$ obtained from (27), is

$$\left|\frac{dV}{dn}\right| = \left\{1 + \left(\frac{R_0}{r}\right)^2 \sin^2\Phi\,(\varepsilon \sin(\xi + \xi_0) + \lambda\, e^{p\,\xi}) + \right.$$
$$\left. + \left(\frac{R_0}{r}\right)^2 (\varepsilon \cos(\xi + \xi_0) + p\,\lambda\, e^{p\,\xi})\right\}^{1/2}. \tag{28}$$

The equation for the radius vector, $\boldsymbol{r}$, is

$$\boldsymbol{r} = \boldsymbol{i}(r\cos\Phi\cos\xi) + \boldsymbol{j}(r\cos\Phi\sin\xi) + \boldsymbol{k}(r\sin\Phi). \tag{29}$$

γ can then be obtained from $\cos\gamma = \boldsymbol{n}\cdot\boldsymbol{r}$, which yiekds the result

$$\cos\gamma = \frac{1}{\left|\frac{dV}{dn}\right|}. \tag{30}$$

The mass can then be calculated using Eqs. (21) and (30)

$$\int\limits_m dm = \int\limits_{\xi=0}^{2\pi} \int\limits_{\Phi=\frac{-\pi}{2}}^{\frac{\pi}{2}} \frac{\varrho\, d\, r^2 \cos\Phi}{\cos\gamma}\, d\Phi\, d\xi. \tag{31}$$

Typical values of λ and ε, as calculated for the Echo II satellite, are $\lambda = 0.0002$ and $\varepsilon = 0.006$. Substitution of these values into Eq. (31) indicates that the approximation $\cos\gamma = 1$ is justified. The mass is then evaluated as

$$\int\limits_m dm = \varrho\, d \int\limits_{\xi=0}^{2\pi} \int\limits_{\Phi=\frac{-\pi}{2}}^{\frac{\pi}{2}} r^2 \cos\Phi\, d\Phi\, d\xi = 4\pi\,\varrho\, d\, R_0^2(1 + 0.005)$$

(the apparent violation of conservation of mass is a result of assuming $\varrho\, d$ to be constant, and assuming $\cos\gamma = 1$). The moment of the mass is evaluated in a consistent manner. The resulting coordinates of the centre of gravity are:

$$\bar{x} = 0.417\,\varepsilon\, R_0, \quad \bar{y} = 0.417\,\varepsilon\, R_0, \quad \bar{z} = 0. \tag{32}$$

4.2 Solar Radiation Force

The solar radiation force can be resolved into two components (assuming spectral radiation) as is illustrated in Fig. 7.

I) components normal to the surface

$$dF_n = m_1 \cos^2\beta\, dA \tag{33}$$

II) components parallel to the incident light rays

$$dF_p = m_2 \cos\beta\, dA \tag{34}$$

where $m_1 = 2\tau\, p_0$; $m_2 = (1 - \tau)\, p_0$; and p_0 is a solar force constant, τ is the reflectivity, and β is the angle between the unit normal $\boldsymbol{n}$ and the incident light rays (the sun vector).

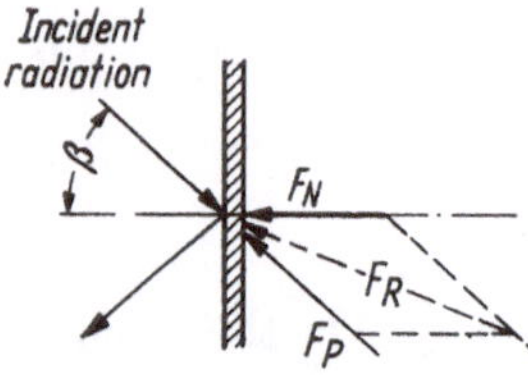

Fig. 7. Parallel and normal radiation pressure forces.

The sun vector is

$$\boldsymbol{S} = \boldsymbol{j}. \tag{35}$$

The shadow-sun boundary is defined by the equation $\boldsymbol{S} \cdot \boldsymbol{n} = 0$ and is

$$\cos\Phi \sin\xi - \frac{R_0}{r} \sin^2\Phi \sin\xi\,(\xi \sin(\xi + \xi_0) + \lambda\, e^{p\xi}) - \frac{R_0}{r} \cos\xi\,(\varepsilon \cos(\xi + \xi_0) + p\,\lambda\, e^{p\lambda}) = 0. \tag{36}$$

Eq. (36) defines the boundary of the area of the sphere which is subjected to solar radiation.

The y component of the resultant F_n is

$$d F_{ny} = d F_n (\boldsymbol{n} \cdot \boldsymbol{j}).$$

Also $\cos\beta = \boldsymbol{n} \cdot \boldsymbol{j}$, and hence

$$F_{ny} = \int_{\Phi_1}^{\Phi_2} \int_{\xi_1}^{\xi_2} \frac{m_1 (\boldsymbol{n} \cdot \boldsymbol{j})^3\, r^2 \cos\Phi}{\cos\gamma}\, d\Phi\, d\xi \tag{37}$$

with the limits of integration being specified by Eq. (36). The moment of the force F_{ny} about the origin is

$$d M_{ny} = r \cos\Phi \cos\xi\, d F_n.$$

Hence

$$M_{ny} = \int_{\Phi_1}^{\Phi_2} \int_{\xi_1}^{\xi_2} \frac{m_1 (\boldsymbol{n} \cdot \boldsymbol{j})^3\, r^3 \cos^2\Phi \cos\xi}{\cos\gamma}\, d\xi\, d\Phi. \tag{38}$$

Similarly, from Eqs. (33) and (34)

$$F_{nx} = \int_{\Phi_1}^{\Phi_2} \int_{\xi_1}^{\xi_2} \frac{m_1 (\boldsymbol{n} \cdot \boldsymbol{j})^2\, (\boldsymbol{n} \cdot \boldsymbol{i})\, r^2 \cos\Phi}{\cos\gamma}\, d\xi\, d\Phi, \tag{39}$$

$$M_{nx} = \int_{\Phi_1}^{\Phi_2} \int_{\xi_1}^{\xi_2} \frac{m_1 (\boldsymbol{n} \cdot \boldsymbol{j})^2\, (\boldsymbol{n} \cdot \boldsymbol{i})^2\, r^3 \cos^2\Phi \sin\xi}{\cos\gamma}\, d\xi\, d\Phi, \tag{40}$$

$$F_{nz} = M_{nx} = 0 \quad \text{by symmetry}. \tag{41}$$

$$F_{py} = \int_{\Phi_1}^{\Phi_2} \int_{\xi_1}^{\xi_2} \frac{(\boldsymbol{n} \cdot \boldsymbol{j})\, r^2 \cos\Phi}{\cos\gamma}\, d\xi\, d\Phi, \tag{42}$$

$$M_{py} = \int_{\Phi_1}^{\Phi_2} \int_{\xi_1}^{\xi_2} \frac{(\boldsymbol{n} \cdot \boldsymbol{j})\, r^3 \cos^2\Phi \cos\xi}{\cos\gamma}\, d\xi\, d\Phi. \tag{43}$$

The shadow-sun boundary, as determined by Eq. (36), is shown graphically in Fig. 8 for the typical values $\xi_0 = \pi/4$, $\varepsilon = 0.00626$, $\lambda = 0.0002$. Clearly, a good approximation of the shadow boundary is $\xi_1 =$ constant and $\xi_2 =$ constant (i.e. ξ_1 and ξ_2 are not dependent upon Φ).

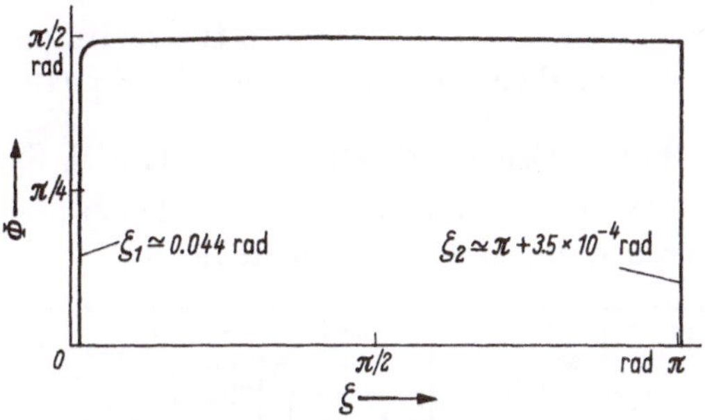

Fig. 8. Position of the shadow-sun boundary.

Substitution of $\xi = 0.044$, $\xi_2 = \pi$, $\Phi_1 = -\frac{\pi}{2}$, $\Phi_2 = \frac{\pi}{2}$, neglecting the higher order terms in ε and λ, and assuming $\cos\gamma = 1$, results in the following equations for the forces and moments:

$$\left.\begin{aligned}
F_{ny} &= m_1 R_0^2\left(\frac{\pi}{2} - 0.019\varepsilon + 10.0\lambda\right),\\
M_{ny} &= m_1 R_0^3(1.46\varepsilon - 7.5\lambda),\\
F_{nx} &= m_1 R_0^2(0.093\varepsilon + 4.9\lambda),\\
M_{nx} &= m_1 R_0^3(0.259\varepsilon - 12.5\lambda),\\
F_{py} &= m_2 R_0^2(0.999\pi + 0.040\varepsilon + 4.8\lambda),\\
M_{py} &= m_2 R_0^3(-\tfrac{2}{3}\xi_1^2 + 2.26\varepsilon - 24.2\lambda).
\end{aligned}\right\} \tag{44}$$

Examination of the dependence of the forces and moments on the shadow-sun boundary, ξ_1 and ξ_2, reveals that equations (44) are valid for $\varepsilon < 0.035$ and $\lambda < 0.001$ (i.e. the error in Eqs. (44) will be less than 1%). The resultant F_n is then approximately F_{ny}, since $F_{nx} \approx 0$ for small values of ε and λ. The center of pressure of the F_n forces and the F_p forces is $\frac{M_{ny} + M_{nx}}{F_{ny}}$ and $\frac{M_{py}}{F_{py}}$ respectively, and hence the resultant torque may be calculated,

$$G = F_{ny}\left(\frac{M_{ny} + M_{nx}}{F_{ny}} - \bar{x}\right) + F_{py}\left(\frac{M_{py}}{F_{py}} - \bar{x}\right). \tag{45}$$

Eq. (45) becomes, upon substitution of the appropriate values of F and M_1 and modifying p_0 to S_{Fp_0}, where S_F is the fraction of sunlight per orbit,

$$G = R_0^3\,\pi S_F p_0 \times$$
$$\times\left[\tau(0.348\varepsilon + 3.20\lambda) + (1-\tau)(0.304\varepsilon - 7.70\lambda) - \frac{2}{3}\frac{\xi_1^2}{\pi}\right]. \tag{46}$$

5. Equation of Spin Motion

The equation of spin motion of the sphere, assuming rotation with the spin axis perpendicular to the ecliptic plane, can be written directly as

$$I\,\dot{\omega} + p_1\,\omega - G = 0\,, \tag{47}$$

where I is the moment of inertia, $p_1\,\omega$ is the retarding torque due to eddy currents in the shell, and G is given by Eq. (46).

6. Comparison with Observed Echo II Spin

To illustrate the theory, a predicted Echo II spin period is compared with the observed data, and is shown in fig. 9.

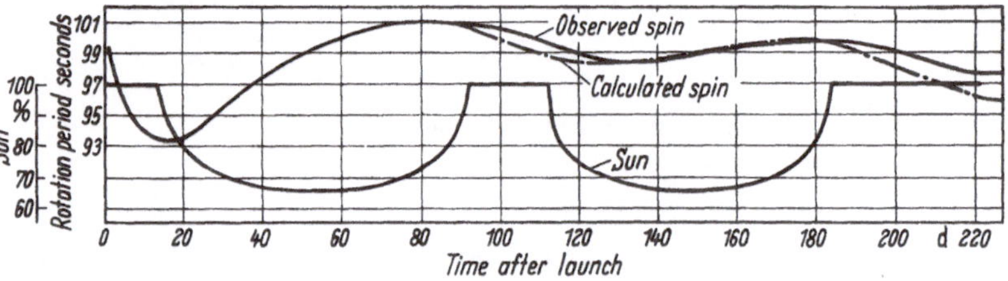

Fig. 9. Echo II spin.

Numerical magnitudes of the parameters used in the calculation are as follows:

C 0.33 cal/gm/°C thermal capacity of the skin,
ν 0.10 Poisson's ratio,
α_T 3.75×10^{-5}/°C coefficient of linear thermal expansion,
ϱd $4.76(10^{-3})$ gm/cm² skin density per unit area,
σ $1.355(10^{-12})$ cal/cm² sec °K⁴ Stefan-Boltzmann's constant,
α 0.434 solar absorptivity,
e 0.3 infrared emissivity,
S 0.0333 cal/cm² sec solar constant.
P_1 $24.7(10^4)$ dyne-cm sec,
P_0 $4.7(10^{-5})$ dynes/cm² solar radiation pressure,
τ 0.7 reflectivity of the skin,
R_0 $2.1(10^3)$ cm. radius of Echo II.
I $7.8(10^{11})$ gm-cm² moment of inertia of Echo II,
S_F percent sun versus time calculated for Echo II.

The numerical calculations were carried out with the parameters K_1 and K_2 being determined at a reference value of $\omega = 0.063$ rad/sec. It was observed that for small variations in ω, K_1 and K_2 are approximately constant, enabling equation (47) to be integrated readily.

The magnitude of the eddy current damping does not include a possible variation of 2.5 times [8] caused by the temperature change of the satellite between the sunlight and shadow.

7. Conclusions

It has been demonstrated that a spherically shaped flexible body may attain a passive spin motion in orbit as a result of static instability caused by interaction between the thermally induced body asymmetry and the solar radiation pressure field. This mechanism appears to explain the observed spin behaviour of the Echo II communication satellite.

In the case of Echo II, it is found that the major variation in the driving torque is caused by the variation in the "percent sun" exposure in orbit.

By the same reasoning, a large flexible thin-walled cylinder with longitudinal axis normal to the solar field may also attain a passive spin in orbit. It is further suggested that thermal distortion and radiation pressure interaction may well apply in some form to the prediction of motion of extended gaseous masses in outer space.

8. Acknowledgements

The authors gratefully acknowledge the assistance of Messrs H. L. Eaker, Project Manager Echo II, E. D. Nelsen, Project Manager of ISIS, and H. Horiuchi, Project Echo II, of the Goddard Space Flight Centre, NASA, Greenbelt, Md., U.S.A., for providing information on Echo II without which this work could not have been completely carried out. In addition, valuable discussions of this work with Dr. T. Garrett, Defence Research Telecommunications Establishment, are acknowledged.

References

[1] Warren, H. R., and J. Mar: Structural Design of the Topside Sounder Satellite, Canad. Aeronaut. Space J. 8, No. 7 (1962).

[2] Mar, J.: Spin Decay of Alouette I, Canadian Aeronautics and Space Institute Astronautics Symposium, Ottawa, Ontario, March 1964 (to be published).

[3] Kemper, A., and K. Farrell: Temperature Gradients and Profile Changes in Long Tubular Elements Due to Incident Radiation, De Havilland Aircraft of Canada Ltd. DHC-SP-TN 164, December 1962.

[4] Mar, J.: Some Thermal Bending and Elastic Effects of Tubular Structures on Spacecraft, CASI Symposium, Toronto, Ontario, Feb. 1965 (to be published).

[5] Etkin, B., and P. C. Hughes: Spin Decay of a Class of Satellites Caused by Solar Radiation, University of Toronto, Institute for Aerospace Studies Report 107 (to be published).

[6] Charnes, A., and S. Raynor: Solar Heating of a Rotating Cylindrical Space Vehicle, American Rocket Society, J., p. 479, May 1960.

[7] Landau-Lifshitz: Theory of Elasticity, Addison-Wesley Publishing Company Inc., 1959.

[8] "Passive Communications Satellite Echo II", Post Launch Analysis Final Report, Goddard Space Flight Center, National Aeronautics and Space Administration, Greenbelt, Maryland, Sept. 1964.

Dynamical Considerations Associated with the Geometric Determination of Position from Satellites

By

Bernard H. Chovitz

U.S. Coast and Geodetic Survey, Washington Science Center, Rockville, Maryland, U.S.A.

Abstract. The author describes the principles of a geometric satellite triangulation system. In order to ascertain the needed accuracy in position determination, it is necessary that the trail of the satellite passes within 0.5 mm of the center of the plate. Prediction problems arising from such requirements are presented. Further, the relations between the geometric and the dynamical coordinate systems are discussed.

Résumé. L'auteur rappelle les principes d'un système géométrique de triangulation par satellites. Afin d'assurer la précision nécessaire dans la détermination des positions, il faut que la trace du satellite passe à moins de 0.5 mm du centre de la plaque. Les problèmes de prédiction liés à cette nécessité sont présentés. Ensuite, l'auteur discute les relations liant les systèmes de coordonnées géométriques et dynamiques.

In a paper presented to the tenth Congress of the International Society of Photogrammetry in September 1964 [*1*], Schmid has covered in some detail the objectives, methods, and expected accuracies of a worldwide spatial triangulation system established by simultaneous observations on light-reflecting satellites. The principal objective is geodetic: to obtain the relative positions of a group of ground stations distributed about 4000 km apart worldwide, which will serve as a super first-order network. The methods employed are photogrammetric: that is, the basic condition imposed on the observations is collinearity of the observed object, the photographic image, and the perspective center of the camera. The expected accuracy is 1:500,000 to 1:1,000,000 among points on the worldwide net relative to each other.

It is pointed out in [*1*] that this first-order triangulation system will establish the necessary geometric fidelity for the tracking stations which observe satellites for orbit analysis. For if the relative positions of these stations can be justifiably held fixed with respect to each other, much more accurate and powerful results can be obtained in studies of the geopotential and perhaps other environmental phenomena like solar radiation pressure, atmospheric density, etc., because a source of correlation among parameters will have been removed. However, although the geometric method can orient the net properly with respect to an astronomic reference frame, it cannot satisfy the ultimate geodetic objective of positioning the geometrically obtained worldwide net in an absolute coordinate system, that is, referred to the Earth's center of mass. This must be accomplished by dynamical methods. Thus, the geometric and dynamic approaches are strongly interrelated and dependent on each other for satisfying their goals.

We shall consider two aspects of this interrelationship here. One is the determination of an absolute reference system by dynamical means in order to complete the geodetic system obtained by the geometric method. The other is the rather pedestrian but very basic use of the dynamic method for preparing predictions of satellite positions for the observers of the geometric phenomena. But first the geometric satellite triangulation system will be briefly described.

Geometric Determination

The basic principle is simple: the use of the satellite as a beacon which is intervisible from two or more ground stations. The observations can be referred unambiguously to an inertial reference frame by taking photographs of the satellite tracks against a stellar background, thus eliminating any connection to the deflections of the vertical at the stations. The result is a three-dimensional triangulation network, the exactitude of whose coordinate system depends only on the quality of the star catalogs used. Of course, this applies only to directional components of the coordinates and not to scale, which is a separate problem.

The idea above is not new, being employed for geodetic purposes long ago by Markowitz and O'Keefe, for example, using the moon as the satellite in question, by Vaisala, using the flare method, and presently by the Smithsonian Astrophysical Observatory in making observations with its Baker-Nunn cameras. What gives the method discussed here its power is the large number of observations available for each photograph. In this sense, it is analogous to the strength of the Doppler method in orbital analysis. The procedure employed is to chop by means of a timed shutter the continuously photographed trail of a sun-illuminated satellite. By this method, an average of 600 observations of the satellite are obtained per plate. The satellite path is construed to be a high-order polynomial function of time, the coefficients of which are obtained from a least squares fit of the measurements. From this a single point is computed as a function of time, which represents the "point of simultaneity", since the same time is chosen for the photograph taken from another station. This method enables synchronization of the observing stations to be carried out smoothly. The satellite plate coordinates are transferred to the stellar

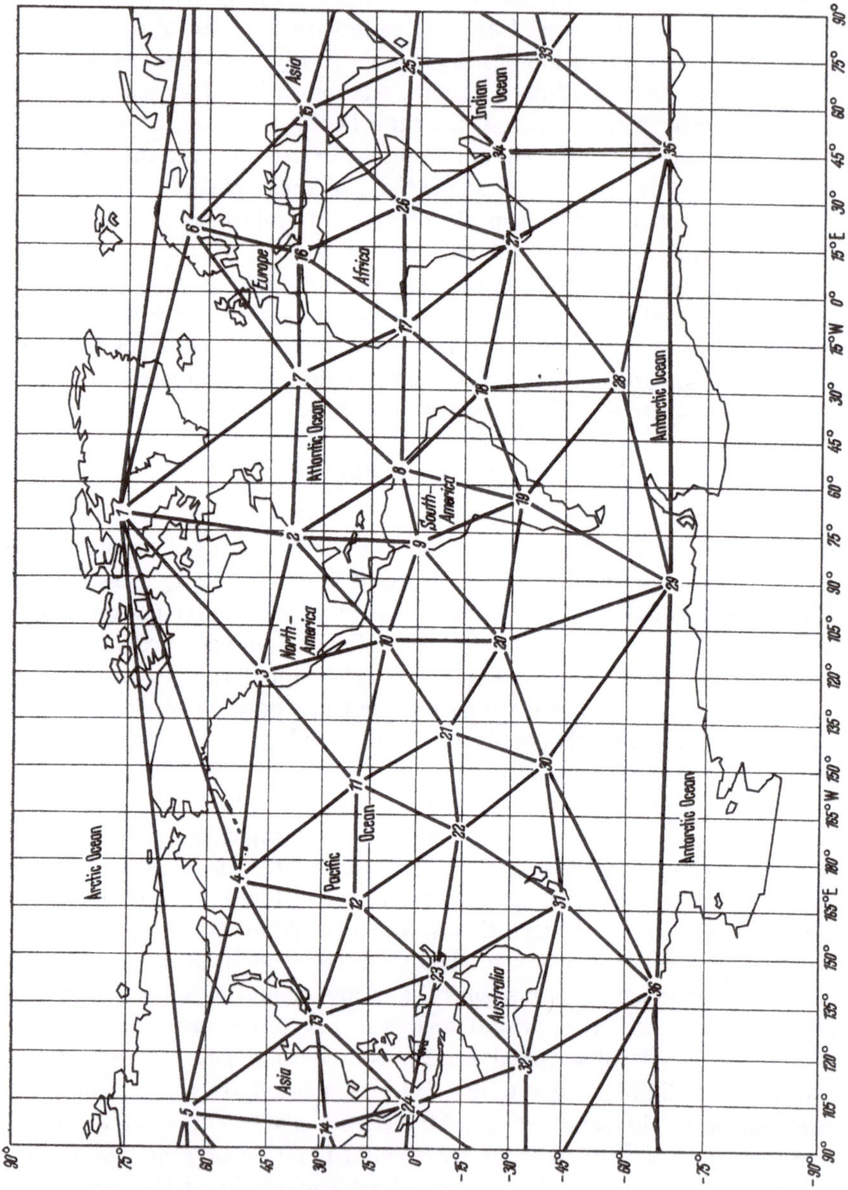

Fig. 1

coordinate system by relating them to measurements of about 200 stars on each plate. Refined analytical photogrammetric procedures are employed to link the plate measurements to the inertial frame.

Considerations which are discussed in detail in [1] connecting the accuracy of the star catalogs, the measurement procedure, instrument quality, scatter of observations, etc., yield a resultant accuracy of about $0''\!.3$ to $0''\!.5$ for the direction of the ray from the camera station to the satellite referred to an inertial coordinate system.

This operation is already in progress, and several triangles of about 1500 km on a side have been observed in North America and adjacent islands. The satellites utilized have been Echo I and Echo II. The first-order worldwide network is designed to consist of 36 stations about 4000 km apart, sighting on the planned PAGEOS satellite with an elevation of 4000 km, near-polar inclination, and circular shape. Fig. 1 is a sketch of the 36 station network.

Prediction

The prediction problem in this context refers to a short-term prediction of up to two weeks, as opposed to long-range predictions of some months. The requirement of the geometric observers on the Echo-type satellites is not satisfied by present prediction services. In carrying out procedures which have as their goal an accuracy of 1:500,000 to 1:1,000,000 it is necessary to set up criteria which may appear unduly stringent with respect to an individual operation but must be enforced to this degree in considering the potential accumulation of error over the entire range of operations. One such criterion in the reduction of data from the geometric observations is that the trail of the satellite should pass within 0.5 mm of the center of the plate. The instrument employed for observing in the worldwide first order geometric network is the BC-4 camera with a focal length of 300 mm. This implies that the allowable across-track error in satellite prediction should be about 5′ of arc. But current predictions of Echo I and Echo II for a week in advance indicate discrepancies in height of as much as 100 km. For a height of 2000 km and a sighting altitude of 45°, this corresponds to an error in altitude of about $1\frac{1}{2}$°, or about 1° projected on the plate. This deviation can be considered random with respect to the satellite, and hence statistically expected to be about 40′ across-track. But this is equivalent to 4 mm. Records of satellite trails photographed up to now roughly confirm discrepancies of this order. Although there are reasons other than inadequate orbit prediction for these discrepancies, as for example, deficient operating procedures and inaccurate station locations, no other cause could conceivably yield errors of this magnitude. It is necessary, therefore, to consider means to improve current orbit predictions for balloon-type satellites by a factor of 5 or so.

There are two aspects to the problem of obtaining precise predictions. One is the choice of orbit analysis; the other is the acquisition of sufficiently accurate observations adequate in number and distribution, available soon enough for prediction purposes. The latter problem is manifestly the more crucial one. Remedying this inadequacy is expensive, however, since it implies more tracking stations of the BAKER-NUNN type. It will pay to determine how much can be squeezed out of the present data network.

An orbit analysis method which appears to be ideally suited for precise prediction is the Smithsonian Astrophysical Observatory Differential Orbit Improvement Program (DOI) [2]. The basic program is empirical in nature, that is, the mathematical formulas for satellite position are designed to flexibly adjust to the observational data without being rigidly tied to a preconceived physical model. It is simply assumed that most phenomena of interest can be expressed by polynomial and trigonometric functions of time. If E represents an orbital element,

$$E = \sum_{i=0}^{N} a_i t^i + \sum_{i=1}^{M} b_i \cos(c_i t + d_i),$$

where a_i, b_i, c_i, d_i are parameters which may be either variable or constant in the solution. This works out very well, for example, in accounting for the secular effects of drag. In this case it is well known that the main effects are given by the coefficient a_1 in the expression for the eccentricity, and by a_2 in the expression for the mean anomaly. Since secular drag effects can be much better accounted for empirically than from theory, these two parameters are treated as variables. On the other hand, since optical observations on passive satellites are relatively few per pass, the determination of short-period perturbations cannot be handled well empirically. Hence, in this case, for $c_i \geqq n$, where n is the mean motion, the corresponding parameters b_i, c_i, d_i are fixed in the solution, their values being based on an a priori model. Another example is perturbations due to tesseral harmonics. Although these can be accounted for from the observations (and have been done successfully in this respect by IZSAK [3]), the number of periodicities involved increase the computer load tremendously if the corresponding parameters are variables. For prediction purposes it suffices to insert values of b_i, c_i, d_i as fixed based on the best values available at the moment. Thus, it is seen that if confidence in an a priori physical model outweighs confidence in the empirical terms (due perhaps to poor or insufficient observations), or if the burden of additional variable parameters is not considered worthwhile, the fixed physical model can easily be made part of the program structure.

The current Smithsonian version of the DOI actually contains physical models which account for short-period perturbations caused by the Earth's flattening, and luni-solar perturbations. Addition of models for tesseral harmonics and for radiation pressure can refine the prediction capability of the DOI even further, especially for tracking balloon-type satellites. The major effort in such an undertaking is the preparation and checking out of computer programs. Fortunately, programs for both of the above models, due to KAULA, are available and can be pressed into service quickly.

Another available model, whose effect is not large or important in this context, will be mentioned here simply because it has not been announced before, and may be of interest otherwise. This is an elaboration if IZSAK's [*4*] model for short-period perturbations due to drag. IZSAK assumed a physical model for the atmosphere which was solely a function of elevation. In order to include the diurnal variation of atmospheric density as part of the atmospheric model, JACCHIA's [*5*] model of the atmosphere was used instead. This substitution yields formulas for short-period perturbations in terms of multiple sines and cosines of the eccentric anomaly somewhat more complicated than IZSAK's, but which have been successfully programmed.

Absolute Positioning

The chief application of the dynamical method to the worldwide geometric network will be to furnish an absolute reference system. The geometric net will have already been oriented with respect to an astronomic coordinate frame at a given epoch; it remains, therefore, only to locate the origin of the reference system, that is, to provide a translation from the geometric origin to the dynamic origin. Shifts of this nature have been accomplished by a number of investigators over the past three years—principally KAULA [*6*], GUIER [*7*], and IZSAK [*3*]. They have placed primary emphasis on the determination of the spectral structure of the Earth's exterior potential, and the deduction therefrom of the spatial variations, the translating of coordinates of observing stations being carried out as an auxiliary operation.

The first degree harmonic terms in the potential are discarded, corresponding to the assumption that the reference system chosen coincides in position with one whose origin is at the Earth's center of mass. However, the reference coordinate system employed by the above investigators for their observations have no direct relationship to the dynamical reference system, being referred to the geometric systems of the Smithsonian Observatory BAKER-NUNN Camera station network or to the TRANET DOPPLER tracking network. A discrepancy thus arises in matching observations with the physical model, due to the elimination of the harmonics of first degree. This difference is recognized and accounted for by a translation of the geometric origin.

If some or all of the stations are referred to the same geometric system. this becomes a uniform change in the rectangular geocentric coordinates of each station [6]. In some cases [3, 7], each station is allowed to shift independently, even if on the same datum as another. This is a potential source of ill-conditioning, but helps to account for internal errors within the geodetic system.

On the other hand, it is usually taken for granted that the geometric and dynamic systems are identically oriented; that is, the polar axis of the geometric system is parallel to a principal axis of inertia of the Earth. The basis for this assumption is that the geometric system has been initially oriented astronomically, and that this orientation is maintained through the correction of azimuths by the LAPLACE azimuth condition. But HOTINE [8] has shown that this is not sufficient. In general, two angles are needed to orient an axis of one system with respect to the corresponding axis of another system. Thus HOTINE has demonstrated that in order for the polar axes of an astronomic and a geodetic coordinate system to be parallel, the following conditions must hold:

$$d\alpha = \sin\varphi\, d\lambda + \cot\beta\,(\sin\alpha\, d\varphi - \cos\alpha\, \cos\varphi\, d\lambda),$$

$$d\beta = -\cos\varphi\, \sin\alpha\, d\lambda - \cos\alpha\, d\varphi,$$

where φ, λ are geographic latitude and longitude and α, β are azimuth and zenith distance. In classical horizontal triangulation, the zenith distance is usually 90°. Hence the first equation reduces to the common form of the LAPLACE condition. The second equation is ignored, but this is not valid because $d\beta$ is not a function of β, and hence independent of whether observations are made in a horizontal plane or not. The implication is that for geodetic systems based on classical triangulation, errors in orientation of this coordinate system should be considered and conceivably allowed for in orbit analysis by the inclusion of two orientation parameters in addition to the three translation parameters. These two added parameters are related to the harmonics of degree 2 and order 1 which represent the deviation of the polar coordinate axis from the principal polar axis of inertia. KAULA [6] carried out solutions in which these harmonics were parameters. They were used, however, strictly as a means of testing the solution, and those solutions for which these harmonics were not vanishingly small were discarded. GUIER [7] states that solutions with these harmonics as parameters yielded values which were effectively zero. There are two reasons why these results appear to satisfy the hypothesis of parallelism of coordinate systems. First, GUIER's method of shifting each station independently tends to allow for a reorientation among stations tied to the same geodetic network. Second, the geodetic systems used as a basis were not strictly classical horizontal systems, GUIER and KAULA both setting forth the initial values of their tracking stations on the system given in [*9*] which combines gravimetric, astrogeodetic and satellite data. These considerations lead to the conclusion that in a solution which emphasizes the positioning of the stations, the choice of parameters and method of procedure should take into account the nature of the provisional coordinates of the tracking station network. In particular, the first-order geometric worldwide triangulation scheme will yield a rigid block of stations oriented properly with respect to the dynamical system to a far greater degree than any other heretofore existing network. The directional rays of this system will, as stated previously, be accurate with respect to an *absolute* reference system to within $0\rlap{.}''3$ to $0\rlap{.}''5$, whereas directions in existing classical first-order nets are accurate to within $0\rlap{.}''2$ to $0\rlap{.}''3$ with respect to a *relative* reference system.

Even so, it will be worthwhile to consider the use of five parameters rather than three in determining the relationship of the geometric to the dynamical coordinate system, the extra two parameters of orientation serving as a check. But the introduction of separate shift parameters for each station is not contemplated as performing any useful purpose, since this will merely serve to load the computation with extra parameters which are essentially functionally interrelated.

In order to obtain the ultimate refinement in accuracy, consideration should also be given to representing the scale by a variable parameter. Thus far the only attempts to do this in the context of satellite observational data have been by ANDERLE and OESTERWINTER [*10*] and VEIS (unpublished). The former set the gravitational constant times the mass of the Earth as a variable parameter; the latter did the same for the Earth's semi-major axis. Their solutions are not comparable since Anderle included both tesseral harmonic variations and station shifts in his solution, while VEIS ignored the tesseral terms of the geopotential. ANDERLE obtained a change in the semi-major axis of 1:50,000; VEIS' change was 1:1,000,000. The result of VEIS appears to be more reasonable, but even so, considering the degree of accuracy required for the positioning of the geometric network, scale variation should be included in the solution.

Probably the greatest difficulty in establishing an absolute geodetic network to the accuracy desired will be in determining the degree of correlation between the parameters for the coordinate adjustment and those for the geopotential. The only published discussion of this topic has been by IZSAK [*3*] who states that "the presence of tesseral harmonics influences the computed station coordinates much more than small errors in the latter affect the computed harmonic coefficients." Since the primary purpose of the investigations up to now has been the determination of the geopotential, the coordinate system positioning being carried out essentially as an auxiliary operation, this has not caused undue concern. If the emphasis is reversed, however, it is essential to establish the effect of the choice of geopotential model on the station shifts, and, in particular, the effect of those terms which are necessarily set equal to zero because of space and time limitations in computing machines.

References

[1] SCHMID, H.: Photogrammetry Applied to Three-Dimensional Geodesy, Société Française de Photogrammétrie, Bull. No. 14, July 1964.

[2] GAPOSCHKIN, E.: Differential Orbit Improvement, SAO S.P. 161, August 1964.

[3] IZSAK, I.: Tesseral Harmonics of the Geopotential and Corrections to Station Coordinates, J. Geophys. Res. June 15, 1964.

[4] IZSAK, I.: Periodic Drag Perturbations of Artificial Satellites, Astronom. J. August 1960.

[5] JACCHIA, L.: A Variable Atmospheric-Density Model from Satellite Accelerations, J. Geophys. Res., September 1960.

[6] KAULA, W.: Tesseral Harmonics of the Gravitational Field and Geodetic Datum Shifts Derived from Camera Observations of Satellites, J. Geophys. Res., Jan. 15, 1963.

[7] GUIER, W.: Determination of the Non-Zonal Harmonics of the Geopotential from Satellite Doppler Data, Nature, Oct. 12, 1963.

[8] HOTINE, M.: A Primer of Non-Classical Geodesy, presented to First Symposium on Three-Dimensional Geodesy, May 1959.

[9] KAULA, W.: A Geoid and World Geodetic System Based on a Combination of Gravimetric, Astrogeodetic and Satellite Data, J. Geophys. Res., June 1961.

[10] ANDERLE, R., and C. OESTERWINTER: A Preliminary Potential for the Earth from Doppler Observations on Satellites, presented to COSPAR, June 1963.

Discussion

Dr. KOVALEVSKY remarks that, as the Smithsonian Catalogue has been used, large systematic errors may be expected in different parts of the sky since the reduction to the FK 4 system is not well known for all catalogues. However, Mr. CHOVITZ remarks that the objective of the data reduction program of the Coast and Geodetic Survey is to select only those stars whose combined standard deviation in right ascension and declination is less than $0''.3$. Prof. BROUWER thinks that this may be easy in the northern hemisphere, but it may be a problem to meet this objective in the southern hemisphere.

Mr. KAULA points out the following facts:

(1) It should be desirable to strengthen the world-wide control system by electronic tracking as well as camera tracking and by using the dynamics of the orbit intensively observed over short arcs, up to 30 minutes.

Mr. Chovitz would also use this method whenever simultaneity cannot be achieved because of weather or other problems.

(2) There should be three orientation angles rather than two to completely relate the three dimensional coordinate systems involved. The orientation angle of primary concern in analysing satellite observations is the Greenwich sidereal time at the epoch pertaining to the orbit being calculated. A problem related to this and not yet satisfactorily solved is the relationship in longitude of the principal Doppler and camera tracking systems. The other two orientation angles expressing the direction of the pole would show observable effects only if the inertially-fixed coordinate system with respect to which the orbit is calculated has an equator at an appreciable tilt to the earth-fixed equator.

(3) Clarification would be needed on the relationship between the central term GM, the equatorial radius a_e, the radial component of station positions, length standards as used by tracking systems and the mean motions and semi-major axes of close satellite orbits.

Dr. Merson thinks that radar observations would become extremely useful to increase the accuracy needed to obtain good orbital parameters within a day or two of the making of the observations[1].

Dr. Anderle (1) agrees with Mr. Kaula's recommendation that the GM determined from the lunar probes be accepted, the information on GM obtained from Doppler data being not yet useful. (2) He thinks that classical first order ground triangulation adds nothing to the determination of intradatum station coordinates which can be got in practice on the basis of the Doppler observations alone. It is to be remarked that this is not the case for the geometric world-wide satellite triangulation, whose accuracy is much greater. (3) Lastly, Mr. Anderle feels that to determine the gravity field and the coordinates of the station relative to the center of mass the most elegant way would be to combine both the dynamic data and the geometric data in the same normal equations. In this way any weakness in one set of data can be offset by the strength in the other set. We must anticipate the possibility of giving weights to the data. It is planned that many of the 36 stations in the optical geometric net will be occupied by Doppler equipment as well. Mr. Chovitz agrees with this remark.

[1] Subsequently, Mr. Chovitz has been informed that very precise predictions from radar observations, good to less than 15″ have been obtained on the BEACON satellite, used for laser beam experiments.

Perturbed Motion of Satellites with Small Orbital Eccentricities

By

Yu. V. Batrakov

Institute for Theoretical Astronomy, Leningrad, USSR

Abstract. The method is developed which allows to use the existing analytical expressions of perturbations of the keplerian elements for determining the coordinates of satellite without loss of accuracy in the case of small eccentricities.

Résumé. L'auteur développe une méthode permettant d'utiliser les expressions analytiques existantes des perturbations des éléments képlériens pour déterminer les coordonnées d'un satellite sans perte de précision dans le cas de faibles excentricités.

When deriving the formulae for perturbations in the motion of satellites the keplerian elements (Kozai, 1959; Batrakov, 1963) or Delaunay variables (Brouwer, 1959; Kozai, 1962) are commonly used. Such a choice of variables is convenient by the simplifications being introduced into calculations and by the rather moderate length of resulting formulae. The perturbations of the elements contain, however, the eccentricity in the denominator, and the errors of coordinates can become large when the eccentricity being small. In such special cases the combinations $\lambda = M + \omega$, $e \sin\omega$, $e \cos\omega$ rather than keplerian elements are used. As a result, the singularity in the equations of motion due to the smallness of eccentricity dissappears completely and all the perturbations can be obtained with the accuracy needed. Some inconvenience of these variables consists in the fact that the work on the derivation of the formulae for perturbations in terms of them must be repeated almost from the very beginning (Kozai, 1961; Chebotarev, 1963). In addition, the different formulae for small and moderate eccentricities must be used and the programming for high-speed computers becomes greatly troublesome. It must be also noted that with equal accuracy, the perturbations of the considered combinations are more lengthy than that ones of the Keplerian elements.

Therefore it is of interest to develop such a method of using the perturbations of keplerian elements that the coordinates would not contain singularity connected with small eccentricities. An analogous method for the BROUWER's theory has been developed (LYDDANE, 1963). Here the formulae are given which allow to use the expressions of perturbations given in (BATRAKOV, 1963) for determining the coordinates of a satellite without the loss of accuracy when the eccentricity is small.

Let us introduce the variables as follows:

$$\lambda = M + \omega, \qquad \tilde{\omega} = x(t - t_0) + \omega_0,$$
$$l = e\cos(\omega - \tilde{\omega}), \qquad h = e\sin(\omega - \tilde{\omega}). \tag{1}$$

Here e, ω, M are the osculating eccentricity, the argument of perigee and the mean anomaly respectively, $\tilde{\omega}$ is the mean argument of perigee, ω_0 is the constant of integration and x is some constant, the value of which is being chosen in such a manner that l and h have not secular perturbations at all. For these variables, using the LAGRANGEan equations for osculating elements, the following equations may be obtained

$$\begin{aligned} \frac{dl}{dt} &= f_l = \frac{l}{e} f_e - (f_\omega - x)\,h, \\ \frac{dh}{dt} &= f_h = \frac{h}{e} f_e + (f_\omega - x)\,l, \end{aligned} \tag{2}$$

where f_e, f_ω are the right-hand members of the LAGRANGEan equations for the osculating eccentricity and the argument of perigee (in these equations λ instead of M is being used).

For determining the perturbations, we develop the right-hand members of (2) into series in terms of the perturbations:

$$\frac{dl}{dt} = f_l^0 + \frac{\partial f_l^0}{\partial l}\,\delta l + \frac{\partial f_l^0}{\partial h}\,\delta h + \sum \frac{\partial f_l^0}{\partial \varepsilon_i}\,\delta\varepsilon_i + \cdots, \tag{3}$$

where ε_i are a, i, λ, Ω respectively. For the remaining variables, the differential equations are of an analogous form. The solution of (3) and of the other equations will be found in the form:

$$\begin{aligned} l &= e_0 + \delta_1 l + \delta_2 l + \cdots, \\ h &= \delta_1 h + \delta_2 h + \cdots, \end{aligned} \tag{4}$$

where e_0, ε_i^0 are the constants of integration and $\delta_k l, \ldots$ are the perturbations of the k-th order (approximation) not containing the secular terms. The constant $\varkappa$ is also taken in the form

$$x = x_1 + x_2 + \cdots. \tag{5}$$

For the first order perturbations from (2) and (4) we obtain easily

$$\left.\begin{aligned} &\delta_1 l = (\delta_1 e), \qquad \delta_1 h = e_0(\delta_1\omega), \\ &x_1 = [[f_\omega^0]] = \frac{1}{(2\pi)^2}\int_0^{2\pi}\int_0^{2\pi} f_\omega^0\, d\lambda\, d\omega, \\ &x_1(t - t_0) = [\delta_1\omega]. \end{aligned}\right\} \tag{6}$$

The parenthesis and the square brackets in (6) and further on, designate the short periodic and secular perturbations respectively.

For the second order perturbations we have

$$\frac{d\delta_2 l}{dt} = \frac{\partial f_l^0}{\partial l}\delta_1 l + \frac{\partial f_l^0}{\partial h}\delta_1 h + \sum \frac{\partial f_l}{\partial \varepsilon_i}\delta_1\varepsilon_i = \frac{\partial f_e^0}{\partial e}(\delta_1 e) + \\ + \frac{\partial f_e^0}{\partial \omega}(\delta_1\omega) + \sum \frac{\partial f_l^0}{\partial \varepsilon_i}(\delta_1\varepsilon_i) - e_0(\delta_1\omega)(f_\omega) \tag{7}$$

if the equalities

$$\frac{\partial}{\partial l} = \frac{l}{e}\frac{\partial}{\partial e} - \frac{h}{e^2}\frac{\partial}{\partial \omega}, \qquad \frac{\partial}{\partial h} = \frac{h}{e}\frac{\partial}{\partial e} + \frac{l}{e^2}\frac{\partial}{\partial \omega}$$

are used, (6) are taken into account, and $l = e_0$, $h = 0$,

When integrating the Eq. (7) and these corresponding to h, ε_i we get

$$\left.\begin{aligned} \delta_2 l &= (\delta_2 e) + \{\delta_2 e\} - \tfrac{1}{2} e_0((\delta_1\omega)^2), \\ \delta_2 h &= e_0(\delta_2\omega) + e_0\{\delta_2\omega\} + (\delta_1 e)(\delta_1\omega), \\ \delta_2\varepsilon_i &= (\delta_2\varepsilon) + \{\delta_2\varepsilon\}. \end{aligned}\right\} \tag{8}$$

The braces here and further on stand for the long periodic perturbations. If only the most significant long periodic terms are taken into account, one gets simply

$$\delta_2 l = \{\delta_2 e\}, \qquad \delta_2 h = e_0\{\delta_2\omega\}, \qquad \delta_2\varepsilon_i = \{\delta_2\varepsilon_i\}. \tag{9}$$

For excluding all the secular terms and obtaining (8) in purely trigonometric form we must assume

$$x_2 = \left[\left[\frac{\partial f_\omega^0}{\partial e}(\delta_1 e) + \frac{\partial f_\omega^0}{\partial \omega}(\delta_1\omega) + \sum \frac{\partial f_\omega^0}{\partial \varepsilon_i}(\delta_1\varepsilon_i)\right]\right], \tag{10}$$

$$x_2(t - t_0) = [\delta_2\omega].$$

In quite a similar way it can be easily shown that the third order perturbations contain long periodic terms of the first order with respect to the small parameter (the flattening of the Earth):

$$\delta_3 l = 0, \qquad \delta_3 h = e_0\{\delta_3\omega\}, \qquad \delta_3\varepsilon_i = \{\delta_3\varepsilon_i\}. \tag{11}$$

Let us denote

$$\{\delta_1\varepsilon\} = \{\delta_2\varepsilon\} + \{\delta_3\varepsilon\}$$

where ε is any of elements a, e, i, Ω, ω, $\lambda = M + \omega$. The expressions for $\{\delta_1\varepsilon\}$ along with $(\delta_1\varepsilon)$, $[\delta_1\varepsilon]$, $[\delta_2\varepsilon]$ are given in (Batrakov, 1963)[1].

At the conclusion we give the set of basic formulae which allows to determine the satellites coordinates in case of small eccentricities with

[1] There are the following misprints in the paper mentioned:

(a) in $\{\delta_1 e\}$ the factor $e\lambda^2$ rather than e must be used,
(b) in $[\delta_2\omega]$ the coefficient -2590 before $e^2\lambda^2$ must be replaced by -5180,
(c) in $(\delta_1 a)/a_0$ the coefficient of $\cos M$ must be $e + \frac{9}{8}e^3 + \frac{87}{64}e^5$.

all attainable accuracy, using the perturbations of the elements from (BATRAKOV, 1963).

At first, the following quantities are to be calculated:

$$\begin{aligned}
\tilde{\omega} &= \omega_0 + [\delta_1\omega] + [\delta_2\omega],\\
l &= e_0 + (\delta_1 e) + \{\delta_1 e\},\\
h &= e_0(\delta_1\omega) + e_0\{\delta_1\omega\},\\
\lambda &= \lambda_0 + n_0(t - t_0) + [\delta_1 M] + [\delta_1\omega] + [\delta_2 M] + [\delta_2\omega] +\\
&\quad + (\delta_1 M) + (\delta_1\omega) + \{\delta_1 M\} + \{\delta_1\omega\},\\
\Omega &= \Omega_0 + [\delta_1\Omega] + [\delta_2\Omega] + (\delta_1\Omega) + \{\delta_1\Omega\},
\end{aligned}$$

$$a = a_0 + (\delta_1 a),$$

$$i = i_0 + (\delta_1 i) + \{\delta_1 i\}.$$

Then the auxiliary quantities l_1, h_1, $E_1 = E + \omega$ (E is the eccentric anomaly) are determined by the formulae

$$\begin{aligned}
l_1 &= e\cos\omega = l\cos\tilde{\omega} - h\sin\tilde{\omega},\\
h_1 &= e\sin\omega = l\sin\tilde{\omega} + h\cos\tilde{\omega},\\
E_1 &- l_1\sin E_1 + h_1\cos E_1 = \lambda.
\end{aligned}$$

The position vector of the satellite $\bar{r}$ is then determined by

$$\bar{r} = \bar{C}\,c + \bar{S}\,s,$$

where

$$\begin{aligned}
\bar{r} &= (x, y, z),\\
\bar{C} &= (\cos\Omega, \sin\Omega, 0),\\
\bar{S} &= (-\sin\Omega\cos i, \cos\Omega\cos i, \sin i),
\end{aligned}$$

$$c = r\cos u = a\left[\cos E_1 + \frac{h_1}{1+\xi}(l_1\sin E_1 - h_1\cos E_1) - l_1\right],$$

$$s = r\sin u = a\left[\sin E_1 - \frac{l_1}{1+\xi}(l_1\sin E_1 - h_1\cos E_1) - h_1\right],$$

$$\xi = \sqrt{1 - l_1^2 - h_1^2}$$

(u is the argument of latitude).

References

[1] KOZAI, Y.: Astronom. J. **64**, No. 9, 367 (1959).

[2] BATRAKOV, YU. V.: Dynamics of Satellites, IUTAM Symposium, Paris, May 28—30, 1962, Berlin/Göttingen/Heidelberg: Springer 1963, p. 74.

[3] BROUWER, D.: Astronom. J. **64**, No. 9, 378 (1959).

[4] KOZAI, Y.: Astronom. J. **67**, No. 7, 446 (1962).

[5] KOZAI, Y.: Astronom. J. **66**, No. 3, 132 (1961).

[6] CHEBOTAREV, G. A.: The Use of Artificial Satellites for Geodesy, Amsterdam: North-Holland Publishing Company, p. 8.

[7] LYDDANE, R. H.: Astronom. J. **68**, No. 555 (1963).

Computational Methods Employed with Doppler Observations and Derivation of Geodetic Results

By

Richard J. Anderle

U.S. Naval Weapons Laboratory, Dahlgren, Virginia, U.S.A.

Abstract. The computational model used at the Naval Weapons Laboratory to obtain geodetic data from DOPPLER observations on satellites is discussed. Some selected problems studied with the method, and discussed in this report, include

(1) The effects of bias on the accuracy of the solution and on the validity of statistical tests of the accuracy of the solution.

(2) The results obtained and the methods of accounting for the effects of tesseral gravitational harmonics such as $C_{13,13}$, $S_{13,13}$ which produce resonant effects on the satellite motion, and

(3) the method of determining and representing the significant gravity coefficients.

The accuracy of orbit computation currently being achieved as a result of recent improvements in the gravity field representation is shown to be about 50 meters.

Résumé. On étudie le programme de traitement du "Naval Weapons Laboratory" pour l'obtention de données géodésiques à partir des observations de satellites. On a choisi de discuter les problèmes suivants:

(1) L'effet d'une erreur en fréquence sur la précision de la solution et sur la validité des calculs statistiques sur la précision de la solution.

(2) Les résultats obtenus et les méthodes qui tiennent compte des harmoniques tesseraux du potentiel tels que $C_{13,13}$, $S_{13,13}$ qui produisent des effets de résonance sur le mouvement du satellite.

(3) La méthode qui permet de déterminer et de représenter les coefficients du potentiel les plus significatifs.

La précision obtenue dans la détermination de l'orbite est de l'ordre de 50 mètres grace aux améliorations récentes dans la représentation du potentiel.

Features of Basic Mathematical Model

The computer program used by the Naval Weapons Laboratory to determine geodetic parameters on the basis of DOPPLER observations made on satellites differs from that used by other research activities in

Table 1. *Some Features of Three Satellite Geodesy Computational Methods*

	KAULA	APL	NWL
Method of orbit computation	General theory	Numerical integration	Numerical integration
Partial derivatives of orbit parameters	General theory	Mixed	Numerical integration
Partial derivatives of gravity parameters	General theory	General theory	Numerical integration
Representation of DOPPLER data per pass	Polynomial	3 parameter	30 points
Atmospheric model for tropospheric refraction	Standard	Computed per pass	Standard
Parameters assumed to be decoupled?	No	Yes	No

a number of ways, as shown in Table 1. The NWL program is based on the numerical integration of the equations of motion. A tenth order COWELL process is generally used to integrate for satellite position while a sixth order process is used to obtain the velocities required to determine the contributions of atmospheric drag to the forces on the satellite. The partial derivatives of satellite position and velocity with respect to orbit parameters, atmospheric parameters, and gravity parameters, are obtained by the numerical integration of the perturbation equations. The 500 or so original data points observed on an individual satellite pass are aggregated in a preprocessing program: each contiguous set of about eight points observed during the pass is represented by an aggregated observation fitted to the eight original points; the standard deviation of the residuals of the least-squares fit is used to compute a weight for the observation. As a result of the aggregation and the rejection of observations made at elevation angles below 10 degrees, each pass is represented by 20 to 30 aggregated observations. A standard atmospheric model invariant with geographic position or time of observation, is assumed in order to determine the effects of tropospheric refraction on the observations. The parameters of the solution are:

(1) Bias parameters for each satellite pass including frequency drift of the oscillator,

(2) Arc parameters including six orbit parameters and scaling factors for the atmospheric density structure,

(3) Geodetic parameters including gravity coefficients and station coordinates.

Full account of the correlation among these parameters is considered in the solution. While there are many other differences in the computational methods used by different activities, it is believed that the foregoing include the principal features which distinguish the methods from the viewpoint of the complexity or the cost of operation of the computer

Table 2. *Geoid Features*

Location	Coordinate	Source				
		KAULA [1] Astro-geodetic 5/61	KAULA [2] Satellite 9/63	IZSAK [3] 1/64	APL [4] 12/64	NWL-5E-6 [5]
England	Latitude	80° N	50° N	55° N	55° N	55° N
	Longitude	140° E	10° E	35° W	5° W	15° W
	Height (m)	43	36	35	67	65
South Africa	Latitude	40° S	50° S	45° S	50° S	45° S
	Longitude	70° E	0°	55° E	50° E	15° E
	Height (m)	29	35	37	39	45
Japan	Latitude	0°	10° S	15° S	25° N	5° N
	Longitude	115 °E	150° E	150° E	140° E	140° E
	Height (m)	34	52	54	60	75
India	Latitude	15° N	10° N	5° N	5° N	10° N
	Longitude	75° E	70° E	70° E	75° E	75° E
	Height (m)	−25	−59	−48	−77	−95
East Pacific	Latitude	10° S		10° N	25° N	15° N
	Longitude	125° W		120° W	130° W	115° W
	Height (m)	+34		−50	−59	−70
South Pacific	Latitude	65° S	30° S	40° S	70° S	70° S
	Longitude	160° W	110° W	110° E	180° E	170° W
	Height (m)	−20	−18	−37	−45	−75
North Pacific	Latitude	25° N	30° N		45° N	40° N
	Longitude	170° W	150° W		175° W	175° W
	Height (m)	−51	−19		−22	−35
West Atlantic	Latitude	30° N	30° N	10° N	30° N	30° N
	Longitude	65° W	60° W	50° W	75° W	70° W
	Height (m)	−44	−11	−23	−70	−45

programs. While there are significant differences in the geoid shape determined by different activities, the data base for the computations as well as the method of computation is also different. For example, eight prominent geoid features appear in the NWL solution for geoid height shown in Fig. 1. The position and magnitude of these features obtained by different researchers are compared in Table 2. The satellite solutions reported by KAULA and IZSAK were based upon optical observations while those reported by APL and NWL were based upon DOPPLER satellite observations. Some of the more significant problems being studied at NWL in connection with such solutions are described in the following paragraphs.

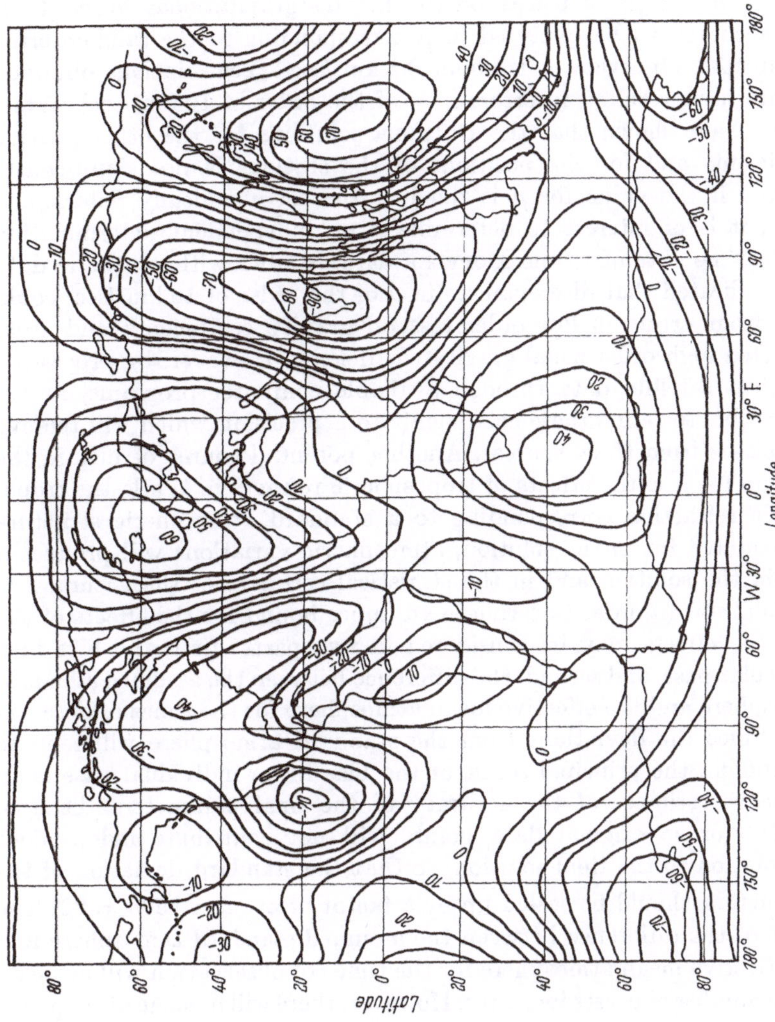

Fig. 1. Contours of equal geoid height in the NWL-SE solution (meters). $1/f_*$ is taken equal to 298.30.

Biases

While considerable effort is expended in equipment and computer program design to remove biases in any system used to determine geodetic parameters, it is inevitable that some bias will remain. For example, first order ionospheric refraction effects are taken into account in the U. S. Navy DOPPLER tracking system at the tracking stations by analog combination of a pair of coherent frequencies transmitted by the satellite. However, second order effects will produce a small bias in the observations. Recently, our attention has been directed toward the effects of biases on the Earth's gravitational constant, μ. The accuracy to be expected in μ corresponding to the random error in DOPPLER observations is about 0.1 km^3/sec^2. Yet solutions obtained to date have resulted in errors in μ of the order of 10 km^3/sec^2. While studies have shown that the DOPPLER solutions for the other gravity coefficients and for the positions of the stations is not significantly affected if the value for μ is fixed at the astronomically determined value, it is of interest to determine if an independent determination of μ can be obtained from DOPPLER data. Studies with synthetic data have indicated that discrepancies in μ of the order of 10 km^3/sec^2 could arise from errors in low order tesseral gravity coefficients and from neglected high order zonal gravity coefficients. However, by processing sufficient satellite data through available computer programs, it will be possible to refine these coefficients to a precision which will remove any biases from these sources. Another potential source of bias in the solution for μ is the effects of tropospheric refraction. While the tropospheric refraction corresponding to a standard atmospheric structure is accounted for in the solution, atmospheric variations will produce a bias in the solution and in the statistical estimate of the accuracy of the solution. In order to estimate an upper bound for the effects of the bias, the effects may be considered in two parts: first, the bias in a particular pass, and second, the difference between the assumed standard atmosphere and the effective mean atmosphere for the times of observation. Since the deviations from the standard atmosphere will tend to be random, the principal result of the bias in an individual pass is an erroneous estimate of the accuracy of the final parameters. That is, the 500 or so original data points will not contribute independent information to the final solution, so that the standard deviation of the parameters should be scaled up by a factor of at most $\sqrt{500} = 22$. The effect of the difference between the assumed standard atmosphere and the effective mean atmosphere for the times of observation will decrease as the number of passes increases. However, there will be some discrepancy between the standard atmosphere and the mean atmosphere for all the passes either due to difficulty in establishing the standard or due to improper sampling of the atmosphere due to the geographic distribution of the DOPPLER stations. Since day to day variations in atmosphere do not affect the tropospheric refraction correction by more than 10% of the correction, one would not expect that the mean correction would be in error by more than 1% for a large number of satellite passes. Studies have indicated that such a bias in the refraction correction would not affect the value of μ by more than 1 km^3/sec^2. Thus it appears certain that Doppler observations will yield a value for μ to an accuracy of at least 2 km³/sec². More detailed studies will have to be conducted in order to determine whether still higher accuracy will be achieved.

While useful information on the value for μ has not yet been obtained on the basis of Doppler observations, tests and analyses have indicated that Doppler observations do yield precise geodetic data if the astronomic value of μ is used in the calculations. The precision is obtained (1) because the solution is not significantly affected by variations in the value of μ which do not exceed the accuracy to which it is known, and (2) because the coupling among the parameters in general is reduced when μ is not included as a parameter of the solution.

Resonance Effects of High Order Tesseral Gravity Coefficients

The effect of a sectorial or tesseral gravity coefficient of order m on the orbit of a satellite is amplified if the nodal period of the satellite is close to the period of Earth's rotation relative the line of nodes of the plane of the orbit divided by m. For example, the nodal period of satellite 1962 $\beta\mu_1$ which is 107.84 minutes gave a five day beat period for the effect of the tesseral gravity coefficients of thirteenth order. The observed and computed effects of the thirteenth order tesseral gravity coefficient on the along-track position of the satellite are shown in Fig. 2 for a sevenday time span. The amplitude of the effect in this case is about 150 meters. Analysis of results such as these have led to values for three pairs of high order gravity coefficients which are listed in Table 3 along with gravity coefficients through the seventh degree and sixth order which were obtained in a separate least-squares solution. One now wishes to know how many additional coefficients must be consi-

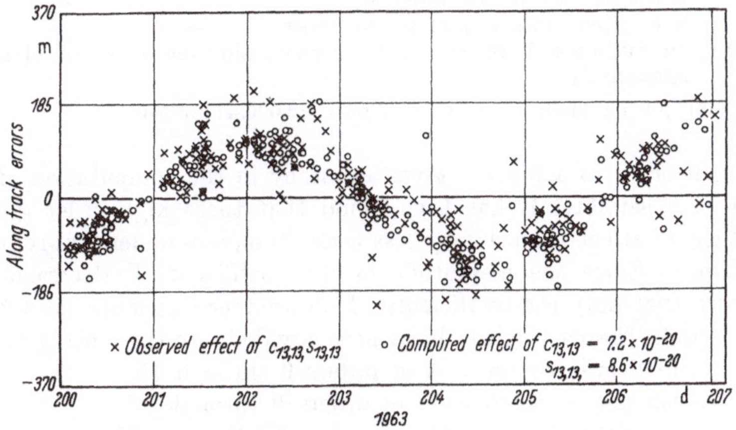

Fig. 2. Satellite 1962 $\beta\mu$ 1 along track errors with NWL-5 A geodetic parameters.

Table 3. *NWL-5 E-6 Normalized Gravity Coefficients*[1]

n	m	$\bar{C}_{nm}$	$\bar{S}_{nm}$	n	m	$\bar{C}_{nm}$	$\bar{S}_{nm}$
2	0	−484.19		6	1	−.09	.19
3	0	.98		6	2	.13	−.46
4	0	.51		6	3	−.02	−.13
5	0	.05		6	4	−.19	−.32
6	0	−.22		6	5	−.09	−.79
7	0	.11		6	6	−.32	−.36
2	1	.02	.06	7	1	.33	.08
2	2	2.45	−1.52	7	2	.35	−.20
3	1	2.15	.27	7	3	.32	.04
3	2	.98	−.91	7	4	−.47	−.24
3	3	.58	1.63	7	5	.06	.02
4	1	−.50	−.58	7	6	−.48	−.24
4	2	.27	.67	13	13	−.03	.11
4	3	1.03	−.25	15	13	−.06	−.06
4	4	−.41	.34	15	14	.01	−.03
5	1	.03	−.12				
5	2	.64	−.33				
5	3	−.39	−.12				
5	4	−.55	.15				
5	5	.22	−.59				

$$V = \mu \Sigma \left[R^n C_{nm} \frac{P_n^m\left(\frac{z}{r}\right)}{r^{n+1}} \cos m\lambda + R^n S_{nm} \frac{P_n^m\left(\frac{z}{r}\right)}{r^{n+1}} \sin m\lambda \right]$$

$$\bar{C}_{n,m} = [(n-m)!\,(2n+1)\,K/(n+m)!]^{-1/2}\, C_{nm}, \quad \text{where } K = 1 \text{ when } m = 0, \quad K = 2 \text{ when } m \neq 0,$$

where

P_n^m is the associated Legendre polynomial,
R is the Earth's radius,
μ is the Earth's gravity constant,
λ is longitude with respect to Greenwich,
z and r are distances above the equatorial plane and from the center of Earth, respectively.

[1] Multiply all coefficients by 10^{-6}. $\mu = 398{,}605.42\ \mathrm{km^3/sec^2}$.

dered in order to achieve a given accuracy in the computation of the orbit of a satellite. It has been found that these high order gravity coefficients affect the orbit by less than 10 meters under non-resonant conditions. Since the magnitude of the coefficients is decreasing, it appears that only the coefficients which produce resonance need to be considered. The coefficients have been divided into three categories:

(1) odd degree coefficients of orders 9 through 15,
(2) even degree coefficients of orders 9 through 15,
(3) even degree coefficients of orders 18 through 30.

Coefficients of orders higher than 15 in categories (1) and (2) or higher than order 30 in category (3) need not be considered because resonance under such conditions would only occur if the nodal period were less than 90 minutes. Coefficients of orders less than 9 in categories (1) and (2) or less than 18 in category (3) correspond to satellites with nodal periods higher than 160 minutes; these periods correspond to satellite altitudes which are high enough so that the effects of gravity field anomalies on the orbit are considerably reduced. The category (1) effects, which are considerably larger than those in the other two categories, are primarily on the along-track position of the satellite. The principal source for this effect is given by the equation [*6*] shown in Table 4 which is applicable to a circular polar orbit. The quadratic divisor is close to zero when the nodal period is close to resonance with a gravity coefficient of order m. The category (1) effects arise when $m\,\omega_E/n_0$ is near unity so that the degree, n, must be odd. The category (2) effects arise from terms in which the small divisor is linear [*7*] rather than quadratic. The principal category (3) effects arise from the terms shown in the figure when $m\,\omega_E/n_0$ is near two, so that the degree, n, is even. Because the coefficients for these terms are so small when the order is 20 to 30, the category (3) effects will be absorbed in two arbitrary parameters in geodetic calculations made at the Naval Weapons Laboratory. The arbitrary parameters will be the phase and amplitude of the effect on the longitude of the satellite in the orbit plane; the period

Table 4. *Principal Effect on Longitude in Orbit Plane of Resonance with High Order Gravity Coefficients*[1]

$$\frac{3}{2} C_{n,m}\left(\frac{R}{a}\right)\sum_{k=0}^{n} P^1_{nmk}\left\{\frac{\sin\left[\left(n-2k+\frac{m\omega_E}{n_0}\right)n_0 t-(n-m)\frac{\pi}{2}\right]}{\left[n-2k+\frac{m\omega_E}{n_0}\right]^2}+\right.$$
$$\left.+\frac{\sin\left[\left(n-2k-\frac{m\omega_E}{n_0}\right)n_0 t-(n-m)\frac{\pi}{2}\right]}{\left[n-2k-\frac{m\omega_E}{n_0}\right]^2}\right\}$$

$C_{n,m}$ is the tesseral harmonic coefficient,
a is the semi-major axis of the orbit,
R is the Earth's equatorial radius,
n_0 is the reciprocal of the nodal period of the satellite,
ω_E is the sidereal rate of the Earth,
P^1_{nmk} is an integer.

---

[1] These formulas are valid only when the small divisor $\left[n-2k-\frac{m\omega_E}{n_0}\right] > 0.03$. B. GARFINKEL of the Ballistic Research Laboratory, Aberdeen, Maryland, has derived formulas applicable to satellites which are closer to resonance.

of the effect will be pre-calculated from the satellite period. The category (1) effects are more difficult to treat because significant effects of gravity coefficients of two or three different orders may exist for a single satellite. The effects of nominal values for resonant tesseral and sectorial coefficients of the lowest degree on the satellite position are shown in Fig. 3. If a satellite is close to resonance with a gravity coefficient of order m, it may have a beat period of close to one day with coefficients of order $(m+1)$ and of order $(m-1)$, which will yield significant effects as can be seen in the figure. (The magnitude of the nominal effects shown in the figure is about right for some of the coefficients found to date, while it is a factor of five too large for other coefficients.) Geodetic calculations made at the Naval Weapons Laboratory will include coefficients of two or three orders and, for each order, two of three degrees in order to account for the resonance effects falling in this category. A similar number of coefficients falling in category (2) will also be considered in calculations. However, the probability of a successful determination of the category (2) coefficients is less certain because the effects are smaller. An example of the computed effects of a pair or these coefficients on the orbit of a satellite is given in Fig. 4. The nominal orbit used is for satellite 1962 $\beta\mu_1$ which, as mentioned previously, has a five day beat period with the thirteenth order gravity coefficient. The figure, however, shows the nominal effects of the twelfth order 14th degree coefficient which yields only an 18 hour beat period. The symbols indicate the effect of the coefficients on the position of the satellite (—), and on the components of the effects on satellite position parallel to the velocity vector (Δ), parallel to the position vector (*), and normal to the orbit plane (+). The high requency effect, has a period of about 110 minutes which corresponds

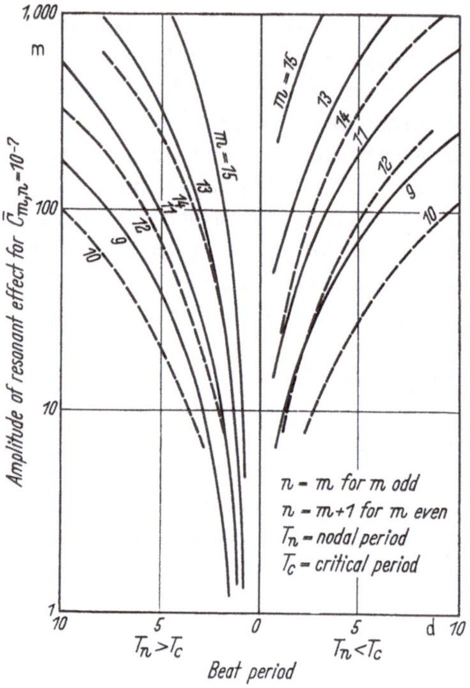

Fig. 3. Resonant effects for polar satellites.

to the orbit period of the satellite. The effect is modulated at a frequency corresponding to the resonant period of about 18 hours.

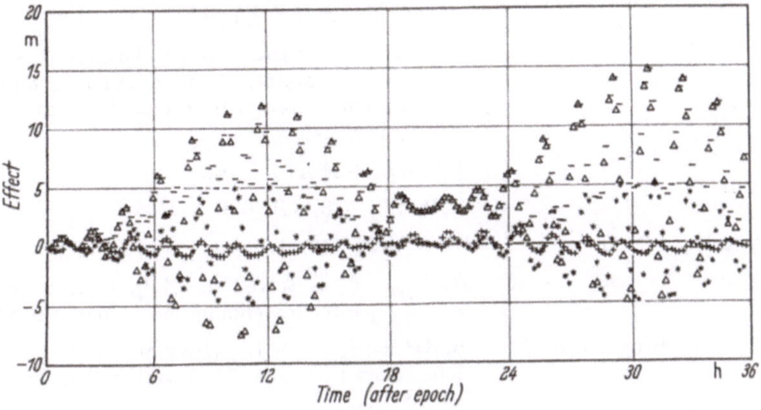

Fig. 4. Effect of $\bar{C}_{14,12} = 10^{-7}$ on the orbit of satellite 1962 $\beta\,\mu$ 1.

Determination of Significant Parameters

The rapid progress in satellite geodesy has resulted in a rapid increase in the number of parameters which must be considered. In the Naval Weapons Laboratory solution made in the Fall of 1964, 58 gravity coefficients and the coordinates of 29 tracking stations were determined. In the solution now in progress, over 200 gravity coefficients and the coordinates of 50 to 100 stations will be considered. However, it is doubtful that a unique solution for the complete set of parameters will be possible. In the past, the solution and the covariance matrix for the solution has been inspected in order to attempt to determine the set of parameters which are well determined. However, this is largely a subjective procedure which may lead to distortion of the results. An alternative procedure developed by C. J. Cohen of the Naval Weapons Laboratory is now under study. Before starting the process, the bias, orbit and station coordinate parameters are eliminated from the normal equations by the process outlined in Table 5. (The parameters eliminated are always certain to be well determined if the number of observations are counted for each pass, orbit, and station.) The eigenvalues and eigenvectors are then found for the normal matrix of gravity parameters which resulted from the parameter elimination. The matrix of eigenvectors which diagnolizes the normal matrix is considered a transformation matrix from a gravity parameter "P space" to "Q space". The reciprocal roots of the eigenvalues, which are the standard deviations of the gravity parameters in Q space can be tested for significance

Table 5. *Parameter Elimination*

Given normal equation	$B\,\Delta P = E.$
Partitioned normal equations	$\begin{pmatrix} B_s & A \\ A^* & B_e \end{pmatrix} \begin{pmatrix} \Delta P_s \\ \Delta P_e \end{pmatrix} = \begin{pmatrix} E_s \\ E_e \end{pmatrix},$ where P_e are the parameters to be eliminated and P_s are the parameters to be determined directly with full consideration to the coupling of the P_s and P_e parameters.
That is,	$\begin{cases} B_s\,\Delta P_s + A\,\Delta P_e = E_s. \\ A^*\,\Delta P_s + B_e\,\Delta P_e = E_e \end{cases}$
Solving for ΔP_s	$(B_s - A\,B_e^{-1}\,A^*)\,\Delta P_s = (E_s - A\,B_e^{-1}\,E_e).$
That is,	$B_s^1\,\Delta P_s = E_s^1$ is solved for ΔP_s, either directly or after parameter elimination is performed.
Back-substituting for ΔP_e^*	$B_e\,\Delta P_e = E_e - A^*\,\Delta P_s$ (Bias parameters are determined by reintegration rather than by back substitution.)

against arbitrary tolerances, and the solution for the Q parameters can be tested against the standard deviations to determine their significance. Unlike corresponding tests using the correlation matrix in P space, Q parameters which are not significant may be eliminated individually since they are completely decoupled. The validity of the tests against arbitrary tolerances presumes that the parameters are satisfactorily normalized. The normalization adopted by KAULA [8] is used; as a result of the normalization, unit perturbations in the normalized coefficients yield equal mean squared contributions to the gravitational potential at the surface of the Earth. An example of the eigenvalues obtained for 58 gravity coefficients on the basis of 37 weeks of observations made on satellites having four distinct orbital inclinations is shown in Table 6. The largest eigenvalue corresponds to an $r\,m\,s$ contribution to the potential of 3 parts in 10^{11} and the smallest to a contribution of 4 parts to 10^9. The components of the eigenvector corresponding to the smallest eigenvalue are shown in Table 7 along with the gravity parameters to which the components refer. The coefficients with the largest components are those which would be the worst determined in this solution. For this case the most poorly determined coefficients are S_{53}, C_{54}, C_{65}, and S_{33}. Now if the Q parameters with the smallest eigenvalues are suppressed, the solution in P space yields a linear combination of parameters rather than a unique set of parameters. For the sake of uniqueness and convenience, additional conditions are then imposed on the solution to remove the extra degrees of freedom. The conditions chosen are that the number of parameters evaluated

Table 6. *Eigenvalues for 37 Weeks of Doppler Observations Made on Satellites Having four Orbital Inclinations*

No.	Value × 10^{-17}	No.	Value × 10^{-17}	No.	Value × 10^{-17}
1	9702.20	21	13.20	41	2.57
2	2354.96	22	12.33	42	2.39
3	1157.65	23	11.87	43	2.21
4	274.01	24	11.25	44	1.94
5	202.68	25	9.15	45	1.89
6	106.57	26	8.62	46	1.81
7	102.39	27	7.85	47	1.56
8	57.67	28	7.48	48	1.42
9	35.85	29	7.09	49	1.34
10	35.30	30	6.53	50	1.27
11	31.15	31	6.17	51	1.24
12	29.74	32	5.25	52	1.06
13	28.32	33	4.16	53	1.02
14	22.59	34	3.81	54	.97
15	21.88	35	3.59	55	.78
16	19.05	36	3.39	56	.76
17	17.83	37	3.10	57	.61
18	15.49	38	2.88	58	.57
19	14.08	39	2.69		
20	13.38	40	2.69		

Table 7. *Eigenvector for Smallest Eigenvalue Obtained for 37 Weeks of Doppler Observations Made on Satellites Having four Orbital Inclinations*

Component	Coefficient	Component	Coefficient	Component	Coefficient
.0004	C_{20}	—.1091	C_{53}	.0485	S_{42}
.0005	C_{30}	—.0162	C_{63}	.0130	S_{52}
.0002	C_{40}	—.0523	C_{73}	.0346	S_{62}
—.0004	C_{50}	—.0259	C_{44}	.0970	S_{72}
—.0020	C_{60}	—.0127	C_{54}	.3363	S_{33}
.0003	C_{70}	.1071	C_{64}	—.1436	S_{43}
—.0080	C_{21}	.1072	C_{74}	—.4283	S_{53}
.0435	C_{31}	.1095	C_{55}	—.0716	S_{63}
.0548	C_{41}	.0663	C_{65}	—.0028	S_{73}
.0922	C_{51}	—.1655	C_{75}	—.0276	S_{44}
.0232	C_{61}	—.0367	C_{66}	.4043	S_{54}
.0379	C_{71}	—.0654	C_{76}	.0573	S_{64}
—.0185	C_{22}	.0472	S_{21}	—.3820	S_{74}
—.0797	C_{32}	—.0164	S_{31}	—.1697	S_{55}
—.0401	C_{42}	—.0163	S_{41}	—.3511	S_{65}
.0086	C_{52}	.0118	S_{51}	.1673	S_{75}
.0374	C_{62}	—.0190	S_{61}	.0216	S_{66}
.0190	C_{72}	—.0287	S_{71}	.0207	S_{76}
.0515	C_{33}	—.0948	S_{22}		
.0867	C_{43}	—.1842	S_{32}		

in P space shall be the same as in Q space and that they shall be the set which yields minimum values for the P parameters. This solution is found by linear programming techniques.

Accuracy of Orbit Computation

As a result of improvements in the representation of the gravity field achieved in the past year a dramatic improvement has been achieved in the accuracy of orbit computations. One measure of the improvement

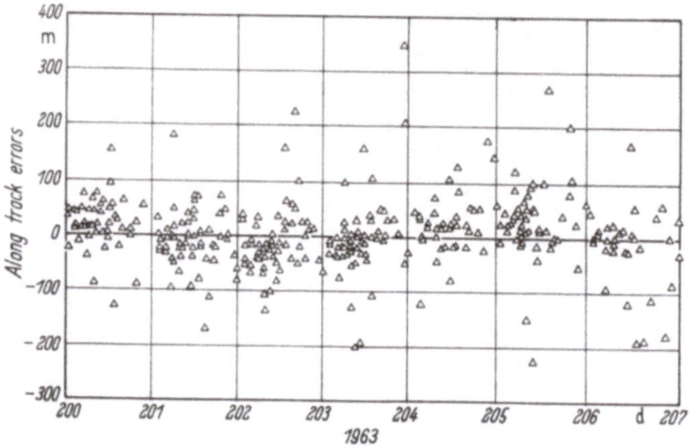

Fig. 5. Satellite 1962 β p 1 along track errors with NWL-5 E-6 geodetic parameters.

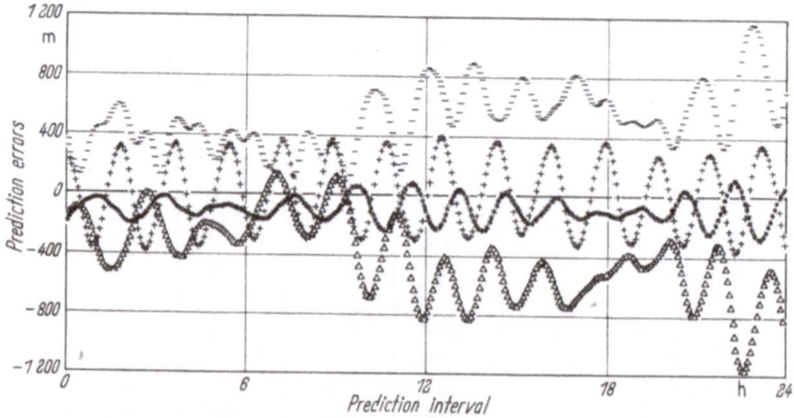

Fig. 6. Components of errors of prediction of satellite orbit corresponding to parameters used from July 1963 to August 1964 (NWL-3 B Parameters).

is the reduction in the residuals of observation. The residuals given in Fig. 2 were reduced by about a factor of about five after the introduction of the resonant gravity coefficients as shown in Fig. 5. An even more striking example of the effects of improvements in geodetic parameters is seen if a satellite orbit predicted on the basis of observations made by a few stations is then compared with an orbit fit to world wide

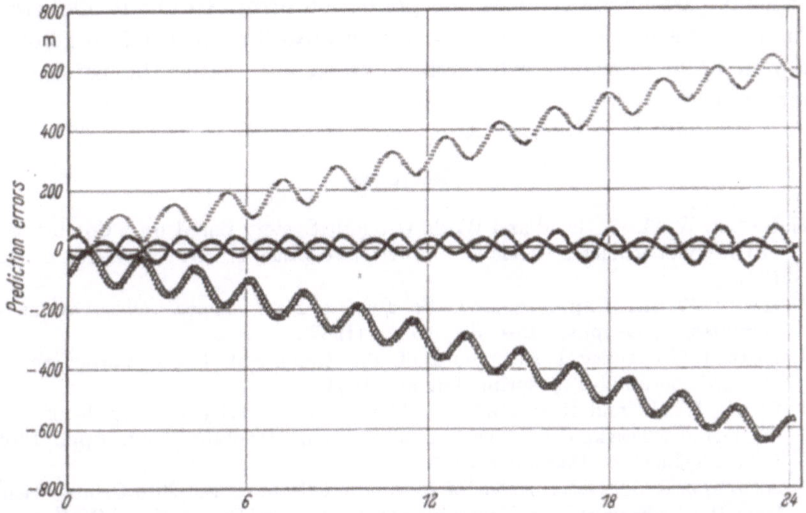

Fig. 7. Components of errors of prediction of satellite orbit corresponding to geodetic parameters used from August 1964 to September 1964 (NWL-5 E parameters).

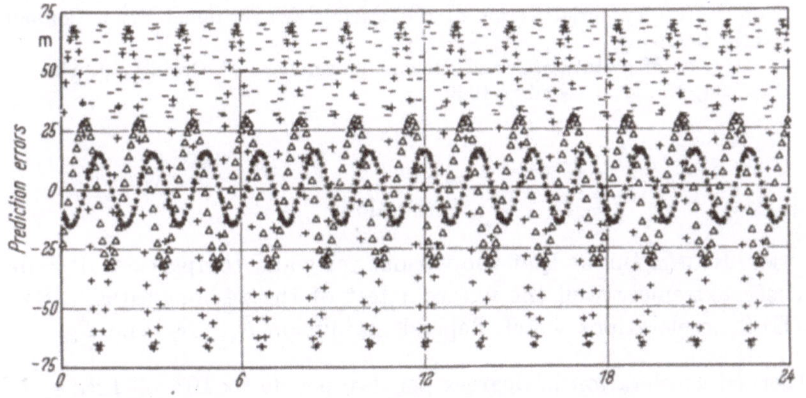

Fig. 8. Components of errors of prediction of satellite orbit corresponding to geodetic parameters used from September 1964 to the present time (May 1965): NWL-5 E-2 parameters.

observations made during the prediction interval. The next figures give such results for a case where the orbit was predicted for a 24-hour time span on the basis of four stations located within a longitude span of 90 degrees and a latitude span of 25 degrees; Fig. 6 shows prediction errors of about 1000 meters which correspond to the gravity parameter set in use at the Naval Weapons Laboratory prior to September 1964; Fig. 7 shows that a small reduction in the errors would result from the use of all gravity parameters given in Table 3 except the resonant gravity coefficients; and Fig. 8 shows that the introduction of the resonant parameters reduced the prediction errors to about 50 meters. Further gains in accuracy are expected as a result of the geodetic solution now in progress which will further extend and refine the parameters now in use.

References

[1] KAULA, W. M.: A Geoid and World Geodetic System Based on a Combination of Gravimetric, Astro-Geodetic, and Satellite Data, NASA T-N D-702 of May 1961.

[2] KAULA, W. M.: Improved Geodetic Results from Camera Observations of Satellites, J. Geophys. Res. 68, No. 18 (1963).

[3] IZSAK, I. G.: Tesseral Harmonics of the Geopotential and Corrections to Station Coordinates, preprint January 1964.

[4] GUIER, W. H., and R. R. NEWTON: The Earth's Gravity Field Deduced from the Doppler Tracking of Five Satellites, Johns Hopkins Univ. Appl. Phys. Labor. TG-634 of December 1964.

[5] ANDERLE, R. J.: Observations of Resonant Effects on Satellite Orbits Arising from the Thirteenth and Fourteenth Order Tesseral Gravitational Coefficients, J. Geophys. Res. in press.

[6] COHEN, C. J.: private communication.

[7] SHEFFIELD, C.: Solutions for the Circular Orbit Perturbations Produced by Non-Central Potential Terms of the Earth's Gravitational Field, Computer Usage Company.

[8] KAULA, W. M.: Statistical and Harmonic Analysis of Gravity, J. Geophys. Res. 64, No. 12, December 1959.

Discussion

Mr. KAULA thinks that the various tests and comparisons described satisfy extremely well the external test of the 24 hour satellite SYNCOM 2 accelerations which depend mainly on $\bar{C}_{22}$, $\bar{S}_{22}$ and $\bar{S}_{33}$.

observed acceleration in degrees per day per day $\times 10^3$	-1.27	-1.32
calculated accelerations from coefficients of ANDERLE	-1.26	-1.26
calculated accelerations from coefficients of GUIER	-1.25	-1.22
calculated accelerations from coefficients of IZSAK	-1.05	-1.09

The $\bar{S}_{3,3}$ harmonic has less effect on a small eccentricity orbit than most terms of the 4th degree, thence the above results indicate that the satellite determinations are good through at least $J_{4,4}$.

Mr. KAULA adds that the effects on the orbit of 1962 $\beta\,\mu$ due to $J_{14,12}$ must depend on terms in the disturbing function which have a coefficient of $O(e)$ where e is the eccentricity since the argument must be:

$$\begin{pmatrix} 0 \\ \text{or} \\ 2 \end{pmatrix} \omega + M + 12(\Omega - \text{G.S.T.})$$

Since the eccentricity of 1962 $\beta\,\mu$ is small these perturbations should be small.

This calculation suggests that resonance perturbations could be found in orbits of greater eccentricity; for example the U.S.S.R. satellite Cosmos 41 has a period very close to 12 hours and an eccentricity of 0,73; it should be appreciably perturbed by all harmonics of order 2 of even as well as odd degree. Furthermore the inclination of Cosmos 41 is very near critical inclination, and the perigee height is low enough for cross action between the drag and the tesseral harmonic acceleration of be perceptible.

A New Determination of the Even Zonal Harmonics of the Earth's Gravitational Potential

By

D. G. King-Hele and **G. E. Cook**

Royal Aircraft Establishment, Farnborough, Hants, England

Summary. The coefficients of the even zonal harmonics in the Earth's gravitational potential are evaluated by analysing the precession of the orbital planes of seven satellites. The most satisfactory representation of the potential is found to be in terms of four coefficients, and their values, in the usual notation, are $10^6 J_2 = 1082.64 \pm 0.02$, $10^6 J_4 = -1.52 \pm 0.03$, $10^6 J_6 = 0.57 \pm 0.07$, $10^6 J_8 = 0.44 \pm 0.11$. The standard deviations quoted refer to a four-coefficient representation of the potential. A six-coefficient solution is also given.

A full version of this paper has been published in the Geophysical Journal of the Royal Astronomical Society. Vol. 10, p. 17.

Discussion

Answering a question by Professor Brouwer, Mr. King-Hele remarks that if you change the number of coefficients evaluated, the values will change more than the standard deviations would suggest. This is due to the fact that it is the total potential that one is trying to estimate. The individual values of the coefficients depend on how many one considers.

On a question by Professor Duboshin about the effects of the odd harmonics, it is answered that by averaging ω over one revolution of the perigee they are almost completely eliminated.

A New Determination of Non-zonal Harmonics by Satellites

By

Imre G. Izsak

Smithsonian Astrophysical Observatory and
Harvard College Observatory, Cambridge, Massachusetts, U.S.A.

Abstract. Determination of non-zonal harmonics of the Earth's gravitational field, using 26,000 precisely reduced BAKER-NUNN camera observations of ten objects. A process of successive approximation is used, which has permitted to improve mean residuals down to a value of 7.″6.

Résumé. Détermination des harmoniques non zonales du champ de gravitation terrestre à l'aide de 26 000 observations effectuées à l'aide de la chambre de BAKER-NUNN sur dix objects différents. Une méthode d'approximations successives a été employée; elle a permis de réduire les résidus moyens à 7″,6.

It is apparent now that the determination of the non-zonal part of the outer gravitational field of the Earth, inasmuch as it affects the motion of artificial satellites, amounts to a process of successive approximations, in the broad sense of the word. Iterative mathematical schemes are widely used at various stages of the work, of course. More interesting perhaps is the tendency on the part of the authors engaged in this kind of investigation to change and presumably improve their methods of data analysis every year or so. In comparison, the derivation of zonal harmonics coefficients shows much less variety. Still nobody seems to have been able to design an approach to the problem in question that would not be objectionable from some point of view (see the critical remarks in (KAULA 1964)). The reason for this lies mainly in the complex statistical nature of the inference called for. The enormous capabilities of modern electronic computers notwithstanding, one is compelled to allow a number of never quite justifiable simplifications of the problem. As the work progresses, investigators learn by experience, mostly their own, but sometimes each other's. For some time to come new, more sophisticated, and necessarily more expensive procedures of analysis will be developed and applied to larger and better observational material. It is encouraging to see that independent determinations carried out by different authors already show reasonable agreement at least as far as the total contribution of the non-zonal terms to the geopotential is concerned. In other words, the main features of the geoid seem to be well established.

The mathematical method of analysis adopted and the corresponding computer program as developed at the Smithsonian Astrophysical Observatory have been described previously in sufficient detail (IZSAK 1964), together with some preliminary results obtained from about 15,000 precisely reduced BAKER-NUNN camera observations of 10 objects. Since then our computer program has undergone a number of changes, the most notable being an extension of its dimensions and an increase of its speed by about 40%. All recent runs of the program, the results of which I will describe presently, utilized over 26,000 observations of 11 satellites, with inclinations ranging from 33° to 96°.

On the basis of former experience, we decided to solve for the unknowns in the following two combinations: coordinates of 12 stations and harmonics coefficients up to the 4th degree (52 quantities); and harmonics coefficients only, but up to the 6th degree (38 quantities). Naturally enough, least squares adjustments of the second kind are essentially more time consuming. Let me call an initial solution of the first kind S_1, and one of the second kind H_1. These adjustments reduced the mean orbital residuals of the observations from 15″.5 to 12″.0, and 12″.1, respectively. Out of curiosity, the results of solution S_1 were fed back into our differential orbit improvement program; that is, the 179 orbits, covering on the average an interval of two weeks, were recomputed, using the estimates of station coordinates and harmonics coefficients just obtained. We noticed with satisfaction that in this process the mean residuals from the new orbits went down to 10″.7.

Although mean residuals from orbits with different perigee and apogee distances are admittedly only rough indicators of the validity of a least squares adjustment, especially when even the use of such an adjustment is somewhat questionable, they are nevertheless quantities to be looked at carefully. In the present case they indicated to us that the whole estimation process should be repeated. To insure greatest flexibility in the final analysis of the observations, and also for a number of purely practical reasons, we considered the estimation of orbital and geodetic parameters as two independent problems. This oversimplification sacrifices apparently too much rigor.

Therefore we started all over again. Using the new residuals of the observations, the physical form of which is fourteen boxes of punched cards, we performed solutions S_2 and H_2. The corresponding mean residuals turned out to be 9″.4 and 9″.3. The additional corrections to station coordinates and harmonics coefficients were in general much smaller than in the initial solutions S_1 and H_1, but often of the same algebraic sign. Once more the iterative cycle was repeated. We recomputed all the orbits involved on the basis of solution $S_1 + S_2$. The mean residual from the new orbits was 8″.3, still decreasing. Again, we analysed the new residuals to see whether additional small corrections should be applied to geodetic parameters. Thus we arrived at solutions S_3 and H_3, with mean residuals 7″.9 and 7″.6. One is inclined to say that the process seems to converge. At this point we felt that another iteration would be rather artificial. To make it meaningful, we would have at least to include further observational material in the future, observations of relatively low orbiting objects in particular. In fact, plans are already under way.

Table 1. *From 26,244 Baker-Nunn Observations of 11 Satellites*

$\bar{C}\,22 \times 10^6$	$\bar{S}\,22 \times 10^6$	2.04	−1.08		2.08	−1.25
C 31	S 31	1.74	.03		1.60	−.04
C 32	S 32	.44	−.88		.38	−.80
C 33	S 33	.18	1.38		−.17	1.40
C 41	S 41	−.39	−.37		−.38	−.40
C 42	S 42	.18	.76		.20	.58
C 43	S 43	.70	−.12		.69	−.10
C 44	S 44	−.13	.82		−.11	.43
C 51	S 51				−.14	−.04
C 52	S 52				.24	−.27
C 53	S 53				−.67	.05
C 54	S 54				−.13	.16
C 55	S 55				.08	−.41
C 61	S 61				−.02	.12
C 62	S 62				.05	−.23
C 63	S 63				.05	.00
C 64	S 64				.07	−.39
C 65	S 65				−.28	−.38
C 66	S 66				−.12	−.59
dX dY dZ						
1 NMEX		−35	40	−46		
2 SAFR		29	−16	−40		
3 AUST		−117	−69	13		
4 SPAIN		0	−16	62		
5 JAPAN		−160	−134	13		
6 INDIA		90	−104	75		
7 PERU		20	−1	−141		
8 IRAN		17	−24	4		
9 CUR		−29	7	−88		
10 FLA		−31	16	−60		
11 ARG		−68	−73	−23		
12 HAW		49	−225	−244		
Mean residuals		7″.9			7″.6	

The total estimates $S = S_1 + S_2 + S_3$ and $H = H_1 + H_2 + H_3$ of geodetic parameters obtained by the method sketched in the preceding paragraphs are exhibited in Table 1. The initial station coordinates are the same as those listed in Table 2 of (IZSAK, 1964). Since standard errors computed in the usual way are completely illusory, we refrain from stating them. Fig. 1 shows a map of geoid heights referred to an oblate spheroid of flattening 1/298.25. This value and the zonal harmonics

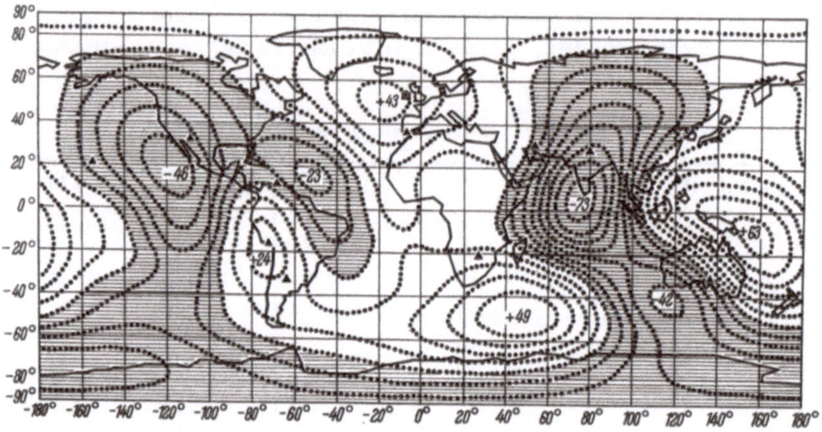

Fig. 1

coefficients from J_3 through J_{14} used in the construction of the figure were borrowed from (KOZAI, 1964). It should be noted that oscillations in geoid height computed from his latest set of coefficients, and those based on his previous set $J_2 - J_9$ (KOZAI, 1962) do not differ by more than 3 meters, and that only near the two poles. The individual coefficients in the third double-column agree best with unpublished results of GUIER and NEWTON (APL determination of July 13, 1964), who analysed a large volume of DOPPLER observations of 4 satellites, and estimated non-zonal coefficients up to the 8th degree. In their latest work (GUIER and NEWTON, December 1964), however, these authors arrived at considerably larger estimates than ours, in several instances. Still, the corresponding maps of geoid heights show striking similarities.

As a check on the validity of our present results, we computed a number of orbits from observations not used yet in least squares adjustments in four different ways:

(a) with original station coordinates, without non-zonal harmonics;
(b) with improved station coordinates, without non-zonal harmonics;
(c) with original station coordinates, with non-zonal harmonics;
(d) with improved station coordinates, with non-zonal harmonics.

Table 2. *Mean Residuals of Observations Not Used in a Simultaneous Determination of Station Corrections and Harmonics Coefficients through the 4th Degree*

1959 η, Vanguard 3: $I = 33^\circ\!.4 \quad e = 0.189 \quad a = 8496$ km

Interval 1963	Observations	σ''_A	σ''_B	σ''_C	σ''_D
4/18— 5/ 1	98	18.0	17.0	11.7	7.8
5/ 6— 5/19	159	19.0	18.4	9.2	7.8
5/22— 6/ 4	195	11.4	8.6	7.2	4.2
7/17— 7/30	172	14.2	13.2	8.9	6.5
8/ 6— 8/19	181	13.7	11.3	8.8	6.5
9/ 7— 9/20	188	13.1	10.9	10.1	7.9
10/ 7—10/20	161	10.6	9.8	8.1	5.8

1962 $\alpha\, \varepsilon_1$, Telstar 1: $I = 44^\circ\!.8 \quad e = 0.242 \quad a = 9672$ km

Interval 1963	Observations	σ''_A	σ''_B	σ''_C	σ''_D
3/12— 3/29	227	8.2	7.1	5.2	2.8
5/14— 5/31	216	4.8	4.4	3.7	3.1
6/13— 6/30	341	8.1	6.9	6.0	3.0
8/ 7— 8/24	205	8.1	7.4	5.9	4.3

1960 ι_2, Echo 1, rocket: $I = 47^\circ\!.2 \quad e = 0.012 \quad a = 7972$ km

Interval 1963	Observations	σ''_A	σ''_B	σ''_C	σ''_D
6/14— 6/25	168	17.1	15.8	6.8	3.3
6/30— 7/11	167	23.6	22.8	9.1	4.8
7/14— 7/25	210	21.8	21.4	8.5	5.4
8/17— 8/28	125	13.0	12.3	7.2	4.4
9/14— 9/25	177	14.6	14.2	5.5	4.0
10/ 2—10/13	208	21.0	18.2	8.8	4.6
11/ 7—11/18	158	17.3	15.4	7.7	4.8

1962 $\beta\, \mu_1$, Anna 1 B: $I = 50^\circ 1 \quad e = 0.007 \quad a = 7508$ km

Interval 1963	Observations	σ''_A	σ''_B	σ''_C	σ''_D
1/20— 1/29	118	21.4	23.0	13.0	8.6
4/17— 4/26	155	32.3	30.3	13.7	10.5
7/11— 7/20	105	38.8	37.9	15.0	9.7
8/19— 8/28	115	27.9	26.1	17.6	13.0

1961 $\alpha\, \delta_1$, Midas 4: $I = 95.9 \quad e = 0.011 \quad a = 10005$ km

Interval 1963	Observations	σ''_A	σ''_B	σ''_C	σ''_D
6/ 1— 6/30	676	7.1	6.3	4.8	3.6
7/ 1— 7/31	659	8.5	7.4	6.5	5.3

For this purpose, the station corrections and harmonics coefficients were those listed in the second double-column of Table 1. The outcome of this experiment is represented in Table 2. At the same time, this table summarizes the present status of orbit computation for high orbiting artificial satellites, and shows the dependability of the BAKER-NUNN optical tracking system.

Acknowledgment. This work was supported in part by grant No. NsG 87/60 from the National Aeronautics and Space Administration.

References

GUIER, W. H., and R. R. NEWTON: The Earth's Gravity Field, Applied Physics Laboratory Techn. Rept. TG-634, 1964.

IZSAK, I. G.: Tesseral Harmonics of the Geopotential and Corrections to Station Coordinates, J. Geophys. Res. **69**, 2621–2630 (1964).

KAULA, W. M.: Determination of Variations in the Gravitational Field for Calculation of Orbit Perturbations, Aerospace Corporation Techn. Rept. TDR-269 (4922-10)-2, 1964.

KOZAI, Y.: Numerical Results from Orbits, Smithsonian Astrophys. Obs. Spec. Rept. 101, 1962.

KOZAI, Y.: New Determination of Zonal Harmonics Coefficients of the Earth's Gravitational Potential, Smithsonian Astrophys. Obs. Spec. Rept. 165, 1964.

Estimation Problems in the Determination of the Even Zonal Harmonics of the External Potential of the Earth

By

A. H. Cook

Standards Division, National Physical Laboratory, Teddington, Middlesex, England

Abstract. The essential difficulty is that the number of parameters (harmonic coefficients) required to describe the effect of the potential on a satellite exceeds the number of independent observations; accordingly, the principle of least squares must be supplemented by some additional hypothesis if unique estimates of the coefficients are to be obtained. Estimates of the parameters are required for three purposes: the description of the existing observations, the prediction of future observation and the specification of the physical state of the Earth and the implications for these three aims of the indeterminancy of the problem are considered.

Another consequence of the indeterminancy is that the estimates of the harmonic coefficients are quite strongly correlated.

A feature of the problem is that the weights of the observations vary in a systematic way with the inclinations of the orbits, a situation not usually met with in least squares estimation problems.

In principle the harmonic coefficients could be separated more efficiently if orbits with a range of semi-major axes were available but the scope of this procedure is rather limited; however, by applying it, the coefficients J_2, J_4 and J_6 can be determined with some assurance.

Résumé. La difficulté essentielle vient de ce que le nombre de paramètres (les coefficients des harmoniques) exigés pour décrire l'effet du potentiel sur un satellite est plus grand que le nombre d'observations indépendantes. De ce fait il faut adjoindre à la méthode des moindres carrés des hypothèses supplémentaires si l'on veut obtenir des valeurs uniques pour les coefficients. Des estimations des paramètres sont nécessaires pour trois raisons: la description des observations existantes, la prédiction des observations futures et la connaissance de l'état physique de la Terre. On a considéré les implications de ces trois buts sur l'indétermination du problème.

Une autre conséquence de l'indétermination est que les valeurs des coefficients des harmoniques sont étroitement liées.

Un des aspects du problème est que les poids des observations varient d'une façon systématique avec les inclinaisons des orbites phénomène inhabituel dans les problèmes d'estimation par les moindres carrés.

En principe les coefficients des harmoniques pourraient être séparés d'une façon efficace si l'on pouvait disposer d'orbites présentant une grande répartition en demi-grands axes. Ce procédé est assez limité, il permet cependant de déterminer J_2, J_4, J_6 avec une bonne certitude.

1. Introduction

The purpose of this note is to indicate some of the problems that have arisen in the determination of the even zonal harmonics from the secular motions of artificial satellites. While the material on which it is based relates to the even zonal harmonics, the same problems are met with in-determinations of odd zonal and tesseral harmonics.

In the determination of the even harmonics, it is supposed that the observed secular motion of the node ($\dot{\Omega}$) or perigee ($\dot{\omega}$) may be represented as the linear combination of components proportional to each even zonal harmonic:

$$\dot{\Omega} = a_2 J_2 + a_4 J_4 + \cdots + a_{2n} J_{2n} + \cdots,$$

$$\dot{\omega} = b_2 J_2 + b_4 J_4 + \cdots + b_{2n} J_{2n} + \cdots.$$

The a's and b's are functions of the elements of the orbit.

(It is supposed that the contribution proportional to J_n^2 has been removed from the observed values of $\dot{\Omega}$ and $\dot{\omega}$).

Similar equations arise in the determination of the odd zonal harmonics, where the observed data are now the amplitudes of terms in the elements e, i, ω or Ω that have the speed of the secular motion of perigee; and in the determination of tesseral harmonics where the observations are the amplitudes of various short period terms. In all cases, the observed rates or amplitudes are theoretically due to an infinite set of harmonics and the essential problem with which this note is concerned is, how much information can be obtained from the data in a situation which, fundamentally, is indeterminate?

2. Validity and Effectiveness of the Method of Least Squares

The usual method of determining parameters from observed linear combinations is by least squares, the sum of squares of residuals being minimised with respect to the parameters to be determined. The principle of the method depends on the statement that the posterior probability of the actual observations, $P(O)$, is proportional to the likelihood of the observations given the parameters, $L(O|p)$, multiplied by the *a priori* probability of the parameters, $P(p)$:

$$P(O) = L(O|p) \cdot P(p).$$

If, as is usual, the likelihood varies rapidly with the parameters, as compared with the variation of $P(p)$, then the maximum value of $P(O)$ is obtained when L is a maximum. In particular, this is strictly the case if there is no *a priori* information about the parameters, that state being represented by writing

$$P(p) = \Pi\, P(p_i)\,,$$

each parameter, p_i, being permitted an infinite range.

To derive the method of least squares, it is further supposed that the observations are uncorrelated and normally distributed, with a standard derivation σ, so that

$$L \propto \exp[-\Sigma(O - C)^2/\sigma^2]\,,$$

where O is an observed datum and C is the value calculated with a specified set of parameters. Then the maximum value of L is obtained when $\Sigma\,(O - C)^2/\sigma^2$ is a minimum.

The conditions for the validity of the method of least squares are therefore:

irrelevance of prior probabilities of parameters;

un-correlated data; data normally distributed.

In the present circumstance, the data may be correlated and the prior probabilities of the J_n's cannot necessarily be ignored; indeed they have been deliberately introduced by some investigators as a means of limiting the indeterminancy of the problem.

It is the indeterminancy which, directly and indirectly, presents the most serious difficulties.

The method of least squares is effective if the variational equations for $\Sigma(O - C)^2/\sigma^2$ (normal equations) are well-conditioned so that the solutions are stable with respect to changes in the sampling of the data. A minimum requirement for this is that the number of observations should exceed the number of parameters to be fitted but this is not of itself sufficient. The data must be formally distinct, that is to say, the linear combination of parameters representing one datum must not be merely a multiple of that representing another, or must not differ from it merely slightly.

These conditions are not in general fulfilled in the present problem.

If the method of least squares is effective, it is possible to determine in addition to estimates of the parameters, the uncertainties of the estimates and the correlations between then, or what is equivalent, linear combinations of the parameters with independent uncertainties. The great advantage of the method of least squares is that one can obtain this information if the postulates of the method are satisfied, as in many cases in physics and astronomy they are near enough. The present problem is unusual in that the postulates are not satisfied but this does not mean that some other method of estimation will give more certain results than least squares; it means that the information that can be derived from the data is limited and it is therefore important to be clear just what the limitations are.

Consider the purposes for which parameters such as the J_n's are estimated. In the first place, it may be possible to describe a large amount of data in a compact way with a few parameters; secondly, such a parametric description of data may enable new observations to be predicted and lastly, the parameters may have more direct physical significance in some problem than do the observed quantities.

The two latter objects require realistic estimates of the uncertainties of the estimates of the parameters especially when, as in the present case, physical ideas such as those of the condition of the interior of the Earth, depend on whether parameters have particular values or no.

3. Data and Parameters

Although many satellites have been placed in orbit there are few with good determinations of the variations of the elements and of these the number having distinct orbits, that is, orbits such that the corresponding observational equations are not nearly identical, is less than twelve (for determination of the even zonal harmonics). The coefficients a_n, b_n, of the equations relating $\dot{\Omega}$ and $\dot{\omega}$ to the J_n's, vary with e, a, i and the order n of the harmonic. The dominant function of e is usually $(1 - e^2)$ and as e is usually quite small it has little effect on the coefficients, a enters as $(R/a)^n$ where R is the radius of the Earth but R/a is only slightly less than 1 for most satellites hitherto used, so a also has little effect on most coefficients. It is the variation of i that induces the greatest variation of the coefficients but owing to launching and other considerations, i is restricted effectively to 7 distinct values.

Now if a is about 7000 km, R/a is about 0.9 and the factor $(R/a)^n$ decreases only very slowly with n, being about 0.3 for $n = 14$. Furthermore, in the equations for perigee, the part of b_n that depends on i may become quite large for large n and in some cases, b_{10} or b_{12} may exceed b_2. This, if harmonics of order, say, 10 to 20 are of comparable magnitude, they will make comparable contributions to the values of $\dot{\Omega}$ or $\dot{\omega}$ for close satellites and with the distinct values of i limited to some 7, the number of distinct observation equations to be derived from observations of close satellites is less than the number of harmonics that may be effective. The problem of estimating those harmonics is thus indeterminate.

If however, a exceeds about 8000 km, the only harmonics that need be considered, at least for the node, are J_2, J_4 and J_6.

Thus it is crucial to consider whether there is any evidence for or against the supposition, that harmonics higher than J_{14} are negligible.

4. Consistency of Data and Sufficiency of Few Parameters

This problem has been discussed in detail elsewhere (COOK, 1965) for the case of the even zonal harmonics. Another way of posing the problem is that it is desired to test the hypothesis that the harmonics of order 10 to 30 are of comparable magnitude.

The method adopted was, in effect, to see if one set of data could be predicted, within the uncertainties of the observations, from another set. There are four sets of nodal data, with more or less overlapping, that have been used by different investigators, and two sets of data from perigee. If a set of parameters is fitted by least squares to one of these sets, the effectiveness with which it predicts another set may be tested by the value of χ^2 for the fit of the parameters to the second set, χ^2 being based on the observational uncertainties of the data. Large values of χ^2 indicate incompatibility of the two sets on the basis of the uncertainties of the data. It is found that the values of χ^2 for most pairs of sets of data are large and sometimes very large. In particular, the perigee data seem highly inconsistent with nodal data and this is important because, as previously mentioned, the higher harmonics often have a larger effect on the motion of perigee than they do on that of the node.

However, there are factors that complicate the analysis and make the conclusions less firm than could be desired. The indeterminancy itself introduces difficulties: to obtain a good description of one set of data, many harmonics must be fitted and the number of degrees of freedom is therefore low, and in consequence the uncertainties of the fitted harmonic coefficients are ill-determined. Again, on account of the few degrees of freedom, the results depend strongly on the weights allotted to the data. But the weights themselves are difficult to estimate and probably vary in a systematic way with the inclination of the orbit (especially for nodal data). The latter feature, a correlation between the weights and coefficients in the observation equations, is rarely met in estimation problems and its formal effect has not been examined.

It may be mentioned here that KING-HELE's various procedures (see KING-HELE and COOK, 1965, for instance) are equivalent to allotting weights in different ways in a least-squares adjustment.

Despite these difficulties it seems that some sets of data, and especially node and perigee data, are inconsistent and indicate the effects of high harmonics.

5. Use of High Satellites

The obvious way of reducing the effect of harmonics of high order is to use data from satellites with semi-major axes great enough to make the effects of those harmonics negligible. At present very few such satellites are suitable but there are four with sufficiently great semi-major axes and sufficiently well separated in inclination to enable J_2, J_4 and J_6 to be determined. The fit of the parameters to the data has two satisfactory features: the value of χ^2 is small and the values of the parameters are not very dependent on weights. These features both indicate that it is the effects of higher harmonics that cause the difficulties in fitting data from closer satellites. The values are

$$\begin{aligned} 10^6 J_2 &= 1082.65 \pm 0.1, \\ J_4 &= -1.60 \pm 0.15, \\ J_6 &= +0.73 \pm 0.4; \end{aligned}$$

they are very close to the recent results of Kozai (1964) and King-Hele and Cook (1965) and it seems that secure values for these coefficients are now available. Even so, they do not fit well the perigee data for some of the same high satellites.

If the remaining close satellite data are used to try to determine J_8 to J_{14}, treating J_2, J_4 and J_6 as known, the fits remain unsatisfactory. Two points should be made here. First, it is not readily possible to extend the use of high satellites to include successively in the data the effects of higher harmonics, for example, choosing data from a set of satellites that would just give J_2 to J_8, then another to give J_2 to J_{10}, and so on, because the choice of the ranges of a to achieve this would be too delicate to be practical.

The second point is that these attempts show up particularly clearly the correlation between estimates of certain harmonics; in particular, with the present distribution of inclinations, J_8 and J_{14} are closely correlated.

6. Additional Remarks

In view of the essential indeterminancy of the problem, it is impossible to make a solution without some restrictions on the prior probability of the harmonics and of course a limit on the number of harmonics determined is equivalent to making the prior probability of the remaining ones zero. Additional restrictions have sometimes been applied to the harmonics that are determined, such as that their magnitudes should not exceed a certain limit. Such restrictions have seemed to be necessary to limit large fluctuations of the estimated parameters that may occur as the samples of data are changed, either because of the poor conditioning of the normal equations or because of correlation between data. Correlation between data could arise particularly if one or two of the neglected higher harmonics were relatively large.

Restrictions on the *a priori* probability of harmonic coefficients do however seem dangerous, since they may operate to conceal both the real indefiniteness of the problem and the possibility of an exceptional value of a coefficient which, if real, might be of geophysical interest. In these circumstances it is very desirable to derive some idea of the probable order of magnitude of neglected harmonics and this is one of the objects of the study referred to (Cook, 1965).

It has sometimes been suggested that functions other than spherical harmonics might be more apt to describe the external potential of the Earth and more tractable in the estimation problem. To make use of such an alternative description, the analytical theory of orbits would have to be re-cast in terms of the preferred functions, but it is doubtful if in fact there would be any advantage. The essential point is that the value of gravity at the surface of the Earth varies quite rapidly from place to place and that a description of the potential at moderate heights that will account for the behaviour of close satellites with adequate accuracy, requires a certain number of distinct items of information. For a satellite at a given height one may for example, specify a certain size of surface area within which the variation of gravity will have no effect on the satellite. It is then sufficient to use the mean value of g for this area and the number of such mean values needed to describe the field at the satellite will depend on the degree of correlation between one area and another. Sufficient is known of the correlation between areas of even $10° \times 10°$ to say that a fairly large number of independent parameters is needed to describe the field experienced by a close satellite and the particular class of functions chosen to represent the field cannot therefore greatly affect the number of distinct functions that must be determined. Of such possible functions outside a sphere, spherical harmonics have the advantage of smoothest variation.

Acknowledgement

This paper has been prepared as part of the programme of general research of the National Physical Laboratory and is published with the permission of the Director of the Laboratory.

References

Cook, A. H.: Geophys. J. Roy. Astronom. Soc. **10** (2), 181–209 (1965).

King-Hele, D. G., and G. E. Cook: Paper in Proceedings of this Symposium, p. 194.

Kozai, Y.: Research in Space Science, Special Report No. 165, Smithsonian Institution, Astrophysical Observatory, 1964, Cambridge, U.S.A.

Discussion

In regard to the question of how rapidly the higher spherical harmonics diminish, Mr. Kaula remarks that we now have three entirely separate estimates of the order of magnitude of coefficients of degree 12 to 15.

(1) The root mean square value of the 3 zonal harmonics $\bar{C}_{12,0}$, $\bar{C}_{13,0}$, $\bar{C}_{14,0}$ determined by Kozai is $\pm 0.045 \times 10^{-6}$.

(2) The root mean square value of the 6 tesseral harmonics of 13th to 15th degree determined by Anderle is $\pm 0.045 \times 10^{-6}$.

(3) The root mean square value of coefficients of degrees 12 to 15 obtained from autocovariance analysis of gravimetry by Kaula is $\pm 0.05 \times 10^{-6}$.

Comment on a Comparison Study Now Underway of Minimum Variance vs. Conventional Least Squares Differential Correction

By

Robert M. L. Baker, Jr.

Computer Sciences Corporation, El Segundo, California, U.S.A.

Comparative analyses of the generalized weighted least squares technique and the approaches which employ the methods of minimum variance, maximum likelihood, KALMAN-SCHMIDT, BAYES, etc., will form principal results of the study effort. A few representative questions that will be answered in the course of the study are:

(1) Which method provides for the smallest residuals in the represented data, and the lowest degree of uncertainty in the elements?

(2) Which method is more vulnerable to error as a result of neglecting second order terms, e.g., what is the influence of non-linearity?

(3) Which method gives better results in the presence on biased data, imperfect representation of forces, etc.? For example, what is the influence of unpredictable atmospheric density variation, and what is the influence of the deviation of the actual atmosphere from standard atmospheric density models in medium and low altitude satellite orbits, and how does such variation influence the orbit determination?

(4) What are the relative adverse effects of ignoring data correlation in the two methods?

(5) Which method is best equipped to handle matrix singularities?

(6) What class of orbit determination problems is best suited to each method? For example, which method is best for use in connection with a transfer orbit extending from the neighbourhood of the Moon?

(7) Which method is more efficient and requires less computer time? For example, how do the methods compare with respect to rapidity of convergence?

(8) Which method accommodates new data points more efficiently and effectively?

Using identical data sequences as inputs, the results using each method will be subjected to a detailed analysis and comparison to determine the accuracy of each method in general, as well as its reliability, effectiveness, and suitability for varying types of orbit missions. It is anticipated that some qualitative evaluations will be made quantitative during the course of the numerical analysis portion of this study. Results of this analysis as the study proceeds might easily be expected to include improvement in the Generalized Weighted Least Squares Technique and or the other methods.

The analysis of methods for weighting observational data will include consideration of geometrical factors, as well as of sensor-dependent weights based on accurate error models for each specific sensor.

In addition it will include a study of time-dependent weighting. The analysis will deal with types of problems which arise, for example, when it is necessary to utilize observations taken over many arcs or over one or two long arcs. The influence of imperfections in the model of the environment will be investigated.

The analysis will also include consideration of procedures for utilizing and weighting correlated data within the framework of the generalized weighted least squares technique.

Investigation of the Conditions for Launching a Cometary Probe

By

J. G. Porter

Formerly at the Royal Greenwich Observatory, Herstmonceux, England

Abstract. The requirements for attaining the necessary accuracy in a cometary orbit are investigated in the case of the launching of a cometary probe. Only a few known comets are possible objects of such probes. A very early recovery of the comet proves to be necessary, and large telescopes should be used by skilled observers in order to achieve this goal.

Résumé. L'auteur donne les résultats d'une recherche sur les conditions de précision avec laquelle on doit connaître l'orbite d'une comète pour réussir le lancement d'une sonde cométaire. Seul un petit nombre de comètes connues peuvent servir de but à ces expériences. Il est nécessaire de retrouver très tôt la comète, et des observateurs expérimentés utilisant des grands télescopes devront être employés pour atteindre ce but.

This paper gives the general conclusions presented in the report of a small committee set up by commission 20 of the IAU in Hamburg (1964).

The conditions for launching a space probe to make contact with a comet have been under investigation for several years, and some reports have already appeared in print. It is the object of this note to study more particularly the requirements for attaining the necessary accuracy in the cometary orbit. It will be assumed that closest approach to the comet will occur at perihelion, which, for technical reasons, will lie within the orbit of Mars. It is desirable that the approach should be within 10,000 km, and this implies that it must be possible to predict the position of the comet at perihelion within this limit. This, in cometary orbits, represents a very high standard of accuracy, and the conditions are quite different from those involved in a planetary mission.

Tables of the motions of the planets are of a high order of accuracy because they are based on some 300 years of observations which, in general, have been made with meridian circles over every part of the orbits. The small eccentricity of a planetary orbit allows its motion to be expressed as a series of terms, from which tables can be constructed and the position of the planet at any future time can be obtained at once. The orbits of comets, on the other hand, are too eccentric to allow this method to be used, and predictions must be made by the method of special perturbations, which involve step-by-step integrations, and are therefore much less accurate. Moreover, comets are generally faint and vague in outline, and their positions are determined by photographic methods. Their orbits may be inclined at any angle, and, unfortunately, only a small part of the orbit nearest the Sun is ever observed, so that the remainder of the orbit must be deduced by calculation.

As a result, cometary orbits are not at all well determined. Of more than 800 cometary orbits in the catalogues, only a mere handful have ever been so thoroughly investigated as to have the required accuracy.

The main difficulty is always in the mean daily motion of the comet, and only very rarely in the orientation of the orbit in space. Thus the uncertainty in a prediction is almost always in the position of the comet in its orbit, and this is generally expressed as an error in the time of perihelion passage. If it is realised that a comet at the Earth's distance from the Sun will have a velocity of the order of 40 km/sec, it will be seen that the required accuracy of 10,000 km calls for an error in the time of perihelion passage of only 250 seconds. This is an almost impossible standard of accuracy for a predicted orbit. It follows at once that the predicted orbit must be corrected by observations made just before launching the probe; that is, the comet must be recovered and observations made well before the time of perihelion passage—perhaps as much as six months beforehand.

We should, of course, prefer to use one of the bright comets, but these are always unexpected visitors, so that they are seldom discovered as much as six months before perihelion. In any case, their orbits are often quite unsuitable, having high inclination, or even retrograde motion. Thus we are forced to make use of the much fainter short-period comets, whose returns can be predicted. These all have orbits of small inclination, with direct motion, and they have the further great advantage that their perihelia always lie near a node, and therefore close to the ecliptic. There are nearly a hundred periodic comets in the catalogues, and 54 of these have made more than one return to the Sun, so that their orbits are more or less well determined. If we omit those of long period, which cannot return at the chosen time, together with those of large perihelion distance, we are left with about 30 comets, of which only a few can meet the other essential conditions—for example, the comet must be well placed for observations throughout the flight of the probe, and especially so at perihelion. The main requirement above all others is that the orbit shall be well determined.

The mean daily motion of a comet can only be found with any accuracy by linking successive returns of the comet. If one return is missed (for example, if the comet is badly placed), then the prediction of the next return becomes less reliable. The best orbits are those in which a number of good observations made at two or more successive returns have been used. There are few orbits of periodic comets which attain even this standard of accuracy. In most cases the calculations have merely aimed at representing the observations with sufficient accuracy to provide an adequate prediction for the next return. The highest accuracy has rarely been achieved and is rarely necessary.

In the few cases in which really accurate work has been attempted, the resulting orbits have proved capable of giving extremely good predictions. Work of this standard will, for example, give an orbit which will represent the observations with a mean error of $\pm 1''.5$. This is the error of one observation in either R.A. or declination, but since the orbital plane is well represented, it is, in practice, the error of position in space. It corresponds to an error of ± 1150 km at unit distance, or to ± 2300 km at 2 a.u. This small error cannot, of course, be carried through to a predicted orbit; the calculation of special perturbations always involves a build-up of errors in the integration schemes, and this cannot be avoided. It is therefore essential that the comet should be recovered before launch takes place, and a few good observations secured. Very few observations would be needed to correct a good orbit of this kind, and to carry the mean error of $\pm 1''.5$ through to perihelion. The selection of a suitable comet thus becomes a matter of finding one which is well placed for recovery some weeks before the date of launching the probe, and is also well placed at the previous return, when an accurate orbit and a good prediction can be calculated.

This prediction must be made with all possible precision. Ephemeris Time must be employed throughout, and perturbations by all the planets included, no approximations being made unless it can be shown that they are thoroughly justified. It need hardly be said that this is laborious and exacting work which calls for judgment and experience. There have been claims that the errors in a prediction are due to loss of mass or to non-gravitational forces, but these claims must be regarded as still open to question. The fact that faulty predictions do occur is almost always due to the use of an inadequate orbit at the commencement of the integration schemes, and there are many other sources of error in this work. Planets may be neglected (particularly Uranus and Neptune) and other approximations may be made; physical constants, especially the masses of the planets, could be improved, and any of these may be responsible for small discrepencies in work which is, in any case, of border-line accuracy. ROEMER (1961) has discussed this matter in some detail.

The success of this cometary project depends essentially on the recovery and subsequent observations of the comet, and since a short-period comet is always a faint object on recovery (magnitude 19—20, perhaps), this calls for large telescopes and long exposures. In the past, a large proportion of the observations have proved to be in error, and this is probably due to the use of an inadequate instrument. The position is better today, but there are still far too few large instruments employed in this work, particularly in the southern hemisphere. The essential measurement is the position of the centre of mass of the comet (i.e., the nucleus) and if this is not visible, as may well happen in a small instrument, then it is not sufficiently accurate to take the centre of light as being the same as the centre of mass. A large instrument is necessary, and since direct guidance on the comet is impossible, it is necessary to make appropriate movements of the cross-wires or of the photographic plate; these movements are calculated from the predicted ephemeris of the comet. Details of the procedure, and of the subsequent measurements of the plate are described by ROEMER (1963).

Plates taken with small telescopes may be reduced using reference stars taken from standard star-catalogues; the errors in this process are then almost entirely due to the difficulty of deciding on the position of the centre of mass of the comet. The greater part of the error when using long-focus instruments seems to come from the star positions in the Astrographic Catalogue, which is the only source for positions of the faint stars photographed in the restricted field of these telescopes. The Astrographic Catalogue gives the positions of several millions of stars down to magnitude 12, in the form of measured coördinates on photographic plates. The plate constants which are provided were obtained many years ago, and the use of modern values of proper motions would effect a considerable improvement in these values. (See, for example, HECKMANN, DIECKVOSS and KOX, 1954). Since only a small part of the sky need be explored for any particular comet, it would be a matter of no great difficulty to recalculate the plate-constants for the selected areas, and this work might well be done beforehand.

The most important point in all such observational work is the attainment of consistent accuracy in these positional measurements. The ideal would be to have observations made at equal intervals of time, but observations may be interrupted by moonlight, bad weather, low altitude of the comet, etc., and hence it is essential that more than one observatory should be involved in the work. Unfortunately there are few observatories in which time could be made available for the use of really large instruments in studying comets. Moreover, the science of astrometry—particularly the astrometry of solar system objects—is badly neglected, and there are few observers today who have the necessary skill. Clearly what is needed is an international network of observatories with long-focus reflectors, and with trained observers who are familiar with the problems of astrometry. To send a space probe to a comet is a very diffeıent proposition from that of a planetary probe.

It may not be out of place to emphasise some of these arguments by giving some statistics of cometary observations. In the ten years 1955 to 1964, inclusive, 80 comets were detected, of which 34 were new ones and 46 were recoveries of periodic comets based on predictions. Of the new comets, only three were discovered more than 170 days before perihelion passage (this figure is taken merely as a convenient dividing-line). Of the periodic comets, no less than 22 were recovered at least 170 days before perihelion, and of these, one was found with the 200-inch telescope at Mt. Palomar, two with the 82-inch McDonald reflector, three with the Crossley 36-inch reflector at Lick, and the remaining 16 were all recovered by Dr. Elizabeth Roemer, using the 40-inch reflector of the U.S. Naval Observatory at Flagstaff. Thus it is clearly possible to recover many of the periodic comets in good time. The 22 comets referred to above include 17 individual comets, five of them appearing twice during these ten years. However, such early recovery requires large instruments—the smallest mentioned is the Crossley 36-inch reflector. The unfortunate feature of modern cometary recovery is that the majority of the searches are made by one observer; Miss Roemer was directly responsible for 17 of these 22 recoveries of periodic comets. We need more telescopes, more observers, and more computers if any real progress is to be made.

References

Heckmann, Dieckvoss and Kox: Astronom. J. 59, 143 (1954).

Roemer, E.: Astronom. J. **66**, 368 (1961).

Roemer, E.: Comets; Discovery, Orbits, Astrometric Observations. The Moon, Meteorites and Comets (Solar System, Vol. IV), Univ. of Chicago Press, 1963

Determination of the Masses of the Moon and Venus, and the Astronomical Unit from the Radio Tracking of Mariner II[1]

By

John D. Anderson and **Michael R. Warner**

Jet Propulsion Laboratory, Pasadena, California, U.S.A.

Abstract. DOPPLER data from the Mariner II space probe to Venus are used to obtain independent determinations of the masses of the Moon and Venus and the astronomical unit.

A number of sources of systematic error are investigated. Particularly emphasized are uncertainties in the Earth and Venus ephemerides and the contribution of lowthrust forces from the attitude-control system.

The inclusion of these effects, along with an improvement in the computation of the Mariner II orbit in the vicinity of Venus, results in more reliable values for the constants than have been published previously.

The mass ratio of the Earth to the Moon is unchanged and is consistent with a value of 81.30. A value of 408600 for the reciprocal mass of Venus satisfies the improved determination. Finally, a provisional value of the astronomical unit is in agreement with other radar determinations.

Résumé. On utilise les données fournies par la sonde spatiale Mariner II envoyée vers Vénus pour obtenir des déterminations indépendantes des masses de la lune de Vénus et de la valeur de l'unité astronomique. On passe en revue un certain nombre des sources d'erreurs systématiques. En particulier les incertitudes des éphémérides de la Terre et de Vénus et la contribution des faibles poussées dues au système de commande d'attitude.

L'introduction de ces effets ainsi qu'une amélioration dans le calcul de l'orbite Mariner II au voisinage de Vénus conduisent à des valeurs des constantes plus sûres que celles publiées jusqu'à présent.

Le rapport de la masse de la Terre à celle de la Lune n'est pas changé et semble se fixer à la valeur 81,30. La valeur 408600 pour l'inverse de la masse de Vénus ressort de la détermination améliorée. Enfin une valeur provisoire de l'unité astronomique est en bon accord avec les autres déterminations par radar obtenues précédemment.

---

[1] This paper presents the results of one phase of research carried out at the Jet Propulsion Laboratory, California Institute of Technology, under Contract No. NAS 7-100, sponsored by the National Aeronautics and Space Administration.

I. Introduction

A previous reduction of the Mariner II data [*1*] resulted in improved values for the masses of the Moon and Venus. However, it was necessary to divide the data into two sections so that the data during encounter could be processed independently of the majority of the data taken during the heliocentric phase of the mission. It was impossible to obtain a reasonable fit to all of the data because of three neglected effects.

The first of these was a significant error in the orbit of the spacecraft during the few hours around planetary encounter. This error was caused by an integration of the equations of motion in geocentric rather than Venus-centered coordinates. The second effect was the neglect in the equations of motion of low-thrust forces which resulted from a nonstandard operation of the attitude-control system. The third effect was the exclusion of the ephemerides of the Sun and Venus as possible sources of error in the solutions.

Since the heliocentric and encounter data could not be combined, it was impossible to realize the full potential of the data to determine, in particular, the astronomical unit. The validity of the value for the mass of Venus as determined from the encounter data was also in doubt. The only really significant result was a value of the Earth-to-Moon mass ratio of 81.30155 $\pm$ 0.0034, which was obtained from the monthly variation in the DOPPLER data during the 87 days of heliocentric flight. After this determination, the successful tracking of four Ranger spacecraft from the Earth to impact on the Moon produced a fairly independent verification of the value [*2*, *3*]. A combination of the Ranger results suggested a value of 81.304 $\pm$ 0.002.

Methods for including the previously neglected effects are described in this article. A model for the low-thrust attitude-control perturbations is discussed, and linear corrections to the ephemerides of the Sun and Venus are developed. Also the technique for including attitude-control parameters as well as orbital elements of the Earth and Venus in the normal equations is explained.

The integration of the equations of motion in Venus-centered coordinates is accomplished. However, the normal equations are still formed in geocentric coordinates. Therefore, after the orbit has been computed in Venus-centered coordinates, it is necessary to transform the answer into geocentric coordinates so that the normal equations can be formed as before.

Two limitations are still present in the solutions for the constants given in this paper. It is not possible to recompute the orbit of the spacecraft after corrections to the new parameters have been determined. Therefore, an iterative solution to the minimization of the sum of the squared residuals cannot be achieved as in the old solutions. Consequently, since nonlinear terms are significant in the solution, reductions of the heliocentric data are not completely satisfactory.

However, a linear correction over the period of planetary encounter does seem valid, and the addition of new parameters and Venus-centered integration produces a definite improvement over previous encounter solutions. As a result, the value of the mass of Venus is changed from the previous value and is preferred as the best determination from the Mariner II data. The new value of the ratio of the mass of the Sun to the mass of Venus (from Solution IV, Sect. III) is

$$\frac{M_S}{M_V} = 408598.1 \pm 25 .$$

The mass ratio of the Earth to the Moon is unchanged by the current solutions:

$$\mu^{-1} = 81.30155 \pm 0.0034 ,$$

A preliminary value of the astronomical unit is obtained in Solution V (Sect. III). This solution represents a fit to all the Mariner II data and hence suffers from the limitation of a linear correction.

The second limitation affecting some of the new solutions is also important. This occurs because the inclusion of heliocentric data requires a geocentric integration of the equations of motion. Therefore, residuals during the few hours of planetary encounter reflect significant errors in the computation of the orbit.

The astronomical unit as determined by Solution V is 149,599,455 $\pm$ 320 km, but due to the two limitations described in the foregoing, the value cannot be considered definitive. The probable error of $\pm$ 320 km is representative of the accuracy to which the astronomical unit can be determined from the Mariner II data. Therefore, when the limitations on the present solution are removed, a definitive value from the data should be available.

II. Improvement of the Normal Equations

A description of additions to the normal equations is given in this Section. A mathematical model for the low-thrust forces is constructed, based on realistic physical processes in the attitude-control system. Differential corrections to the Earth and planetary ephemerides are also developed. Finally, the inclusion of the attitude-control parameters and the ephemeris elements in the equations of condition is discussed.

A. Low-Thrust Forces

The attitude of the Mariner II spacecraft is controlled so that the solar panels are always facing the Sun and the high-gain antenna can be directed at the Earth. This control is produced by the release of cold nitrogen gas from a number of jets. In standard operation the jets are coupled so that there is no significant perturbation on the trajectory of the spacecraft. However, previous least-squares fits to the DOPPLER data (see [*1*], p. 141) indicate that low-thrust forces are present because of a nonstandard operation of the attitude-control system. These forces are assumed unknown and a set of attitude-control parameters is introduced for estimation. An iterative minimization of the sum of squares of the residuals requires that the attitude-control model be incorporated into the equations of motion for the spacecraft.

The low-thrust forces are represented by two physical processes. The first is a slow gas leak from some unknown point in the attitude-control system. The second is a failure of the gas jets to act in couples. With a model of this sort the number of attitude-control parameters can be kept to a minimum. Also, an estimation of physically meaningful parameters can aid in the postflight evaluation of the attitude-control system.

For an ideal slow leak the perturbative acceleration in the equations of motion acts in an unknown direction with a magnitude proportional to the pressure in the gas reservoir. If the leak is slow enough, the pressure can be considered a constant. Because the spacecraft remains attitude-stabilized over the entire Mariner II mission, the leak must not be so fast as to deplete the supply of gas. Rather than assuming a constant pressure, and consequently a constant perturbative acceleration, the leak is represented by a quadratic function in the time from the epoch. The direction of the thrust is assumed fixed with respect to the principal axes of the spacecraft. The reference plane for these axes is the plane containing the Sun, the spacecraft, and the Earth. One of the axes in the reference plane, the roll axis, lies along the Sun-spacecraft line.

Let $\boldsymbol{U}_{\odot S}$ be the unit vector in the direction of the heliocentric position of the spacecraft. Also let $\boldsymbol{U}_{ES}$ be the unit vector in the direction of the geocentric position of the spacecraft. Both unit vectors are simply computed as a function of time from geocentric ephemerides of the Sun and spacecraft. The unit vector $\boldsymbol{N}$, normal to the reference plane, is given by a vector product of $\boldsymbol{U}_{\odot S}$ and $\boldsymbol{U}_{ES}$

$$\boldsymbol{N} = \frac{\boldsymbol{U}_{\odot S} \times \boldsymbol{U}_{ES}}{|\boldsymbol{U}_{\odot S} \times \boldsymbol{U}_{ES}|}, \tag{1}$$

where $|\boldsymbol{U}_{\odot S} \times \boldsymbol{U}_{ES}|$ is the magnitude of the vector product. The third axis $\boldsymbol{T}$ is computed by the following formula:

$$\boldsymbol{T} = \boldsymbol{N} \times \boldsymbol{U}_{\odot S}. \tag{2}$$

Thus $(\boldsymbol{U}_{\odot S}, \boldsymbol{T}, \boldsymbol{N})$ defines an orthogonal right-handed coordinate system which is fixed with respect to a completely attitude-controlled spacecraft. The perturbative acceleration for the leak is therefore given by $(1 - \alpha_1 \tau - \alpha_2 \tau^2)(f_1 \boldsymbol{U}_{\odot S} + f_2 \boldsymbol{T} + f_3 \boldsymbol{N})$, where τ is the time from the epoch t_0. The parameters in the leak model are (α_1, α_2), the coefficients in the assumed quadratic decrease in the thrust, and (f_1, f_2, f_3), the magnitude of the thrust at the epoch multiplied by the respective direction cosines of the thrust vector in the spacecraft-fixed system of coordinates.

The failure of the gas jets to act in couples introduces a perturbative acceleration which depends on the degree of unbalance between opposing jets, the limit-cycle characteristics of the attitude-control system and the disturbing torques acting on the spacecraft. In the normal limit-cycle operation the net average thrust over the duration of the mission is practically zero. However, if there is a significant unbalance in the individual thrust levels of the jets, then on the average a constant low thrust is imparted to the spacecraft. This effect can be absorbed in the model already introduced for the slow-leak thrust. Again the direction of the net thrust is fixed with respect to the principal axes of the spacecraft.

For the situation where the attitude-control system senses the effects of disturbing torques and subsequently opposes them, an unbalance in the jets imparts a thrust which on the average is proportional to the disturbing torque. If this torque is constant then the average thrust produced by the unbalanced jets in opposing the torque is also constant. Therefore the previous model will suffice. Notice that torques produced by the slow gas leaks can also be represented by the previous model and thus the thrust imparted by the unbalanced jets is absorbed in the coefficients of the thrust from the leak itself.

The only other significant time-varying torque arises from the solar-radiation pressure acting on a center of pressure not coincident with the center of gravity. Then the torque is proportional to the inverse square of the distance $r_{\odot S}$ of the spacecraft from the Sun. The perturbative acceleration which results from the solar-radiation pressure itself has previously been included in the equations of motion as a radial perturbation along $\boldsymbol{U}_{\odot S}$ with a magnitude proportional to $r_{\odot S}^{-2}$.

It is impossible to separate this radial acceleration from that produced by the uncoupled jets which react to radiation-pressure torques. However, it is necessary to add a tangential and normal component to the equations of motion in order to represent the general reaction to these torques. Thus the additional term in the acceleration to account for the unbalanced jets is simply $(K/r_{\odot S}^2)(G_T \boldsymbol{T} + G_N \boldsymbol{N})$. The constant K is introduced arbitrarily to make G_T and G_N dimensionless parameters.

It is assigned the value $K = 1.99209464 \times 10^6\ \mathrm{km^3/sec^2}$ and represents the constant of proportionality for the acceleration which results from the incident radiation on the Mariner II spacecraft. The total acceleration for the incident and reflected radiation is given by $K(1 + v)/r^2_{\odot S}$ where v allows for the reflected radiation and errors in the evaluation of K. The adopted value of K is computed with an assumed area and mass of the spacecraft of 3.83 m^2 and 198.22 kg respectively. These values are accurate enough that a nonzero value of v should essentially represent the contribution of the reflected radiation and the radial low-thrust effects.

Formerly, the trajectory of the spacecraft was computed from a set of equations of motion which included the standard n-body terms and the solar-radiation-pressure term. Adding to these equations a low-thrust perturbative acceleration $\dot{\boldsymbol{r}}_t^{\backslash}$, which combines the effects of a slow leak and uncoupled jets, results in

$$\dot{\boldsymbol{r}}_t^{\backslash} = f_1\,\alpha(\tau)\,\boldsymbol{U}_{\odot S} + \left[f_2\alpha(\tau) + \frac{K\,G_T}{r^2_{\odot S}}\right]\boldsymbol{T} + \left[f_3\alpha(\tau) + \frac{K\,G_N}{r^2_{\odot S}}\right]\boldsymbol{N}, \tag{3}$$

where $\alpha(\tau)$ is the quadratic function

$$\alpha(\tau) = 1 - \alpha_1\,\tau - \alpha_2\,\tau^2 .$$

B. Correction to Planetary Ephemerides

In the least-squares solution for orbital elements of the Earth and target planet, the capability to proceed beyond the linear correction to the elements is introduced. As a result it is necessary to allow a variation in the planetary ephemerides in order that a new probe trajectory can be computed based on the new set of planetary orbital elements.

When heliocentric positions and velocities of the Earth-Moon barycenter or the target planet are required in the numerical integration, the values are interpolated from the standard JPL ephemeris tapes [*4* and *5*]. Now if nonzero corrections to the respective orbital elements are available from a previous iteration in the least-squares solution, then the positions and velocities as read from the ephemeris tapes are corrected by two-body differential correction formulas. The correction occurs in two steps.

The first step is to convert corrections to the orbital elements at the epoch into corrections to Cartesian components of position and velocity at the same epoch. Let $\Delta\boldsymbol{E}$ and $\Delta\boldsymbol{T}$ represent corrections to the orbital elements of the Earth and target planet respectively. Also let $\Delta\boldsymbol{X}_0$ represent corrections to the six components of position and velocity for the target planet, and similarly let $\Delta \boldsymbol{X}_0'$ represent the same thing for the Earth.

Symbolically the linear mappings between orbital elements and Cartesian coordinates can be written in matrix form:

$$\Delta \boldsymbol{X}_0 = M \Delta \boldsymbol{T}, \tag{4}$$

$$\Delta \boldsymbol{X}_0' = M' \Delta \boldsymbol{E}. \tag{5}$$

The epochs for the corrections to the Earth and target-planet elements are t_0' and t_{0T} respectively. Hence, including the epoch t_0 for the elements of the probe's orbit, three distinct epochs enter into the least-squares solution.

The second step in the ephemeris correction is to map the corrections $\Delta \boldsymbol{X}_0$ at t_{0T} and $\Delta \boldsymbol{X}_0'$ at t_0' to Cartesian coordinate corrections $\Delta \boldsymbol{X}$ and $\Delta \boldsymbol{X}'$ at an arbitrary time t. Again matrix notation can be used:

$$\Delta \boldsymbol{X} = U \Delta \boldsymbol{X}_0, \tag{6}$$

$$\Delta \boldsymbol{X}' = U' \Delta \boldsymbol{X}_0'. \tag{7}$$

A two-step mapping of the orbital-element corrections into Cartesian corrections at a later time is somewhat cumbersome when compared with a single mapping such as that given in [*6*], Chap. IX. However, the method was chosen so that various orbital-element sets could be selected for correction with a minimum modification to the computer program. For example, the elements of the target planet are selected as the Cartesian components themselves as a preliminary approach to the problem. The reason for selecting Cartesian components is that it is probably the heliocentric position of the planet that must be modified in order to achieve a satisfactory representation of all the data. However, a constraint to hold the mean distance of the planet at a constant value is included in the differential correction. Therefore, the matrix M is not the unit matrix of dimension six but has dimensions 6×5 instead. One degree of freedom is removed by the linear form of the vis-viva integral for a constant mean distance, as follows:

$$\dot{s} \Delta \dot{s} + \frac{\mu}{r^2} \Delta r = 0. \tag{8}$$

In terms of the position $\boldsymbol{r}$ and velocity $\dot{\boldsymbol{r}}$, the speed $\dot{s}$ is given by $\dot{s}^2 = \dot{\boldsymbol{r}} \cdot \dot{\boldsymbol{r}}$ and the radius r by $r^2 = \boldsymbol{r} \cdot \boldsymbol{r}$. Therefore Eq. (8) can be written as a linear combination of position and velocity corrections:

$$\frac{\mu \boldsymbol{r}}{r^3} \cdot \Delta \boldsymbol{r} + \dot{\boldsymbol{r}} \cdot \Delta \dot{\boldsymbol{r}} = 0, \tag{9}$$

The coefficients of the position corrections are simply the negative of the two-body accelerations. Because the position components are retained for an eventual correction, one of the velocity components is eliminated as an orbital element. For both the Mariner II and Mariner IV orbits the component $\dot{x}$ is the largest of the heliocentric velocity coordinates during the period of planetary encounter. Therefore it is eliminated and the matrix M appears as follows:

$$M = \begin{bmatrix} 1 & 0 & 0 & 0 & 0 \\ 0 & 1 & 0 & 0 & 0 \\ 0 & 0 & 1 & 0 & 0 \\ \ddot{x}_0/\dot{x}_0 & \ddot{y}_0/\dot{x}_0 & \ddot{z}_0/\dot{x}_0 & -\dot{y}_0/\dot{x}_0 & -\dot{z}_0/\dot{x}_0 \\ 0 & 0 & 0 & 1 & 0 \\ 0 & 0 & 0 & 0 & 1 \end{bmatrix}$$

Corrections to the orbital elements of the Earth-Moon barycenter appear as in set VI of Brouwer and Clemence (See [6], p. 245). Because only radio tracking data are used, the correction $\Delta\psi_2'$ is assumed equal to zero. This has the effect of fixing the direction of the vernal equinox. The other two corrections to orientation are then $\Delta\psi_1$, a correction to the obliquity of the ecliptic, and $\Delta\psi_3'$, a correction to the longitude of perihelion. Along with corrections to the eccentricity and the mean longitude at the epoch t_0', there are a total of four elements in the ephemeris corrections for the Earth-Moon system. The position part of the 6×4 matrix M' is given by Brouwer and Clemence (op. cit.). By differentiating the position part with respect to time the velocity part is also obtained. The complete matrix is evaluated at the epoch.

Let the ordering of the elements be as follows:

$\Delta E_1 = \Delta e'$ (eccentricity correction),

$\Delta E_2 = \Delta\psi_1$ (obliquity correction),

$\Delta E_3 = \Delta l_0' + \Delta\psi_3'$ (mean longitude correction),

$\Delta E_4 = e'\Delta\psi_3'$ (longitude of perihelion correction).

Then the position $\Delta\boldsymbol{r}_0'$ and velocity $\Delta\dot{\boldsymbol{r}}_0'$ corrections are written in vector form:

$$\Delta\boldsymbol{r}_0' = \sum_{i=1}^{4} \boldsymbol{\alpha}_i \Delta E_i, \tag{10}$$

$$\Delta\dot{\boldsymbol{r}}_0' = \sum_{i=1}^{4} \dot{\boldsymbol{\alpha}}_i \Delta E_i = \sum_{i=1}^{4} \boldsymbol{\beta}_i \Delta E_i. \tag{11}$$

The vectors $\boldsymbol{\alpha}_i$ and $\boldsymbol{\beta}_i$ can be evaluated as follows:

$$
\begin{aligned}
\boldsymbol{\alpha}_1 &= H_1' \boldsymbol{r}_0' + K_1' \dot{\boldsymbol{r}}_0', \\
\boldsymbol{\alpha}_2 &= (1, 0, 0) \times \boldsymbol{r}_0', \\
\boldsymbol{\alpha}_3 &= \frac{1}{n'} \dot{\boldsymbol{r}}_0', \\
\boldsymbol{\alpha}_4 &= \frac{1}{e'} \left[(0, -\sin\varepsilon, \cos\varepsilon) \times \boldsymbol{r}_0' - \frac{1}{n'} \dot{\boldsymbol{r}}_0' \right],
\end{aligned}
\tag{12}
$$

$$
\begin{aligned}
\boldsymbol{\beta}_1 &= H_2' \boldsymbol{r}_0' + K_2' \dot{\boldsymbol{r}}_0', \\
\boldsymbol{\beta}_2 &= (1, 0, 0) \times \dot{\boldsymbol{r}}_0', \\
\boldsymbol{\beta}_3 &= -n' \left(\frac{a'}{r_0'} \right)^3 \boldsymbol{r}_0', \\
\boldsymbol{\beta}_4 &= \frac{1}{e'} \left[n' \left(\frac{a'}{r_0'} \right)^3 \boldsymbol{r}_0' + (0, -\sin\varepsilon, \cos\varepsilon) \times \dot{\boldsymbol{r}}_0' \right];
\end{aligned}
\tag{13}
$$

where ε is the obliquity of the ecliptic. The factors in the eccentricity correction are the following:

$$H_1' = \frac{r_0' - a'(1 + e'^2)}{e' p'}, \tag{14}$$

$$K_1' = \frac{r_0' \dot{r}_0'}{a'^2 n'^2 e'} \left(1 + \frac{r_0'}{p'} \right), \tag{15}$$

$$H_2' = \frac{\dot{r}_0'}{e' p'} \left[1 - \frac{a'}{r_0'} \left(1 + \frac{p'}{r_0'} \right) \right], \tag{16}$$

$$K_2' = \frac{1}{e'(1 - e'^2)} \left(1 - \frac{r_0'}{a'} \right). \tag{17}$$

Orbital elements, when required, are obtained from the ephemeris values of $\boldsymbol{r}_0'$, $\dot{\boldsymbol{r}}_0'$ and the usual two-body formulas. The matrix M' is comprised of elements of the vectors $\boldsymbol{\alpha}_i$ and $\boldsymbol{\beta}_i$ and, in fact, M' can be written in the following partitioned form if $\boldsymbol{\alpha}_i$ and $\boldsymbol{\beta}_i$ are taken as column matrices:

$$M' = \left(\begin{array}{c|c|c|c} \boldsymbol{\alpha}_1 & \boldsymbol{\alpha}_2 & \boldsymbol{\alpha}_3 & \boldsymbol{\alpha}_4 \\ \hline \boldsymbol{\beta}_1 & \boldsymbol{\beta}_2 & \boldsymbol{\beta}_3 & \boldsymbol{\beta}_4 \end{array} \right).$$

In the second step of the differential correction, the mapping of the Cartesian components, the same formulas are used to evaluate both matrices U and U'. Following HERRICK (see [7], Sect. 3. and 6.) a "universal" formulation of the matrix is introduced. Although in the particular case of planetary ephemeris correction the universal formulation is not required, it was chosen so that a general computational program could be made available for other purposes.

One more consideration enters in the application of the $\Delta\boldsymbol{E}$ and $\Delta\boldsymbol{T}$ corrections to the orbit of the spacecraft. It is necessary to apply the final position and velocity corrections to the equations of motion, which are in turn subjected to a numerical integration procedure. This is a straightforward correction for the heliocentric equations of motion. However, depending on the position of the spacecraft, the central body for the integration can be the Earth, the Sun, or the target planet. In the case of Mariner II most of the useful data occurs when the spacecraft is in a heliocentric phase, yet the very important data near encounter are represented by a Venus-centered integration. Therefore, the possibility of using the Earth or target planet as a central body is included in the ephemeris correction program.

When the equations of motion are expressed in geocentric coordinates, the corrections to the solar coordinates and consequently to all the geocentric planetary coordinates are given as the negative of the corrections $\Delta \mathbf{X}'$ to the Earth-Moon barycenter. However, the geocentric corrections to the coordinates of the target planet are given by $\Delta \mathbf{X} - \Delta \mathbf{X}'$ in order to account for the variation in the heliocentric coordinates of both the Earth and target planet. The geocentric ephemeris of the Moon is assumed unaffected.

For the case where the equations of motion are given with respect to the target planet, all coordinates of the other planets and the Sun are corrected by applying the negative of the heliocentric corrections $\Delta \mathbf{X}$ to the target planet. In addition, the coordinates of the Earth and Moon are corrected by $\Delta \mathbf{X}' - \Delta \mathbf{X}$.

It is now possible to assess the effect of small variations in the orbital elements of the Earth and target planet on any trajectory representative of a planetary probe.

C. Augmented Equations of Condition

The equations of condition are obtained by numerically integrating a system of variational equations associated with the equations of motion for the spacecraft (see [1], pp. 136—138). As a result, the sensitivity of the state $\mathbf{X}$ (position and velocity) of the spacecraft at an arbitrary time t is obtained as a function of the initial state $\mathbf{X}_0$ at the epoch t_0 and the various constants $\mathbf{Y}$. The function is given in the linearized form $\delta \mathbf{X} = U\, \delta \mathbf{X}_0 + V\, \delta \mathbf{Y}$ and the differential equations for the state-transition matrix U and parameter-sensitivity matrix V are $\dot{U} = \Phi\, U$ and $\dot{V} = \Phi\, V + \Theta$. The matrices Φ and Θ represent the linearized sensitivity of the equations of motion to $\mathbf{X}$ and $\mathbf{Y}$. Thus Φ and Θ are obtained analytically by forming differentials in the equations of motion according to the formula $\delta \dot{\mathbf{X}} = \Phi\, \delta \mathbf{X} + \Theta\, \delta \mathbf{Y}$. The initial conditions on U and V are that U be the unit matrix and V be null at the epoch t_0.

The complete equations of condition are $\delta \boldsymbol{z} = A\, \delta \boldsymbol{x}$, where $\delta \boldsymbol{z}$ and $\delta \boldsymbol{x}$ are corrections to the data and parameters respectively. As shown in [1], the matrix A can be written in the partitioned form $A = (G\,U \,\vdots\, G\,V + H)$ where G and H represent the explicit dependence of the data on the vectors $\boldsymbol{X}$ and $\boldsymbol{Y}$. In the linear form the equation for this dependence is $\delta \boldsymbol{z} = G\, \delta \boldsymbol{X} + H\, \delta \boldsymbol{Y}$.

The augmentation of the equations of condition involves the attitude-control parameters as well as orbital elements of the Earth and target planet. The previous inclusion of the parameters $\boldsymbol{X}$ has made the matrix G available. In addition, there is no explicit first-order dependence of the doppler data on the attitude-control parameters, the solar ephemeris, or the ephemeris of the target planet. Therefore, nothing need be added to the matrix H. New terms are introduced in the matrix Θ only so that an augmented matrix V can be computed by numerically integrating the augmented differential equations $\dot{V} = \Phi\, V + \Theta$. The dependence of Φ on the new low-thrust forces is neglected as a second-order effect.

Elements are introduced into the matrix Θ by forming differentials in the perturbative accelerations. In the case of the attitude-control parameters the elements are obtained by a straightforward differentiation of Eq. (3) with respect to the parameter set $(\alpha_1, \alpha_2, f_1, f_2, f_3, G_T, G_N)$.

For the orbital elements of the planet and the Earth-Moon barycenter it is necessary to evaluate first the differentials of the equations of motion with respect to the heliocentric planetary positions. Differentials in the planetary positions at the time argument t are then evaluated with respect to the heliocentric planetary orbital elements at the epoch. Designate the variations in the spacecraft's equations of motion with respect to the planetary positions by

$$\Delta \ddot{\boldsymbol{r}} = \left(\frac{\partial \boldsymbol{f}}{\partial \boldsymbol{r}}\right) \Delta \boldsymbol{r} + \left(\frac{\partial \boldsymbol{f}}{\partial \boldsymbol{r}'}\right) \Delta \boldsymbol{r}', \tag{18}$$

where $\boldsymbol{r}$ and $\boldsymbol{r}'$ refer to the coordinates of the planet and the Earth-Moon barycenter respectively. The second variation, that with respect to the orbital elements, is obtained by combining Eqs. (4) and (6) for the planet and Eqs. (5) and (7) for the Earth. Thus the complete variations in the equations of motion are given by

$$\Delta \ddot{\boldsymbol{r}} = \left(\frac{\partial \boldsymbol{f}}{\partial \boldsymbol{r}}\right) U_P\, M\, \Delta\, \boldsymbol{T} \tag{19}$$

$$\Delta \ddot{\boldsymbol{r}}' = \left(\frac{\partial \boldsymbol{f}}{\partial \boldsymbol{r}'}\right) U'_P\, M'\, \Delta\, \boldsymbol{E}, \tag{20}$$

where U_p is the upper 3×6 half of the 6×6 matrix U.

For the evaluation of the matrices $(\partial \boldsymbol{f}/\partial \boldsymbol{r})$ and $(\partial \boldsymbol{f}/\partial \boldsymbol{r}')$, consider the perturbative acceleration $\dot{\boldsymbol{r}}_n^{\backslash}$ on the spacecraft caused by the Earth-Moon system and the planets. Because the computer program used for the least-squares reduction operates in geocentric coordinates, it is also necessary to write the perturbative acceleration in geocentric coordinates:

$$\dot{\boldsymbol{r}}_n^{\backslash} = GS[\boldsymbol{h}(\boldsymbol{r}_{\odot S}) + \boldsymbol{h}(\boldsymbol{r}_{\oplus\odot})] + \sum_i GM_i[\boldsymbol{h}(\boldsymbol{r}_{iS}) + \boldsymbol{h}(\boldsymbol{r}_{\oplus i})]. \quad (21)$$

Following a suggestion by P. R. PEABODY, the vector function $\boldsymbol{h}$ in Eq. (21) is defined by

$$\boldsymbol{h}(\boldsymbol{x}) = -\frac{\boldsymbol{x}}{x^3}, \quad (22)$$

where x is the magnitude of the vector $\boldsymbol{x}$. The various arguments in the function are $\boldsymbol{r}_{\odot S}$, the heliocentric position of the spacecraft; $\boldsymbol{r}_{\oplus\odot}$, the geocentric position of the Sun; $\boldsymbol{r}_{iS}$, the planet-centered position of the spacecraft; and $\boldsymbol{r}_{\oplus i}$, the geocentric position of the planet. A variation in the function $\boldsymbol{h}$ can be written in terms of a variation in the argument $\boldsymbol{x}$:

$$\Delta\boldsymbol{h}(\boldsymbol{x}) = J(\boldsymbol{x})\,\Delta\,\boldsymbol{x}. \quad (23)$$

The matrix $J(\boldsymbol{x})$ is given by

$$J(\boldsymbol{x}) = \frac{3\,\boldsymbol{x}\,\boldsymbol{x}^T}{x^5} - \frac{1}{x^3} I_3, \quad (24)$$

where the superscript T indicates a transpose and I_3 is the unit matrix of order three. With these definitions the two variational matrices of interest are the following:

$$\left(\frac{\partial \boldsymbol{f}}{\partial \boldsymbol{r}}\right) = GM_T[J(\boldsymbol{r}_{\oplus T}) - J(\boldsymbol{r}_{TS})], \quad (25)$$

$$\left(\frac{\partial \boldsymbol{f}}{\partial \boldsymbol{r}'}\right) = GS[J(\boldsymbol{r}_{\odot S}) - J(\boldsymbol{r}_{\oplus\odot})] + \sum_i GM_i[J(\boldsymbol{r}_{iS}) - J(\boldsymbol{r}_{\oplus i})]. \quad (26)$$

The subscript T indicates the particular planet designated as the target. Actually, all planets except the target produce a negligible effect on the matrix $(\partial \boldsymbol{f}/\partial \boldsymbol{r}')$. Therefore, Eq. (26) is replaced by a convenient approximation:

$$\left(\frac{\partial \boldsymbol{f}}{\partial \boldsymbol{r}'}\right) = GS[J(\boldsymbol{r}_{\odot S}) - J(\boldsymbol{r}_{\oplus\odot})] + GM_T[J(\boldsymbol{r}_{TS}) - J(\boldsymbol{r}_{\oplus T})]. \quad (27)$$

Results obtained by numerically differentiating the variational equations $\dot{V} = \Phi\, V + \Theta$ for the attitude-control parameters and the planetary orbital elements are shown in Figs. 1, 2 and 3.

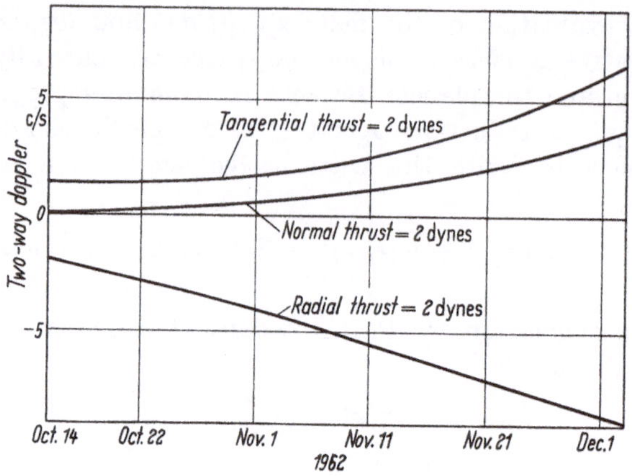

Fig. 1. Sensitivity of Doppler data to low-thrust fores.

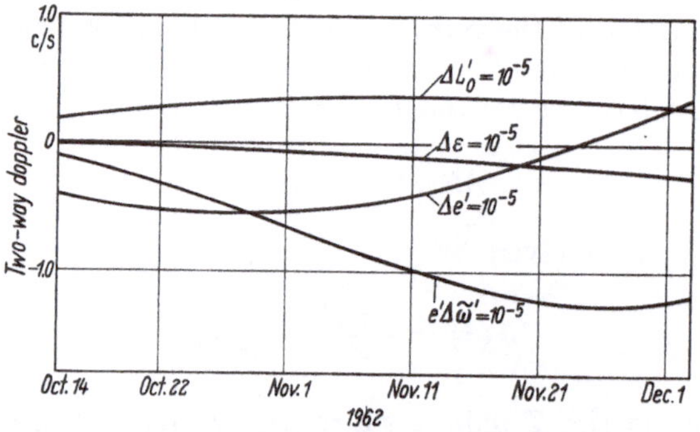

Fig. 2. Sensitivity of Doppler data to Earth orbital elements; epoch of osculation is October 19.0, 1962; spacecraft epoch is September 5, 1962.

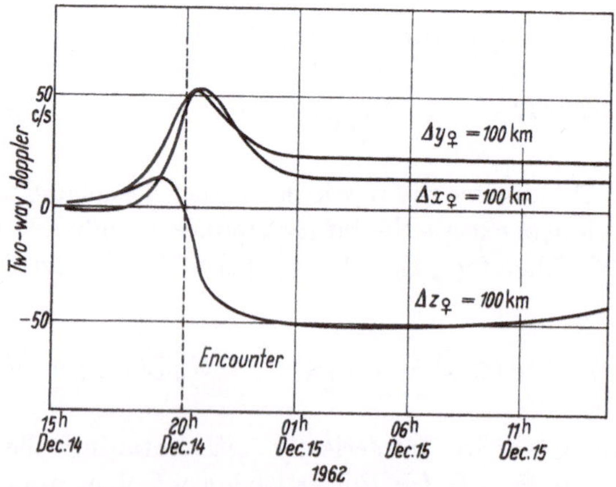

Fig. 3. Sensitivity of Doppler data to heliocentric position of Venus at encounter; spacecraft epoch is December 7, 1962.

III. Numerical Results

Five solutions for various parameter and data configurations are presented in this Section. They are largely experimental in nature and conclusions about the behavior of solutions for the new parameters are emphasized. Solutions I and II use data taken during the heliocentric portion of the mission. Solutions III and IV deal with the encounter data. In Solution V an attempt to use both heliocentric and encounter data is evaluated.

The normal equations for the solutions are used in a form which allows *a priori* values of the parameters to be introduced as additional data (see [*1*], p. 133). Whenever the statistics on these *a priori* values are chosen such that the solution is conditioned by them, the assumptions on the *a priori* parameter covariance matrix $\tilde{\Gamma}_x$ are stated. If for example the *a priori* error on a parameter is so small that the Doppler data are unable to appreciably reduce it, then the effect of including this parameter is as an error source only. The probable errors from the inverse normal matrix on the fitted parameters are larger than they would be if the essentially uncorrected parameter was eliminated from the solution.

The resulting probable errors after the fit depend on the assumed weights on the data. For these solutions a uniform weighting is assumed for all doppler data based on the same cycle count time. The only exception is for points taken at low elevation angles where the data error is increased because of errors in the atmospheric refraction correction.

For data sampled every 60 sec the assumed error is 0.05 cps in the average Doppler frequency. After October 15, 1962 the sample time is increased to a nominal interval of 600 sec. Also, time over which the frequency is averaged, a process mechanized by a cycle-count device, is increased to 600 sec. To account for the fact that more information is contained in an observation sampled after 600 sec of counting than in one sampled after 60 sec, the value of 0.05 cps is multiplied by $\sqrt{60/T_s}$ where T_s is the sample time. Occasionally the time T_s is less than 600 sec even after October 15, 1962. This occurs when the counter misses a count. Thus it is necessary to reset the count at zero and construct the next observation.

The validity of assuming an error of 0.05 cps on the 60-sec data has been investigated by fitting individual passes of data and then by computing the *r m s* value of the residuals for each pass. After reducing the *r m s* residual to a common basis of 60-sec samples, the typical random error on these data appears to be about 0.015 cps. The range of errors for individual passes is from 0.014 to 0.018 cps, although the *r m s* residual for data taken on November 6, 1962, is only 0.009 cps. However, this is an exception. The assumed data error of 0.05 cps does seem justified and, hopefully, the fact that it is almost three times larger than the random error can account for possible systematic errors.

A. Solution I

The sensitivity plots of Fig. 2 indicate that the Mariner II data are not particularly suited to the determination of the Earth's orbital elements. To demonstrate further the accuracy to which these elements can be determined, a solution for the elements along with the initial position and velocity of the spacecraft is presented. Data from September 5 to December 1, 1962 are included. However, only every tenth observation is used. The initial conditions for the spacecraft's orbit are taken from the best previous fit to the heliocentric data. The epoch is September 5, 1962, 00^h, 23^m, $32^s.000$. Values are assigned to the solar-radiation coefficient v, the gravitational constant GM for the Moon, and the astronomical unit A, as follows:

$$v = 0.91340329, \quad GM = 4902.7752\ \text{km}^3/\text{sec}^2, \quad A = 149{,}598{,}850\ \text{km}.$$

Any constant not mentioned in this solution and those that follow is given a value as adopted in [8]. In all the solutions the epoch of osculation for the Earth's elements is October 19.0, 1962, U.T.

The *a priori* probable errors on the parameters given below are chosen large so that the data can dominate the solution.

Parameters	*A priori* error
x_0, y_0, z_0	$\pm 10^6$ km
$\dot{x}_0$, $\dot{y}_0$, $\dot{z}_0$	± 3162 m/sec
$\Delta e'$, $\Delta\varepsilon$, $\Delta L_0'$, $e'\,\Delta\tilde{\omega}'$	$\pm 5 \times 10^{-5}$

The solution itself is of little interest because several important parameters have been excluded. However, the errors computed from the diagonal elements of the inverse normal matrix indicate how well the Earth's elements could be determined from the data under the most ideal condition: that for which the attitude-control forces and all the astronomical constants are known perfectly from independent measurements. In the real situation the errors cannot be less than those listed at the left. Thus the probable errors for the Earth's elements are, at best, of the same order of magnitude as correspon-

Parameters	Probable errors
x_0	± 107 km
y_0	± 142 km
z_0	± 133 km
$\dot{x}_0$	± 0.013 m/sec
$\dot{y}_0$	± 0.009 m/sec
$\dot{z}_0$	± 0.062 m/sec
$\Delta e'$	$\pm\ 1.6 \times 10^{-6}$
$\Delta\varepsilon$	$\pm 22.3 \times 10^{-6}$
$\Delta L_0'$	$\pm\ 3.8 \times 10^{-6}$
$e'\,\Delta\tilde{\omega}'$	$\pm\ 1.6 \times 10^{-6}$

ding tolerable corrections. In fact, the error in the obliquity amounts to almost 5″, and any correction to this parameter could not be considered significant. As a result, the inclusion of the Earth's elements in any real solution can be accomplished by considering them as error sources.

For completeness the corrections computed in this solution are given below. The correlation matrix for the solution is given in Table 1.

$$\Delta x_0 = -781 \text{ km},$$

$$\Delta y_0 = -1047 \text{ km},$$

$$\Delta z_0 = 44 \text{ km},$$

$$\Delta x_0' = 0.0228 \text{ m/sec},$$

$$\Delta y_0' = -0.0294 \text{ m/sec},$$

$$\Delta z_0' = 0.287 \text{ m/sec},$$

$$\Delta e' = -8.0 \times 10^{-6},$$

$$\Delta \varepsilon = 105.8 \times 10^{-6},$$

$$\Delta L_0' = -24.4 \times 10^{-6},$$

$$e' \Delta \tilde{\omega}' = 10.7 \times 10^{-6}.$$

Table 1. *Correlation Matrix, Reduced to its Lower Half, of Estimated Parameters (Heliocentric Data)*

X_0	Y_0	Z_0	$\dot{X}_0$	$\dot{Y}_0$
+1.0000	—	—	—	—
+0.9806	+1.0000	—	—	—
−0.1740	−0.1463	+1.0000	—	—
−0.6285	−0.5426	+0.2808	+1.0000	—
+0.7516	+0.7090	−0.1010	−0.9144	+1.0000
−0.6295	−0.6346	−0.0323	+0.1506	−0.4656
−0.6796	−0.7007	−0.2221	+0.1986	−0.4209
−0.5928	−0.5913	+0.1163	+0.5365	−0.4151
−0.8979	−0.9407	−0.1658	+0.3958	−0.6336
−0.4447	−0.4455	−0.0944	+0.6918	−0.7520

$\dot{Z}_0$	$\Delta e'$	$\Delta \varepsilon$	$e' \Delta \tilde{\omega}'$	$\Delta L_0'$
+1.0000	—	—	—	—
+0.6735	+1.0000	—	—	—
−0.2174	+0.3240	+1.0000	—	—
+0.6555	+0.7602	+0.5097	+1.0000	—
+0.2116	+0.5819	+0.4258	+0.4509	+1.0000

Because of the neglect of other important parameters, it is not surprising that the corrections are several times larger than their associated probable errors.

B. Solution II

In [1], p. 141, a fit to the data from September 5 to December 1, 1962, is shown to yield a value for the Earth-Moon mass ratio of 81.30155 $\pm$ 0.0034. The effect of adding attitude-control parameters to that solution is investigated here. Again, only every tenth observation is used. Also, the elements $\Delta e'$, $\Delta L_0'$ and $e' \Delta \tilde{\omega}'$ are included with an assumed *a priori* error of 5×10^{-7}.

The inability to recompute the spacecraft's orbit after corrections have been applied to the parameters imposes a severe limitation on the solutions presented in this article. Particularly, it is impossible to determine corrections to the attidute-control parameters α_1 and α_2 because the nominal values of the parameters f_1, f_2, and f_3 are zero. Therefore, a variation of the data with respect to α_1 and α_2 is also zero. After corrections have been computed for f_1, f_2, and f_3, it is possible to recompute the orbit and to evaluate corresponding variations with respect to α_1 and α_2. A successive repetition of this process should converge to values of all the attitude-control parameters and provide a least-squares fit to the data. However, for the present it is necessary to remain with a linear correction to parameters which are nominally zero.

Consequently, only the attitude-control parameters f_2 and f_3 are included in the solution. It is futile to attempt to determine f_1 because so long as α_1 and α_2 are zero, it behaves almost exactly like the radiation-pressure parameter v. Similarly, f_2 and f_3 behave much like G_T and G_N respectively. Therefore, the latter two parameters are neglected. The separation that occurs because v, G_T, and G_N are divided by the square of the Sun-spacecraft distance is not significant over the distances involved. The force at Venus is about two times that at the Earth for v, G_T, and G_N.

Telemetry information from the spacecraft indicates that the low-thrust attitude-control forces are not constant over the mission. Therefore, the consideration of only v, f_2 and f_3 is not realistic and certain anomalies are bound to arise in the solutions. In order to keep corrections to f_2 and f_3 at reasonable values, *a priori* errors are assigned compatible with telemetry information on the magnitude of the forces. Two dynes of force, a reasonable upper bound, corresponds to a magnitude of 10^{-10} km/sec^2 in f_2 and f_3. A reflection of two dynes of force in the parameter v amounts to about 0.9. This figure is also reasonable for the solar-radiation contribution. The parameter v should be bounded by zero and unity if no attitude-control forces are present. Therefore, *a priori* errors are assigned as listed below:

Parameters	A priori error
x_0, y_0, z_0	$\pm 10^6$ km
$\dot{x}_0$, $\dot{y}_0$, $\dot{z}_0$	± 100 m/sec
$\Delta e'$, $\Delta L_0'$, $e' \Delta \tilde{\omega}'$	$\pm 5 \times 10^{-7}$
v	± 1.0
f_2, f_3	$\pm 10^{-10}$ km/sec^2
GM	± 50 km^3/sec^2
A	$\pm 500{,}000$ km
R_2, R_3	± 300 m
λ_2, λ_3	$\pm 0\overset{\circ}{.}003$

As in [*1*], the geocentric radii R and longitudes λ of the Goldstone Deep Space Intrumentation Facility transmitter (station 3) and receiver (station 2) are included in the solution. Also, in addition to assigning *a priori* errors to R and λ, the two radii and two longitudes are assumed perfectly correlated. This is done because the relative position of the two stations is known quite accurately. Thus a 4×4 *a priori* covariance matrix associated with the station coordinates is constructed for appropriate location in the matrix $\tilde{\Gamma}_x$, which has the following form:

$$\begin{bmatrix} \sigma_R^2 & 0 & \varrho\,\sigma_R^2 & 0 \\ 0 & \sigma_\lambda^2 & 0 & \varrho\,\sigma_\lambda^2 \\ \varrho\,\sigma_R^2 & 0 & \sigma_R^2 & 0 \\ 0 & \varrho\,\sigma_\lambda^2 & 0 & \sigma_\lambda^2 . \end{bmatrix}$$

For perfect correlation the coefficient ϱ is exactly unity. However to avoid a singular matrix in $\tilde{\Gamma}_x$, the coefficient ϱ is set equal to 0.998.

The solution for the parameters follows. The obliquity of the ecliptic is omitted because of its small effect on the data. All initial conditions for the parameters are the same as in solution I except for the solar-radiation coefficient, which is set equal to 0.83119147.

$$\Delta x_0 = -645 \pm 150 \text{ km},$$

$$\Delta y_0 = -857 \pm 205 \text{ km},$$

$$\Delta z_0 = 1207 \pm 247 \text{ km},$$

$$\Delta \dot{x}_0 = 0.101 \pm 0.043 \text{ m/sec},$$

$$\Delta \dot{y}_0 = -0.125 \pm 0.036 \text{ m/sec},$$

$$\begin{aligned}
\varDelta \dot{z}_0 &= 1.215 \pm 0.177 \text{ m/sec},\\
\varDelta e' &= 1.31(\pm 5) \times 10^{-7},\\
\varDelta L_0' &= -0.55(\pm 5) \times 10^{-7},\\
e'\varDelta \tilde{\omega}' &= -0.22(\pm 5) \times 10^{-7},\\
\varDelta \varepsilon &= -0.32 \pm 0.19,\\
\varDelta f_2 &= -0.274(\pm 0.057) \times 10^{-10} \text{ km/sec}^2,\\
\varDelta f_3 &= 0.886(\pm 0.105) \times 10^{-10} \text{ km/sec}^2,\\
\varDelta GM &= 0.84(\pm 0.20) \text{ km}^3/\text{sec}^2,\\
\varDelta A &= -9797 \pm 5787 \text{ km},\\
\varDelta R_2 &= -85 \pm 20 \text{ m},\\
\varDelta \lambda_2 &= 0^\circ\!.00085 \pm 0^\circ\!.00046,\\
\varDelta R_3 &= -85 \pm 20 \text{ m},\\
\varDelta \lambda_3 &= 0^\circ\!.00085 \pm 0^\circ\!.00046.
\end{aligned}$$

The correlation matrix for this solution is given in Table 2.

At least in one respect the new solution is better behaved than that in [*1*]. Consider the two values of the astronomical unit:

$$\text{A} = 149{,}615{,}890 \pm 4640 \text{ km} \quad \text{(no attitude-control forces)},$$

$$A = 149{,}589{,}050 \pm 5787 \text{ km} \quad \text{(linear correction, attitude-control and Earth elements present)}.$$

In the new solution the only significant correlation of A with other parameters is in the tangential low-thrust parameter f_2. In the old solution, A was significantly correlated with the initial velocity of the spacecraft and the solar-radiation parameter v. It appears that all of these parameters attempted to compensate for the neglected normal and tangential forces. The new value of A deviates from that obtained from radar-bounce experiments by about 1.7 times its probable error. Compare this with a deviation of about 3.6 times for the old A. The factor of 1.7 could arise because of the nonlinear behavior of the solution with respect to variations in f_2. The question of whether A can be brought into consistency with other radar determinations by means of the heliocentric data alone must wait for future iterative solutions. It seems apparent now that the large discrepancy of A in the old solution was caused by a neglect of important low-thrust forces.

Table 2. *Correlation Matrix, Reduced to its Lower Half, of Estimated Parameters (Heliocentric Data)*

X_0	Y_0	Z_0	$\dot{X}_0$	$\dot{Y}_0$	$\dot{Z}_0$
+1.0000	—	—	—	—	—
+0.9823	+1.0000	—	—	—	—
−0.6616	−0.6579	+1.0000	—	—	—
−0.4445	−0.4731	+0.5311	+1.0000	—	—
+0.4645	+0.5057	−0.5000	−0.9798	+1.0000	—
−0.6740	−0.7109	+0.5224	+0.7020	−0.8124	+1.0000
−0.0237	−0.0049	+0.0079	−0.0532	+0.0375	+0.0335
−0.0298	−0.0282	−0.0355	−0.0296	+0.0561	−0.1240
−0.4386	−0.4454	+0.3275	+0.2910	−0.3302	+0.4232
−0.0345	+0.0179	−0.2681	−0.8550	+0.8606	−0.5345
+0.0051	−0.0048	−0.0006	+0.0310	−0.0264	−0.0006
−0.4362	−0.4141	−0.0270	−0.4022	+0.4509	−0.2805
−0.6483	−0.6733	+0.2126	+0.0563	−0.1931	+0.6470
−0.2449	−0.2536	+0.1415	+0.2388	−0.0941	−0.2626
+0.5903	+0.6069	−0.6890	−0.5812	+0.6197	−0.7024
−0.3301	−0.4316	+0.3291	+0.7827	−0.7737	+0.5513
+0.5906	+0.6071	−0.6892	−0.5810	+0.6194	−0.7023
−0.3303	−0.4319	+0.3300	+0.7828	−0.7736	+0.5511

$\Delta e'$	$e'\Delta\tilde{\omega}'$	GM	v	$\Delta L_0'$	f_2
+1.0000	—	—	—	—	—
+0.0004	+1.0000	—	—	—	—
+0.0151	−0.0083	+1.0000	—	—	—
−0.0048	+0.0414	−0.1183	+1.0000	—	—
+0.0008	+0.0002	−0.0025	−0.0028	+1.0000	—
−0.0379	+0.0308	+0.0102	+0.7852	+0.0028	+1.0000
+0.0853	−0.2880	+0.4551	+0.1789	+0.0076	+0.2579
+0.0669	+0.0467	+0.0628	+0.0472	+0.0093	+0.5171
−0.0155	+0.0562	−0.3476	+0.4000	+0.0018	+0.1784
−0.1185	−0.0290	+0.2181	−0.6710	+0.0642	−0.2069
−0.0155	+0.0562	−0.3477	+0.3996	+0.0017	+0.1781
−0.1185	−0.0290	+0.2182	−0.6710	+0.0642	−0.2067

f_3	A	R_2	λ_2	R_3	λ_3
+1.0000	—	—	—	—	—
−0.1887	+1.0000	—	—	—	—
−0.3710	+0.0702	+1.0000	—	—	—
+0.1053	+0.3027	−0.4008	+1.0000	—	—
−0.3710	+0.0701	+0.5353	−0.3998	+1.0000	—
+0.1051	+0.3031	−0.4018	+0.9134	−0.4008	+1.0000

As for the mass of the Moon, it appears that the old solution yields a better value. A comparison of the correlations of GM given below, with parameters common to the two solutions makes this understandable.

Correlation Coefficients for GM

Parameter	Old solution	New solution
x_0	-0.201	-0.439
y_0	-0.192	-0.445
z_0	0.208	0.328
$\dot{x}_0$	0.224	0.291
$\dot{y}_0$	-0.200	-0.330
$\dot{z}_0$	0.081	0.423
v	-0.170	-0.118
GM	1.000	1.000
A	0.289	0.063
R_2	-0.135	-0.348
λ_2	0.137	0.218
R_3	-0.115	-0.348
λ_3	0.151	0.218
$\Delta e'$	—	0.151
$e' \Delta \tilde{\omega}'$	—	-0.826
$\Delta L_0'$	—	-0.247
f_2	—	0.102
f_3	—	0.455

In the old solution, even though the corrections to most of the parameters are erroneous, the low correlations with the mass of the Moon indicate an independent determination. In the new solution the determination is not independent and, as a result, the inclusion of the attitude-control parameters must be accomplished carefully. This conclusion is emphasized by the correlation of 0.455 between the mass of the Moon and the normal-force parameter f_3. Note that the corrections to f_2 and f_3 are large with respect to their probable errors. Therefore, until the complete attitude-control model can be included in a nonlinear solution, the mass of the Moon is best determined by neglecting the low-thrust forces entirely. This is an example of the generalization that it is often better to omit an important parameter in a least-squares solution than to include it improperly.

The Mariner II value of the mass ratio of the Earth to the Moon remains, as before,

$$\mu^{-1} = 81.30155 \pm 0.0034 \text{ (Mariner II)}.$$

This agrees quite well with independent determinations from the Ranger VI and VII data (see [*3*], p. 47).

$$\mu^{-1} = 81.30362 \pm 0.0023 \quad \text{(Ranger VI)},$$

$$\mu^{-1} = 81.30439 \pm 0.0028 \text{ (Ranger VII)}.$$

C. Solution III

Unlike the sensitivity plots for the Earth elements, those for Venus (Fig. 3) indicate that it should be possible to determine the position of Venus during the encounter period. Therefore, a least-squares fit to all the encounter data from December 7 to December 20, 1962, is presented with the initial conditions for the orbit of the spacecraft and the five Venus elements ($x_{♀}$, $y_{♀}$, $z_{♀}$, $\dot{y}_{♀}$, $\dot{z}_{♀}$) as free parameters. The spacecraft epoch is December 7.0, 1962, U.T., while the epoch for the elements of Venus is the closest approach time December 14, 1962, $20^{\mathrm{h}}.0$, U.T. No significant limit is placed on the *a priori* errors. The linear corrections to the best previous encounter solution (see [*1*], p. 143) are as follows:

$$\Delta x_0 = -75 \pm 80 \text{ km}, \qquad \Delta y_0 = 84 \pm 146 \text{ km},$$

$$\Delta z_0 = -457 \pm 186 \text{ km}, \qquad \Delta \dot{x}_0 = -0.039 \pm 0.111 \text{ m/sec},$$

$$\Delta \dot{y}_0 = -0.122 \pm 0.220 \text{ m/sec}, \qquad \Delta \dot{z}_0 = 0.634 \pm 0.262 \text{ m/sec},$$

$$\Delta x_{♀} = -4.79 \pm 6.06 \text{ km},$$

$$\Delta y_{♀} = 3.93 \pm 10.63 \text{ km},$$

$$\Delta z_{♀} = -1.91 \pm 6.06 \text{ km},$$

$$\Delta \dot{y}_{♀} = -0.96 \pm 1.91 \text{ m/sec},$$

$$\Delta \dot{z}_{♀} = 0.70 \pm 1.82 \text{ m/sec}.$$

The correlation matrix for this solution is given in Table 3.

Table 3. *Correlation Matrix, Reduced to its Lower Half, of Estimated Parameters (Encounter Data)*

X_0	Y_0	Z_0	$\dot{X}_0$	$\dot{Y}_0$	$\dot{Z}_0$
+1.0000	—	—	—	—	—
−0.9891	+1.0000	—	—	—	—
+0.9231	−0.9611	+1.0000	—	—	—
−0.9938	+0.9967	−0.9526	+1.0000	—	—
+0.9857	−0.9982	+0.9727	−0.9968	+1.0000	—
−0.9410	+0.9752	−0.9979	+0.9657	−0.9831	+1.0000
+0.1953	−0.1217	−0.1323	−0.1118	+0.0664	+0.0702
+0.1680	−0.2180	+0.4469	−0.2352	+0.2750	−0.3895
+0.1239	−0.2153	+0.4469	−0.2161	+0.2627	−0.3912
+0.3534	−0.4055	+0.6139	−0.4188	+0.4578	−0.5631
−0.2040	+0.2947	−0.5175	+0.2947	−0.3410	+0.4660

$x_{♀}$	$y_{♀}$	$\dot{y}_{♀}$	$z_{♀}$	$\dot{z}_{♀}$
+1.0000	—	—	—	—
−0.8811	+1.0000	—	—	—
−0.8289	+0.9061	+1.0000	—	—
−0.7905	+0.9788	+0.9065	+1.0000	—
+0.8031	−0.9070	−0.9965	−0.9229	+1.0000

As expected, the velocity of Venus cannot be determined from the Mariner data. However, the computed errors on the position are small enough to encourage their incorporation in solutions involving the encounter data. The velocity elements are neglected in the next two solutions.

D. Solution IV

The encounter solution of [*1*] is now repeated with three improvements. The first of these is a computation of the Mariner II orbit in Venus-centered rather than geocentric coordinates. The second is the snclusion of the attitude-control parameters f_2 and f_3. Because of the thort 13-day interval of time involved, these parameters, along with the solar-radiation coefficient, should provide a satisfactory representaiion of the low-thrust forces. The third improvement is the addition of the three position components of Venus in the solution along with the constraint [Eq. (9)] that the semimajor axis remain constant.

When the astronomical unit is also included in the solution, the component $\dot{x}_{♀}$ should depend on the correction to A as well as the corrections $\Delta x_{♀}$, $\Delta y_{♀}$, and $\Delta z_{♀}$. This is because the mean distance in astronomical units is constant. The effect of this variation is implicit in the matrix M.

It can be concluded that apparently all difficulties with the previous encounter solution of [*1*] have been removed. Furthermore, a linear correction to the parameters should be sufficient. Therefore, this solution is the most significant of those presented so far.

The initial values of the parameters and their *a priori* probable errors are tabulated in the following:

$$x_0 = -36{,}494{,}851 \pm 10^6 \text{ km},$$
$$y_0 = -29{,}910{,}920 \pm 10^6 \text{ km},$$
$$z_0 = -9{,}454{,}576.4 \pm 10^6 \text{ km},$$
$$\dot{x}_0 = -7.3400224 \pm 0.1 \text{ km/sec},$$
$$\dot{y}_0 = -10.825375 \pm 0.1 \text{ km/sec},$$
$$\dot{z}_0 = -5.4165860 \pm 0.1 \text{ km/sec},$$
$$\Delta x_{♀} = \Delta y_{♀} = \Delta z_{♀} = 0 \pm 500 \text{ km},$$
$$f_2 = f_3 = 0 \pm 0.707 \times 10^{-10} \text{ km/sec}^2,$$
$$v = 0.83119147 \pm 10,$$
$$M_V = 2.4477798 (\pm 0.00632) \times 10^{-6} \text{ (solar mass units)},$$
$$A = 149{,}598{,}500 \pm 500 \text{ km},$$
$$R_2 = 6371.8620 \pm 0.3 \text{ km}, \quad \lambda_2 = 243^\circ\!.14997 \pm 0^\circ\!.003,$$
$$R_3 = 6371.8478 \pm 0.3 \text{ km}, \quad \lambda_3 = 243^\circ\!.19367 \pm 0^\circ\!.003.$$

The initial conditions are geocentric equatorial coordinates of date. The epochs for the spacecraft and Venus are the same as in Solution III.

Corrections to the parameters are given below and the correlation matrix is given in Table 4.

$$\Delta x_0 = -20 \pm 152 \text{ km},$$

$$\Delta y_0 = 94 \pm 122 \text{ km},$$

$$\Delta z_0 = -336 \pm 140 \text{ km},$$

$$\Delta \dot{x}_0 = -0.009 \pm 0.059 \text{ m/sec},$$

$$\Delta \dot{y}_0 = -0.130 \pm 0.081 \text{ m/sec},$$

$$\Delta \dot{z}_0 = 0.496 \pm 0.198 \text{ m/sec},$$

$$\Delta x_♀ = -7.39 \pm 2.59 \text{ km},$$

$$\Delta y_♀ = 5.90 \pm 2.50 \text{ km},$$

$$\Delta z_♀ = -0.14 \pm 0.75 \text{ km},$$

$$\Delta f_2 = 0.0002 (\pm 0.685) \times 10^{-10} \text{ km/sec}^2,$$

$$\Delta f_3 = 0.0129 (\pm 0.704) \times 10^{-10} \text{ km/sec}^2,$$

$$\Delta v = -0.335 \pm 0.619,$$

$$\Delta M_V = -3.87 (\pm 1.49) \times 10^{-10} \text{ (solar mass units)},$$

$$\Delta A = 359 \pm 499 \text{ km},$$

$$\Delta R_2 = -28 \pm 10 \text{ m},$$

$$\Delta \lambda_2 = 0\overset{\circ}{.}00077 \pm 0\overset{\circ}{.}00014,$$

$$\Delta R_3 = -28 \pm 10 \text{ m},$$

$$\Delta \lambda_3 = 0\overset{\circ}{.}00077 \pm 0\overset{\circ}{.}00014.$$

The corrections to the position of Venus can be transformed into corrections to geocentric right ascension and declination:

$$\cos\delta\,\Delta\alpha = 0\overset{''}{.}033 \pm 0\overset{''}{.}012,$$

$$\Delta\delta = 0\overset{''}{.}0007 \pm 0\overset{''}{.}0021.$$

These corrections are to be applied to the position as computed with the current JPL Venus ephemeris tape [4]. The uncertainties are remarkably small and deserve careful scrutiny in the future.

The uncertainty in the attitude-control parameters and the astronomical unit are not reduced appreciably from their *a priori* errors. However, the corrections to the attitude-control parameters imply about 0.0004 dynes of force in the tangential direction and about 0.03 dynes in the normal direction.

Table 4. *Correlation Matrix, Reduced to its Lower Half, of Estimated Parameters (Encounter Data)*

X_0	Y_0	Z_0	$\dot{X}_0$	$\dot{Y}_0$	$\dot{Z}_0$
+1.0000	—	—	—	—	—
+0.8253	+1.0000	—	—	—	—
+0.3943	+0.0996	+1.0000	—	—	—
−0.4308	−0.0393	+0.1158	+1.0000	—	—
+0.3472	−0.1147	+0.5100	−0.7642	+1.0000	—
−0.0648	+0.1934	−0.9384	−0.2740	−0.4094	+1.0000
−0.1599	−0.3715	+0.7583	+0.2300	+0.3496	−0.8553
+0.1119	+0.2080	−0.6569	−0.2837	−0.2006	+0.7176
−0.1825	−0.1633	+0.3728	+0.6934	−0.2918	−0.5126
+0.2879	+0.0276	+0.9818	+0.1384	+0.4813	−0.9334
+0.0379	−0.1825	+0.2006	−0.1254	+0.2773	−0.2206
−0.0463	−0.0259	+0.2172	+0.5707	−0.3089	−0.3079
−0.0376	+0.0577	+0.0228	+0.1183	−0.0159	−0.1360
−0.9775	−0.9053	−0.4074	+0.2544	−0,2836	+0.0916
−0.0163	−0.0085	−0.0118	+0.0099	−0.0148	+0.0082
+0.0210	+0.0097	−0.0410	−0.0657	+0.0335	+0.0537
−0.0162	−0.0087	−0.0115	+0.0096	−0.0143	+0.0079
+0.0212	+0.0089	−0.0383	−0.0657	+0.0353	+0.0509

$x_♀$	$y_♀$	v	M_V	$z_♀$	f_3
+1.0000	—	—	—	—	—
−0.8911	+1.0000	—	—	—	—
+0.2813	−0.1910	+1.0000	—	—	—
+0.8264	−0.7298	+0.3005	+1.0000	—	—
−0.0545	+0.4532	+0.2117	+0.1630	+1.0000	—
+0.0344	+0.0354	+0.9565	+0.1102	−0.2176	+1.0000
−0.0061	+0.0142	+0.0375	−0.0576	−0.0533	+0.0102
+0.1356	−0.0511	+0.1235	−0.3191	+0.0338	+0.0098
+0.0152	−0.0040	−0.0036	−0.0072	−0.0050	−0.0065
−0.0394	+0.0195	−0.0705	−0.0412	−0.0326	−0.0573
+0.0155	−0.0041	−0.0035	−0.0070	−0.0048	−0.0064
−0.0365	+0.0171	−0.0697	−0.0384	−0.0323	−0.0571

f_3	A	R_2	λ_2	R_3	λ_3
+1.0000	—	—	—	—	—
−0.0042	+1.0000	—	—	—	—
−0.0068	+0.0148	+1.0000	—	—	—
−0.0134	−0.0123	+0.0194	+1.0000	—	—
−0.0070	+0.0147	−0.7247	+0.0347	+1.0000	—
−0.0136	−0.0125	+0.0005	+0.0833	+0.0158	+1.0000

Besides the corrections to the position of Venus the only other correction of astronomical interest is to the mass of Venus. The mass ratio of the Sun to Venus is

$$\frac{M_S}{M_V} = 408{,}598.1 \pm 25\,.$$

The value given in [*1*] is 408,533.5 ± 30. Because the new value differs from this by about two times the previously quoted probable error, it is concluded that the previous error was overly optimistic. The old error was a factor of three larger than that given by the diagonal element in the inverse normal matrix. It appears that a factor of six would have been more realistic in allowing for the obvious systematic errors inherent in the old solution.

There is no apparent reason to suspect the new value of the mass or to doubt its assigned probable error. Of course, it is possible that computational errors are introducing undetectable systematic effects into the solution. However, unlike the old solution, there is no real basis for increasing the computed probable error of ±30. It is possible to obtain a fairly independent verification of the solution by comparing the station locations for station 3 with those obtained by the Ranger VII data (Ref. [*3*], p. 22). Station 2 was not used by Ranger VII so it cannot be compared.

Only two coordinates are determined by Doppler tracking data. These are $R \cos\varphi'$ and λ. The third component $R \sin \varphi'$ cannot be determined. A comparison given below reveals that the two solutions agree in the $R \cos \varphi'$ direction to 5.9 m and in the longitude direction to 5.4 m.

Station 3 coordinates	Ranger VII	Mariner II
$R \cos\varphi'$	5212.0388 km	5212.0447 km
λ	243°19438	243°19444

E. Solution V

Finally a least-squares fit to the combined data from September 5 to December 20, 1962 is performed. As in Solutions I and II, every tenth observation is used, and the trajectory is computed in geocentric coordinates. Again, an inability to consider nonlinear effects makes it impossible to achieve a meaningful fit to all the data. The linear correction is presented here to demonstrate the capability of the Mariner II data. A similar solution was presented in [*1*], but at that time it was impossible to assess the effects of the attitude-control forces and the planetary ephemerides on the computed uncertainties.

The probable errors on the parameters are listed below. The first column indicates the parameters considered in the new solution. The second column gives the assumed a priori error on each parameter for the new solution. The probable errors from the inverse of the normal matrix are given in the third column, and finally, the errors from the old solution are given in the fourth column. The difference between the third and fourth columns is an indication of the contribution to error by the attitude-control parameters and the position coordinates of Venus. The error on the data is essentially the same for both columns.

Parameter	A priori error	Error New solution	Error Old solution
x_0	$\pm 10^6$ km	± 73	± 25
y_0	$\pm 10^6$	± 101	± 7
z_0	$\pm 10^6$	± 75	± 18
$\dot{x}_0$	± 10 m/sec	± 0.0121	± 0.003
$\dot{y}_0$	± 10	± 0.0083	± 0.001
$\dot{z}_0$	± 10	± 0.0240	± 0.002
$\Delta x_{♀}$	± 500 km	± 3.4	± 0.0 (neglected parameter)
$\Delta y_{♀}$	± 500	± 8.7	± 0.0 (neglected parameter)
$\Delta z_{♀}$	± 500	± 4.8	± 0.0 (neglected parameter)
v	± 10	± 0.1	± 0.013
f_2	7.07×10^{-11} km/sec²	$\pm 3 \times 10^{-12}$	± 0.0 (neglected parameter)
f_3	7.07×10^{-11}	$\pm 6 \times 10^{-12}$	± 0.0 (neglected parameter)
GM	0.2 km³/sec²	± 0.13	± 0.18
M_V	6.3×10^{-9}	$\pm 1.6 \times 10^{-11}$	$\pm 1.4 \times 10^{-11}$
A	4000 km	± 316	± 55

The correlation matrix for this solution is given in Table 5.

An a priori error of ± 0.2 km³/sec² was assigned to the gravitational constant for the Moon in the expectation that the solution for the constants might have more significance. Without this constraint a value of GM similar to Solution II would result. The values of the constants from the new solution are given below:

$$\Delta x_{♀} = -37.2 \text{ km},$$

$$\Delta y_{♀} = -371.4,$$

$$\Delta z_{♀} = -218.0,$$

$$v = 1.05, \qquad GM = 4902.9669 \text{ km}^3/\text{sec}^2,$$

$$f_2 = -9.6 \times 10^{-12} \text{ km/sec}^2, \quad M_S/M_V = 408{,}606.7,$$

$$f_3 = 6.89 \times 10^{-11}, \qquad A = 149{,}599.455 \text{ km}.$$

The corrections to the position of Venus are large. This can be explained by recognizing that the changes to the position are compensating for errors in the model for the low-thrust forces and the geocentric integration of the trajectory during planetary encounter.

Table 5. *Correlation Matrix, Reduced to its Lower Half, of Estimated Parameters (All Data)*

X_0	Y_0	Z_0	$\dot{X}_0$	$\dot{Y}_0$
+1.0000	—	—	—	—
+0.9604	+1.0000	—	—	—
−0.8851	−0.9043	+1.0000	—	—
+0.7936	+0.9239	−0.7508	+1.0000	—
−0.8913	−0.9622	+0.7922	−0.9685	+1.0000
−0.8676	−0.9310	+0.9320	−0.8516	+0.8406
−0.2208	−0.2356	−0.0287	−0.3555	+0.3627
−0.0721	−0.0751	−0.1150	−0.1103	+0.1193
−0.9665	−0.9869	+0.8433	−0.9143	+0.9747
−0.2917	−0.2899	+0.2322	−0.2392	+0.2431
+0.6211	+0.6412	−0.5941	+0.5837	−0.6237
−0.1226	−0.1271	−0.0878	−0.1686	+0.1798
−0.9227	−0.9554	+0.7957	−0.9142	+0.9815
−0.8951	−0.9662	+0.8814	−0.9274	+0.9219
−0.9703	−0.9869	+0.8618	−0.8957	+0.9523

$\dot{Z}_0$	$x_♀$	$y_♀$	v	GM
+1.0000	—	—	—	—
+0.1455	+1.0000	—	—	—
−0.0086	−0.1436	+1.0000	—	—
+0.8681	+0.3180	+0.1144	+1.0000	—
+0.2476	+0.0511	−0.0296	+0.2925	+1.0000
−0.5538	−0.1764	+0.2149	−0.6393	−0.2969
+0.0346	−0.0509	+0.9951	+0.1729	−0.0160
+0.7907	+0.3052	+0.1493	+0.9818	+0.2681
+0.9776	+0.2947	+0.0471	+0.9311	+0.2674
+0.8946	+0.2364	+0.2090	+0.9889	+0.2827

M_V	$z_♀$	f_2	f_3	A
+1.0000	—	—	—	—
+0.1848	+1.0000	—	—	—
−0.6355	+0.2050	+1.0000	—	—
−0.5833	+0.1035	+0.8712	+1.0000	—
−0.6003	+0.2616	+0.9600	+0.9406	+1.0000

The situation with respect to the attitude-control parameters is the same as in Solution II. An iterative minimization of the sum of the squared residuals is required for stability of the solution. In Fig. 4 the mean residual from each pass of tracking data is plotted as a function of the date in order to evaluate the total fit to the data. The mean residual is defined as the arithmetic mean of residuals computed after the fit by the linear relation $\delta\hat{z}_{\text{new}} = \delta\hat{z}_{\text{old}} - A\,\delta x^*$. The vector δx^* represents the corrections to the parameters.

It is apparent that the linear correction is unable to produce a satisfactory fit to the data during the heliocentric portion of the orbit. The encounter data is probably satisfied because of the corrections to the position coordinates of Venus. However, the only reasonable method for handling both the encounter and heliocentric data is to introduce

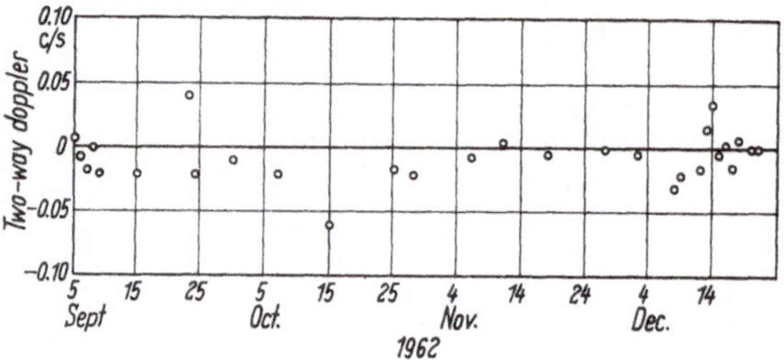

Fig. 4. Mean residuals from Solution V.

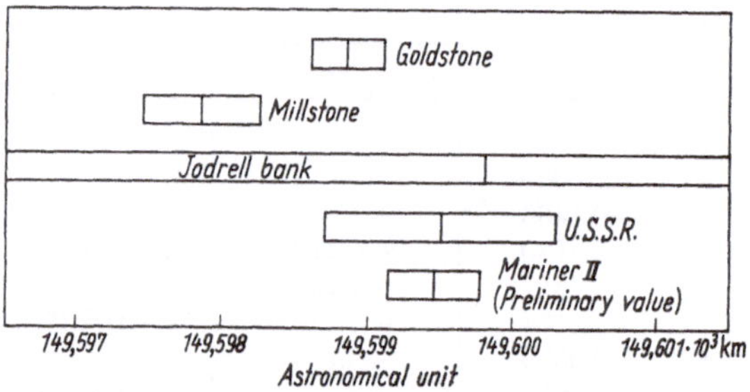

Fig. 5. Comparison of Mariner II value of astronomical unit with other radar values.

the additional low-thrust parameters and to allow an accurate computation of the orbit by changes of phase.

It should be pointed out that this solution is the only one produced to date that is able to fit both sets of data at all. It is interesting that the values of the mass of Venus and the astronomical unit are quite reasonable, although this could be fortuitous. The mass of Venus falls within the error range assigned to the adopted Mariner II value from Solution IV. Notice, however, that when all the data are used the uncertainty in the mass is reduced by a factor of ten. Thus in the future a further improvement in the mass of Venus should be possible.

The value of the astronomical unit is consistent with other radar-determined values [*9, 10, 11* and *12*]. Fig. 5 is included to illustrate this point. The width of the rectangles in the figure indicate the probable errors on each value as assigned by the respective experimenters. The probable error assigned to the Mariner II value is from the current solution. Of course, the value itself must be considered a preliminary one. Finally, when the difficulties with the combined solution are removed there is no assurance that the corresponding value of A will not change by several times the probable error.

Acknowledgments

The solutions described in this paper were accomplished with the aid of the orbit-determination computer program developed by the Jet Propulsion Laboratory (Ref. [*13*] and [*14*]). Modifications to this program for the Mariner II solutions were made by J. A. Flynn, E. H. Larson, and M. W. Nead of the JPL computing section. The authors gratefully acknowledge the assistance given by T. W. Hamilton and G. W. Null of the JPL analysis section, and P. R. Peabody and D. K. Rose of the computing section, in obtaining these solutions.

References

[*1*] Anderson, J. D., G. W. Null and C. T. Thornton: "The Evaluation of Certain Astronomical Constants from the Radio Tracking of Mariner II," in AIAA Series on Progress in Astronautics and Aeronautics. Vol. 14: Celestial Mechanics and Astrodynamics, edited by V. Szebehely, New York/London: Academic Press 1964.

[*2*] Sjogren, W. L., D. W. Curkendall, T. W. Hamilton, W. E. Kirhofer, A. S. Liu, D. W. Trask, R. A. Winneberger and W. R. Wollenhaupt: The Ranger VI Flight Path and Its Determination from Tracking Data, Technical Report No. 32-506, Jet Propulsion Laboratory, Pasadena, Calif., December 15, 1964.

[*3*] Wollenhaupt, W. R., D. W. Trask, W. L. Sjogren, E. G. Piaggi, D. W. Curkendall, R. A. Winneberger, A. S. Liu and A. L. Berman: Ranger VII Flight Path and Its Determination from Tracking Data, Technical Report No. 32-694, Jet Propulsion Laboratory, Pasadena, Calif., December 15, 1964.

[*4*] Peabody, P. R., J. F. Scott and E. G. Orozco: JPL Ephemeris Tapes E 9510, E 9511, and E 9512, Technical Memorandum No. 33-167, Jet Propulsion Laboratory, Pasadena, Calif., March 2, 1964.

[*5*] Peabody, P. R., J. F. Scott and E. G. Orozco: Users Description of JPL Ephemeris Tapes, Technical Report No. 32-580, Jet Propulsion Laboratory, Pasadena, Calif., March 2, 1964.

[*6*] Brouwer, D., and G. M. Clemence: Methods of Celestial Mechanics, New York/London: Academic Press 1961.

[*7*] Herrick, S.: Universal Variables, Astronom. J. **70**, 309—315 (May 1965).

[8] Clarke, V. C.: Constants and Related Data for Use in Trajectory Calculations, Technical Report No. 32-604, Jet Propulsion Laboratory, Pasadena, Calif., March 6, 1964.

[9] Muhleman, D. O., D. B. Holdridge and N. Block: The Astronomical Unit Determined by Radar Reflections From Venus, Astronom. J. **67**, 191—203 (May 1962).

[10] Pettengill, G. H., H. W. Briscoe, J. V. Evans, E. Gehrels, G. M. Hyde, L. G. Kraft, R. Price and W. B. Smith: A Radar Investigation of Venus, Astronom. J. **67**, 181—190 (1962).

[11] Thompson, J. H., G. N. Taylor, J. E. B. Ponsonby and R. S. Roger: A New Determination of the Solar Parallax by Means of Radar Echoes from Venus. Nature **190**, 519 (1961).

[12] Kotelnikov, U. A.: Radar Contact With Venus, Paper presented at the 12th International Astronautical Congress, Washington, D.C., October 1961.

[13] Warner, M. R., M. W. Nead and R. H. Hudson: The Orbit Determination Program of the Jet Propulsion Laboratory, Technical Memorandum No. 33—168, Jet Propulsion Laboratory, Pasadena, Calif., March 18, 1964.

[14] Warner, M. R., and M. W. Nead: SPODP-Single Precision Orbit Determination Program, Technical Memorandum No. 33-204, Jet Propulsion Laboratory, Pasadena, Calif., February 15, 1965.

Discussion

Prof. Brouwer believes that this solution would be improved again, if it was possible to combine the observational data from Mariner II with informations from radar-echo experiments or from other sources.

Plans for Analysis of Lunar Satellite Orbits[1]

By

William M. Kaula

Institute of Geophysics and Planetary Physics,
University of California, Los Angeles, U.S.A.

Abstract. Two satellites are scheduled to orbit around the Moon in 1966. Their orbits are fixed by non-selenodetic considerations and there are engineering limitations on their range and range-rate tracking.

On the assumption of stress implications equal to those of the Earth, the lunar gravitational field is expected to cause a rich variety of perturbations of more than ± 100 meters amplitude. The principal orbital effects expected to distort determination of lunar gravitational variations are third body perturbations and lunar ephemeris errors of monthly and semi-monthly period; in addition, one tracking system will suffer considerable ionospheric refraction.

The program of observations to optimize results within engineering limitations is recommended to be a combination of deliberate distribution over the month and randomization within sub-intervals of a month.

Data analysis should be a minor modification of Earth satellite—orbit analysis.

Résumé. Deux satellites doivent être envoyés autour de la Lune en 1966. Les orbites ne sont pas fixées suivant des critères sélénodésiques et des considérations techniques limitent le repérage en distance et en variation de distance.

En supposant que les conditions de tension de la Lune sont identiques à celles de la Terre, on devra s'attendre à ce que le potentiel lunaire crée une grande variété de perturbations de plus de 100 mètres d'amplitude. Les principaux effets qui gêneront la détermination du potentiel lunaire seront les perturbations du problème des trois corps et les erreurs de périodes un mois et un demi-mois dans les éphémérides de la Lune. De plus, un système de poursuite sera soumis à des réfractions ionosphériques.

On recommande un programme d'observations qui, compte tenu des limitations techniques, optimisera les résultats par combinaison d'une distribution délibérée au cours du mois et d'une distribution au hasard d'intervalles inférieurs au mois.

L'analyse des données sera la même que celle utilisée pour les satellites terrestres à quelques modifications près.

Introduction

In 1966, it is planned to start launching at least two series of satellites around the Moon. A geophysical by-product of these projects

[1] Publication No. 430 of the Institute of Geophysics and Planetary Physics.

should be information about the variations of the Moon's gravitational field deduced from orbital perturbations. We therefore wish to examine in turn five principal topics:

(1) The characteristics of the satellites, their orbits, and their tracking, and thence the limitations they may impose on the by-product application;

(2) the anticipated nature of the perturbations we hope to measure in order to determine the lunar gravitational variations;

(3) the anticiped nature of other perturbations of the satellite orbit, and of the Moon's orbit itself, which may distort determination of the gravitational variations;

(4) the appropriate observing program in view of the conclusions from items (1), (2), and (3) above; and

(5) the technique of data analysis.

1. Satellite Characteristics

The two satellites series planned by NASA are the Lunar Orbiter and the Anchored IMP.

The NASA Lunar Orbiter is a relatively heavy satellite designed to photograph possible astronaut landing areas on the Moon. Its attitude stabilization is by gas jets. Its orbital specifications will be approximately:

Semi-major axis: 1.5 lunar radii,
eccentricity: 0.34,
inclination: 15°.

The primary purpose of the Lunar Orbiter will be accomplished within a few days after injection into orbit. However, the orbital specifications indicate it should have an indefinite orbital lifetime unless the J_3 term of the lunar gravitational field turns out to be unexpectedly large. The tracking lifetime of the lunar orbiter may be limited to a few months, however, if extra precautions are not taken to prevent it from becoming too cold when it will be eclipsed by the Earth.

The Jet Propulsion Laboratory Deep Space Network will track the Lunar Orbiter with ranging accuracies on the order of a few meters and range rates on the order of a fraction of a centimeter per second. An instrumental limitation of sorts is the lengthy preparation time required for tracking, which makes it necessary to track in a few long duration blocks rather than many short duration blocks.

The NASA-Goddard Space Flight center satellite series IMP (Interplanetary Monitoring Probe) is designed to measure the magnetic field intensity, energetic particle density, and micrometeorite flux in the region of space influenced by the Earth. Since this region appears to include a "tail" many Earth radii in length, it is appropriate that the IMP series include a satellite which is "anchored" to the Moon. The IMP satellites are spin-stabilized and rather small, weighing less than 50 kg. The rocket system by which the Anchored IMP will be injected into orbit is also relatively small, so as a consequence the design orbit is not very close to the Moon and the scatter around the design orbit of possible actual orbits is considerable. Most of the orbits within this scatter fall within the ranges:

semi-major axis: 1.8 to 9 lunar radii,
eccentricity: 0.3 to 0.7, and
inclination: 150° to 180°.

The retrograde inclination is prefered for reasons of orbital stability. However, perturbations by the Earth limit the lifetimes of some orbits severely, so that there is only about a 70 per cent chance that the satellite will survive more than four months.

The Goddard Space Flight Center range-and-range-rate system will track the IMP with ranging accuracies of some tens of meters and range-rate accuracies on the order of several centimeters per second. Power limitations prevent simultaneous telemetering and tracking. Consequently, tracking will be available only a few minutes at a time, and in an amount which is a compromise with the requirements for magnetic field, energetic particle, and micrometeorite measurements.

2. Perturbations by the Lunar Gravitational Field

There are two questions:

(1) what is the anticipated magnitude of the spherical harmonic coefficients of the lunar gravitational field? and

(2) how would these harmonics perturb the orbit?

To estimate the magnitude of the coefficients, we have three guides:

(1) the physical libration and orbital motion of the Moon, which yield only the second degree harmonics;

(2) the topography of the Moon, application of which entails assumptions as to isostasy or other internal differentiation in the Moon; and

(3) the variations of the Earth's gravitational field, application of which entails the assumption that the Moon sustains density irregularities in some way comparable to those of the Earth.

The physical libration gives for the ratio of the difference of the moments of inertia about the equatorial axes to that about the polar axes,

$(B - A)/C$, the amount $204.9 \pm 0.9 \times 10^{-6}$ (JEFFREYS, 1961). Under the assumption that the Moon is homogeneous, C/Ma^2 is 0.4, which yields

$$J_{22} = \frac{B - A}{3Ma^2} = 0.000027. \tag{1}$$

The contribution of the Moon to the secular motion of its node and perigee leads to a slightly different value, but (1) is adequate for an order of magnitude estimate.

To use the topography of the Moon, there is needed firstly a system of measurements of lunar features and secondly a harmonic analysis of the measurements. Probably the best execution of the first task is that by BREECE, HARDY, and MARCHANT (1964). The second task requires some sort of assumption to compensate for our lack of knowledge of the back side of the Moon. The minimum assumption is stationarity i.e. the back side topography has the same statistical properties as the front side. Assuming stationarity, we can obtain a power spectrum of the topography by autovariance analysis, and then use the spectrum to construct covariance matrices for linear regression analysis, as is being applied to geophysical data on Earth (KAULA, 1966).

The harmonic analysis of the lunar topography should be done in any case in order to interpret more effectively the variations of the gravitational field which will be determined. Meanwhile, to use the topography as a predictor for the gravitational variations requires the not inconsiderable assumption that the topography is a rigidly supported surface load, or, alternatively, that some particular mode of internal differentiation prevails, such as isostatic compensation.

The principal harmonic analysis of the topography published, by GOUDAS (1964), assumed front-back symmetry of the Moon's external form. Assuming the topography to be a rigidly supported surface load yields some harmonics of considerable magnitude; for example:

$$J_3 = -8.3 \times 10^{-5},$$

$$J_4 = 26.4 \times 10^{-5}. \tag{2}$$

However, the Moon's structure may be such that the topography is as poor an indicator of the low degree harmonics of the gravitational field as it is on Earth, where treating the topography as a rigidly supported load would lead to estimates averaging about four times too big and virtually random in location with respect to the actual field. We therefore would prefer a rule that depends on more fundamental properties. The principal quantity that limits the magnitude of the density irregularities a planet can sustain, regardless of the mechanism involved, is the shearing stress differences they entail. Such stresses are the result of the mutual gravitational attraction of the density irregularities and the entire mass of the planet, and the magnitude of irregularities a planet can sustain varies inversely with its mass: a tiny asteroid, such as Eros, can have a shape as elongated as a coffin, while a massive planet, such as Jupiter, must have a shape very close to fluid equilibrium.

If we assume that the stresses which the Moon's material can sustain are equal to those of the Earth, we in effect assume comparable chemical composition and comparable distance of internal pressures and temperatures from melting conditions—both of which do apply for the most plausible models of the Moon (MACDONALD, 1962).

The conventional "dimensionless" potential coefficients C_{lm}, S_{lm}, or J_{lm} would be the same as dimensioned ($L^2\,T^{-2}$) potential coefficients in a system of units where the gravitational constant, the mass of the planet, and the radius of the planet are all unity. The dimension of stress is force per unit area, or $ML^{-1}\,T^{-2}$. Since the gravitational constant (dimension $L^3\,M^{-1}\,T^{-2}$) is the same everywhere, time must scale between two planets as $M^{-1/2}\,L^{3/2}$. Substituting this time scaling in the stress scaling yields M^2L^{-4} as the scaling law for the Moon; i. e., the stress implied in the Moon by a given "dimensionless" coefficient in the Moon's potential is $(M_M^2/L_M^4)/M_E^2/L_E^4) = 1/35.7$ times as much as that implied by an equal term in the Earth's potential. Hence, if it is assumed that the Moon's material is as strong as the Earth's, we would expect the gravitational irregularities to average 35.7 times as much, proportionate to the central term.

The J_{22} of the Earth is about 1.7×10^{-6}; multiplying by the scaling factor of 35.7 yields about 6×10^{-5}, or more than twice as much as that deduced from the physical libration. Similarly, there is obtained 9×10^{-5} for the estimated absolute value of J_3—about the same as from the topography. To make more extensive estimates, however, we should do some statistical averaging. If the coordinate system is rotated, a spherical harmonic of a particular degree l will be transformed into other harmonics all of the same degree l. Geophysical structures in the Earth implying stress do not have any particular orientation with respect to the pole of rotation (except for discrepancy of the oblateness from hydrostatic equilibrium, which has a special explanation). It therefore seems appropriate to use a normalization of the spherical harmonics such that their order of magnitude does not depend on the orientation axes. i.e., such that for normalized spherical harmonics $\bar{S}_{lm}$,

$$\int_{\text{sphere}} \bar{S}_{lm}^2\, ds = K_l^2, \tag{3}$$

K_l^2 is a function of the degree l only. If we choose a normalization such that K_l^2 is the area of the unit sphere, 4π, then the contribution σ_l^2

of a particular degree l to the mean square σ^2 of the variations of the field will be simply the sum of squares of the normalized coefficients $\bar{C}_{lm}$, $\bar{S}_{lm}$ of that degree:

$$\sigma_l^2 = \sum_{m=0}^{l} (\bar{C}_{lm}^2 + \bar{S}_{lm}^2), \qquad \sigma^2 = \sum_{l=2}^{\infty} \sigma_l^2. \tag{4}$$

Estimates of σ_l^2 for the Earth's gravitational field are available from two sources:

(1) autocovariance analysis of gravimetry; and
(2) satellite orbits.

Table 1. *Power Spectrum of the Earth's Gravity Field*
In Non-Dimensional Units: $\sigma_l^2 = \sum_m (\bar{C}_{lm}^2 + \bar{S}_{lm}^2) \times 10^{12}$

Degree l	From gravimetry σ_l^2	From satellite orbits
2	7.0	7.5
3	8.2	9.2
4	2.3	2.5
5	0.1	1.0
6	0.7	1.3
7	0.2	1.2
8	0.2	0.6

Table 1 is a compilation of these results. Taking for the expected magnitude of a gravitational coefficients $\bar{C}_{lm}$ or $\bar{S}_{lm}$:

$$E\{\bar{C}_{lm}\bar{S}_{lm}\} = \pm \sqrt{\frac{\sigma_l^2}{2l+1}}, \tag{5}$$

there is obtained as a rule of-thumb from Table 1 for the Earth:

$$E\{\bar{C}_{lm}\bar{S}_{lm}\} \approx 10^{-5}/l^2, \tag{6}$$

and hence for the Moon:

$$E\{\bar{C}_{lm}\bar{S}_{lm}\} \approx 35 \times 10^{-5}/l^2. \tag{7}$$

To obtain an estimate of the periodic effects on a satellite orbit of coefficients of the magnitude given by (7), integrate the Lagrangian equations of motion under the assumption that the semi-major axis, eccentricity, and inclination a, e, and I are constant on the right side of the equations and that the angles are secularly changing; for example, for a perturbation of the nodal longitude (KAULA, 1961)

$$\Delta\Omega_{lm} = \frac{(\mu/a)(a_e/a)^l}{na^2(1-e^2)^{1/2}\sin I} \sum_{p,q} \frac{(\partial F_{lm}/\partial I)\, G_{lpq}(e)}{\{(l-2p)\dot{\omega} + (l-2p+q)\dot{M} + m(\dot{\Omega}-\dot{\theta})\}} \times$$
$$\times \begin{Bmatrix} C_{lm} \\ \text{or} \\ S_{lm} \end{Bmatrix} \begin{Bmatrix} \cos \\ \text{or} \\ \sin \end{Bmatrix} \{(l-2p)\,\omega + (l-2p+q)\,M + m(\Omega-\theta)\}. \tag{8}$$

The dominant feature of (8) is that the magnitudes of the perturbations depend on the magnitudes of the rates $\dot{\omega}$, $\dot{M}$, $\dot{\Omega}$, $\dot{\theta}$ in the denominator, where θ is the rate of rotation of the planet with respect to an inertially fixed reference. For a close satellite of the Moon, these rates will be on the order of

$$\left.\begin{aligned} O(\dot{M}) &\approx 10 \text{ cycles/day}, \\ \dot{\theta} &= 0.037 \text{ cycle/day}, \\ O(\dot{\omega}) &\approx 0.002 \text{ cycle/day}, \\ O(\dot{\Omega}) &\approx 0.002 \text{ cycle/day}. \end{aligned}\right\} \tag{9}$$

The smallness of the pericenter and nodal rates $\dot{\omega}$ and $\dot{\Omega}$ is due to the smallness of the lunar oblateness, J_2. The interesting feature, however, is the enhanced effect of tesseral harmonics in the lunar gravitational field due to the slowness of the Moon's rotation. $\dot{\theta}$. Consequently a wide variety of perturbations should be expected. Table 2 is the result of calculating the periodic perturbations anticipated from coefficients of the magnitude given by (7). It can be seen from the table that estimates of lunar tesseral harmonics should be obtainable even from retrograde orbits as far out as seven lunar radii. The perturbations are not so big, however, that the simple representation of (8) will not suffice, except perhaps for some non-linearities of the form:

$$\Delta_2(\Omega, \omega)_{lm} = \frac{\partial(\dot{\Omega}, \dot{\omega})}{\partial(e, I)} \int \Delta(e, I)_{lm}\, dt. \tag{10}$$

3. Other Perturbations Anticipated

The principal complication in the perturbations of the satellites own orbit is, of course, the considerable third body perturbation by the Earth: $81.3^2 = 6600$ times as great as for an Earth satellite, in terms of the ratio masses of disturbing to primary body. Comparing the factor $\mu a_e^2 J_2/r^3$ of the oblateness term to the factor $r^2 m^*/r^{*3}$ of the leading third body term, we get that the two perturbations are about equal at 1.9 lunar radii. Hence the rates $\dot{\Omega}$, $\dot{\omega}$, used in calculating perturbations, or partial derivates, as in (8), should include the effect of the Earth. Consideration should also be given to a theoretical development combining J_2 and third body effects; for example, a von Zeipel transformation type theory in effect combining the J_2 theory of Brouwer (1959) and the third-body theory of Hori (1963) should be quite feasible (Kaula, 1965). First attention in this theory should be given to the possibility of long periodic or secular effects arising from the mutual interaction of short periodic terms. If they are not significant, as appears likely, then a feasible solution could be to obtain the long term variations of the orbit by a variation of parameters numerical integration using a disturbing function averaged with respect to the mean anomaly and short periodic perturbations incremented as linear oscillations similar to (8).

Radiation pressure accelerations will be unchanged from Earth satellites, since they are functions of the solar radiation and satellite characteristics, not of the planet properties, except in the sense of shadow effects. Hence the ratio of the acceleration of the central term will be larger by a factor of six, and the shadow effects will be enhanced by about 25 per cent over those for an Earth satellite of the same semi-major axis in planetary radii, due to the longer period. However, by (6) and (7), the ratio of the accelerations to those of the gravitational variations will be smaller by a factor of six.

The commensurability of the rotation of the Moon and its revolution about the Earth suggests that determination of monthly, bimonthly etc. perturbations of the satellite by the lunar gravitational field may be distorted by monthly, semi-monthly, etc. periodic errors in the lunar ephemeris. Comparison of the lunar ephemeris with numerical integration does reveal a discrepancy of several hundred meters in the monthly variation (PEABODY et al., 1964) however it appears that this can be removed by an appropriate scale factor correction dependent on more

Table 2. *Amplitudes of Periodic Perturbations Anticipated due to Variations in Lunar Gravitational Field*

Assuming Normalized Spherical Harmonic Coefficients

$$(\bar{C}_{lm}, \bar{S}_{lm}) = \frac{\pm 35 \times 10^{-5}}{l^2}$$

Semi-major axis lunar radii	Eccen-tricity	Incli-nation radians	Spherical harmonic indices lm for terms expected to cause perturbation of magnitude		
			> 1000 m.	500 – 1000 m.	100 – 500 m.
1.993	0.3028	0.992	22, 30, 31, 40	32, 33, 41, 42, 43	44
2.462	0.4760	2.617	22, 30, 31, 32 40, 41	42	33, 43
2.597	0.4426	1.840	22, 30, 31, 40	32, 33, 41	42, 43, 44
2.691	0.6629	2.199	22, 30, 31, 32 33, 40, 41, 42 43	44	
3.555	0.6706	1.218	22, 30, 31, 32 33, 40, 41	43, 44	42
3.620	0.4686	2.224	22, 30, 40	31	32, 33, 41, 42
6.097	0.6079	1.409	22, 30, 40		31, 32, 33, 41
6.142	0.3993	2.381	30	22	31
10.512	0.5323	1.637	30	22	

modern values of parameters (MARSDEN, 1964), to the extent necessary to determine perturbations of the sort in Table 2, if not to exploit fully the few meters accuracy of the instrumentation.

4. The Observing Program

We must determine the six constants of integration of the orbit plus as many gravitational coefficients of the Moon as practicable, given that we are tracking from the Earth with the time and accuracy limitations stated in section 1, that the perturbations sought are approximately as described in section 2, and that possibly interfering perturbations are as in section 3. Practicability is of course difficult to define, but let us assume that from a single satellite we might try to estimate the twenty or so coefficients corresponding to the harmonic indices in Table 2. The prescription of the observing program may thus defined as obtaining a distribution of observations adequate to

(*1*) yield a sufficient number of effective degrees of freedom to determine the 25 or so unknowns; and

(2) average out any distorting effects.

The perturbations discussed in sections 2 and 3 are all of considerably longer period than the satellite period. Hence, the variations in range and range-rate due to the motion of the satellite around its orbit might be considered as a carrier wave, and the perturbations as modulations of this carrier wave. We thus wish to determine the time series constituted by these modulations. The application of techniques of time series analysis to observations of close Earth satellites has been somewhat discouraging because of the complications of non-uniformly distributed observations and the rapid rate of change of the quantities actually observed with respect to the components of the satellite state vector. In the case of a lunar satellite tracked from the Earth, the time series approach is easier because we can observe whenever we wish, provided the satellite is not behind the Moon; we can observe only one or two components of the satellite state; and the relationship of these components to dynamically convenient orbital elements has a relatively slow rate of change. However, "whenever we wish" does not mean "continuously" in the practical situation: on the contrary, demands such as other vehicles being tracked by the J.P.L. Deep Space Network and the other measurements requiring the same power system on the IMP will require a sound justification for the tracking program requested.

In observing a time series, the change of information from one cycle to the next of the carrier wave is very slight; several successive cycles are generally observed only to reduce noise effects. To estimate the effective number of measurements made in order to calculate the degrees of freedom we thus can consider the information obtained by observing a single revolution. The most favorable situation will be when there is maximum variation in the observations; i.e. when the observer is in the plane of the orbit. If the orbit is circular, a pure sine wave will be observed in both range and range-rate (neglecting parallax), and two effective measurements will be made: the semi-major axis, a, from the amplitude, and the argument of latitude, $\omega + M$, from the phase angle. If the orbit is eccentric, then the sine wave will be distorted, and the eccentricity, e, and the pericenter argument, ω, can be determined from the amplitude and phase angle of distortion. If the nodal angle is changed, the change in range will be undistinguishable from a change in ellipticity and pericenter argument but the change in range rate will be inconsistent with the change in range so that in effect the nodal longitude, Ω, will be measured. Hence, there are five effective measurements in the most favorable situation: all KEPLER elements except the inclination, I. The most unfavorable situation will be when there is minimum variation in the observations, i.e. when the orbital plane is normal to the line of sight. Then all that is measurable is the fact that the orbit is so oriented; i.e., the two orientation angles: the nodal longitude Ω, and the inclination I.

To obtain the fullest information of the six orbital constants of integration we thus want to observe over a full cycle of the orientations possible: i.e. over a complete month. This conclusion has also been reached by other studies of simular situations—e.g. by DEUTSCH (1960). The mean number of effective measurements averaging over all orientations will be closer to the maximum of 5 than to the minimum of 2, since the former will occur for all orbits but the latter only for polar orbits. However, there will be a loss of information due to the satellite going behind the Moon which is maximized for equatorial orbits, so it seems reasonable to take about 3.5 as the mean number of effective measurements per cycle observed.

Hence, for the J.P.L. Deep Space Network which can observe continuously for hours, to obtain, say, 20 degrees of freedom for the 25 unknowns, the minimum program would be to measure $(20 + 25)/3.5 = 13$ full orbital cycles distributed throughout the one month cycle: i.e. roughly a revolution of the satellite every other day. For observations of IMP, the duration of which, due to the power limitation, must be limited to the few minutes necessary to smooth out high frequency noise, the minimum program would be to measure at $(20 + 25)/2 = 23$ times distributed with respect to both the one month cycle and the orbital anomaly.

To accomplish the second purpose of averaging out distortions, the exact distribution of a set of observations O_i to determine a set of parameters of interest P_j in the presence of suspected distortions D_k should be such that the sum of products of partial derivatives—essentially the off diagonal elements of the least-squares normal equations—are minimized for any pair of P_j, D_k:

$$\sum_i \frac{\partial O_i}{\partial P_j} \frac{\partial O_i}{\partial D_k} = minimum. \tag{11}$$

The test (11) will be applied to any proposed distribution of observations, but we are still in need of guides as to what distributions to dropose. As discussed in section 3, the principal distortions D_k of an orbital nature will be associated with the third body effects on the satellite orbit and on the orbit of the moon itself. Hence the periodicities of these perturbations will be various combinations of the periodicities of the Moon's orbit about the Earth and of the Earth's orbit about the Sun. The possibility of the minimum (11) being close to zero thus is reduced to simply:

(1) If only Sun-Earth periodicities are involved, then distribution of observations over the month should suffice;

(2) if only Earth-Moon periodicities are involved, then no distribution will be of much avail, due to the exact coincidence of the Moon's rotation and its revolution about the Earth;

(3) if both periodicities are involved, then the distribution will have to be carried over six months or a year.

The principal perturbation of the lunar orbit, the variation, contains in its argument $2\lambda^*$, twice the longitude of the Sun referred to the Earth. Hence it is desirable that the distribution of observations extend over half a year.

In the case of the range-and-range-rate tracking of IMP, there will also be distortions with 24-hour periodicity, because of ionospheric refraction of the relatively low frequency (136 Mcs) signal, and perhaps because of station position error. Hence a third distribution should be added. The sidereal period of the Moon is about $27\frac{1}{3}$ mean solar days (appropriate for ionospheric effects), which suggest as a minimum program 69 observations distributed uniformly over 82 days—provided, of course, that the mean solar day or the sidereal month are not nearly integer multiples of the satellite period.

The considerations discussed thus far presume that the periodicities of distortions are known. If the periodicities are unknown, then a randomized distribution of observations should be used, i.e., a Poisson distribution, if the unknown periodicities are assumed to be Gaussian white noise (Shapiro and Silverman, 1960). In a modification of this technique, the month of $27\frac{1}{3}$ days would be divided into, say, eight parts. The number of observations in each part would be equal, but their distribution within each part randomized. In view of the possibily of unknown distortions, randomized distributions of observations should be subjected to the test (11) for distortions of known periodicities. If the increase of the sum is slight for a moderate increase above the minimum number of observations, then the randomized distribution should be adopted.

It is to the emphasized that the numbers suggested here in are for a minimum program corresponding to the limited number of harmonics in Table 2. Manifestly, full exploitation of the accuracy of the instrumentation — particularly the J.P.L. Deep Space Network — will require the determination of considerably more coefficients and hence more observations. The essential limitation will be the same as for Earth satellites: the inability to resolve ambiguities because of the lack of variety of orbital specifications. In this connection, the principal value of the IMP with its less accurate tracking is the provision of orbital variety to resolve ambiguities.

5. The Technique of Data Analysis

The methods we propose to use are essentially those applied to Earth satellite orbits (KAULA, 1963 a). The principal parts of the analysis program and the features particularly applicable to the analysis of lunar orbiters of this program are:

a) Data selection and conversion. Aggregation, coordinate transformation, etc. The principal modification of this part will be to add to the information pertaining to each observation the selenocentric inertial position and velocity of the tracking station, thus eliminating subsequent reference to the lunar ephemeris.

b) Preliminary analysis. Determination of preliminary orbits and calculation of arrays to form partial derivatives as in equation (8).

c) Final analysis. Determination of corrections to parameters. All statistical devices—various types of weighting, etc.—will be applied in this part. One device of particular application to early analyses of only one or two lunar satellite orbits is the use of pre-assigned variances for the parameters to be determined. The parameters are treated mathematically as effectively observations with weights inversely proportionate to the variances, and hence the probability of unrealistically distorted solutions due to ill conditioning can be reduced.

Conclusion

The perturbations of lunar satellite orbits should be a rich source of information pertaining to the Moon's interior. Third body perturbations and lunar ephemeris errors appear to be annoying but manageable.

The instrumentation is of adequate accuracy, and the limitation of the system appears to be as for Earth satellites: inability to resolve ambiguities due to lack of variety in orbital specifications.

References

BREECE, S., M. HARDY and M. Q. MARCHANT: Horizontal and Vertical Control for Lunar Mapping (Part II: AMS Selenodetic Control System 1964), Army Map Service Tech. Rep., N. 29 (1964).

BROUWER, D.: Solution of the Problem of Artificial Satellite Motion Theory without Drag, Astronom. J. **64**, 378–397 (1959).

DEUTSCH, A. J.: Orbits for Planetary Satellites from Doppler Date Alone, ARS J. **30**, 536–542 (1960).

GOUDAS, C. L.: A Contour Map of the Moon, Icarus **3**, 476–485 (1964).

HORI, G.: New Approach to the Solution of the Main. Problem of the Lunar Theory, Astronom. J. **68**, 125–146 (1963).

JEFFREYS, H. On the Figure of the Moon, Monthly Not. Roy. Astr. Soc. **122**, 421–432 (1961).

KAULA, W. M.: Analysis of Gravitational and Geometric Aspects of Geodetic Utilization of Satellites, R.A.S. Geophys. J. **6**, 104–133 (1961).

KAULA, W. M.: Tesseral Harmonics of the Gravitational Field and Geodetic Datum Shifts Derived from Camera Observations of Satellites, J. Geophys. Res. **68**, 5183–5190 (1963a).

KAULA, W. M.: The Investigation of the Gravitational Fields of the Moon and Planets with Artificial Satellites, Adv. in Space Sci. and Tech. **5**, 210–226 (1963b).

KAULA, W. M.: Gravitational and Other Perturbations of a Satellite Orbit, Ch. 7 in G. V. GROVES, Ed.: Dynamics of Rockets and Satellites, Amsterdam: North Holland Publ. Co. 1965.

KAULA, W. M.: Global Harmonic and Statistical Analysis of Gravimetry, in H. ORLIN, Ed.: Gravity Anomalies: Unsurveyed Areas, Geophys. Monograph, Vol. 9, Washington: AGU 1966, pp. 58–67.

MACDONALD, G. J. F.: Internal Constitution of the Minor Planets. J. Geophys. Res. **67**, 2945–2974 (1962).

MARSDEN, B. G.: On a Consistent Ephemeris, Jet Prop. Lab. Space Program Summary **4** (37–29) 61 (1964).

PEABODY, P. R., J. F. SCOTT and E. G. OROZCO: User's Description of JPL Ephemeris Tapes, Jet Prop. Lab. Tech. Rep. **32–580**, 1964.

SHAPIRO, H. S. & SILVERMAN, R. A.: Alias-free sampling of random noise, J. Soc. Ind. & Appl. Math. 8, 225–248 (1960).

Discussion

Professor BROUWER remarks that, in most theoretical works done so far on the subject, there have not been accounted for the indirect solar perturbations due to the solar effect on the Moon's motion. It would be interesting to see what periodic perturbations of a moon satellite would be produced in such a case.

Dr. KOVALEVSKY indicates that he has studied analytically the long period behaviour of lunar satellites, care being taken of the second and the third zonal harmonics ($\tau_2 = 0.00021$, $\tau_3 = 0.00009$) as well as the perturbations by the Earth. Very large long period perturbations develop, that in many cases force the satellite to come down on the surface of the Moon before a complete perigee period. The Fig. 1 shows the various types of orbits that exist for a semi-major axis of 2500 km, with a periselene at launch at is maximum latitude ($g = +90°$) or minimum latitude ($g = -90°$). Given eccentricity and inclination are the mean values of these parameters not corrected for long period perturbations, for $g = +90°$ or $g = -90°$.

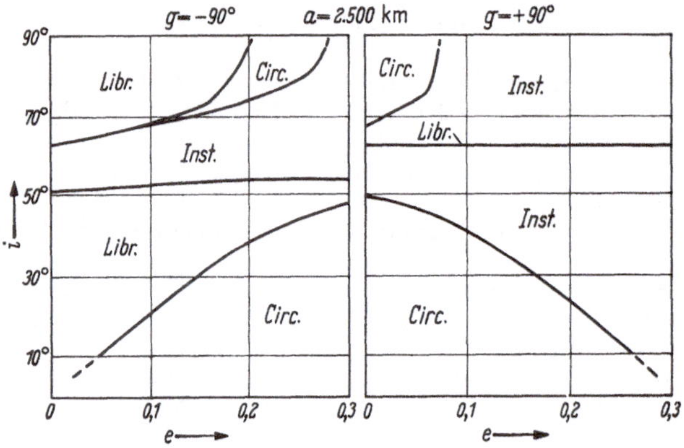

Fig. 1

The three possible types of orbits are

(1) Circulation orbits for which the periselene revolves around the Moon.

(2) Libration orbits for which the periselene oscillated around a stable equilibrium position. Three families of such orbits exist.

(3) Unstable orbits for which the satellite reaches the surface of the Moon before a complete revolution or oscillation of the perigee.

In the given case, only low-inclination and high-inclination with low eccentricity satellites can remain in orbit for a long period.

For semi-major axis larger than 4000 km, the stable family of polar orbits disappear, and, generally speaking, orbits of inclination higher than 55° as well as some other high eccentricity orbits sometimes will fall on the surface of the Moon. This feature will prevent having a good distribution of satellites orbits around the Moon, which would be necessary for a good solution of the determination of its gravitational field.

Trajectories Around Both Masses in the Restricted Three Body Problem

By

Richard F. Arenstorf

Marshall Space Flight Center, NASA, Huntsville, Alabama, U.S.A.

A representative collection of trajectories referred to in the title and belonging to different families of periodic solutions of the so-called 2. kind and higher order have been shown on slides. Their mathematical theory and classification is given in [*1*], [*2*]. Also, a new iterative method for the treatment of perturbation problems in celestial mechanics was presented, which yields accelerated convergence on preassigned finite time ranges and, as an application, shows the existence of synodically periodic solutions of the plane restricted three body problem near the smaller of the primaries, which are close to precessing elliptic orbits of arbitrary eccentricity. This work is to be published in [*3*], [*4*]. By numerical extension of such families of orbits interesting trajectories can be found, which demonstrate the phenomenon of temporary capture with satellite motion about each of the primaries and periodically alternating transitions to the vicinity of the other primary. Examples of such trajectories have been shown on slides, and are being published in [*5*]. Finally, three selected trajectories with the properties of the title have been dynamically projected in physical time from a motion picture film, which has been produced by electronic computers at the NASA-Marshall Space Flight Center.

References

[*1*] Arenstorf, R. F.: Periodic Solutions of the Restr. Three Body Problem Representing Analytic Continuations of Keplerian Elliptic Motions, Amer. J. Math. **85**, 27–35 (1963).

[*2*] Arenstorf, R. F.: Periodic Trajectories Passing Near Both Masses of the Restricted Problem of Three Bodies. Proceedings XIVth Internat. Astronautical Congress, Paris 1963, Vol. IV, pp. 85–97.

[*3*] Arenstorf, R. F.: A New Method of Perturbation Theory and Its Application to Periodic Motions in the Restricted Problem of Three Bodies. Proceedings of the Symposium on Celestial Mechanics, Oberwolfach, March 1964, Hochschultaschenbücher, Mannheim: Bibliographisches Institut.

[*4*] Arenstorf, R. F.: A New Method of Perturbation Theory and Its Application to the Satellite Problem of Celestial Mechanics, J. reine angew. Math. **221**, 113–145 (1966).

[*5*] Davidson, M. C.: Numerical Examples of Transition Orbits in the Restricted Three Body Problem, Astronautica Acta **10**, 308–313 (1964).

Analytical Index

Examples of presentation of references

Brouwer **75—79**: part of the paper by D. Brouwer extending from the beginning page 75 to the end of page 79.
Chovitz **168**, 7—28: part of paper by B. Chovitz extending from line 7 to line 28 of page 168.
Anderson **218**, 26 —**221**, 18: part of the paper by J. D. Anderson and M. R. Warner extending from line 26 of page 218 to line 18 of page 221.
Disc. Kozai **52**, 12: line 12 of the discussion on page 52.

(Papers are referred to by the name of the first author)

I. Celestial Mechanics

1. Variables

a) Sets of variables

b) Formulation of equations and theory with new variables

2. Orbit Determination

a) Generalities

b) Methods in orbit determination

3. Radiation Pressure Effects

a) Description of solar and earth reflected radiation

b) Direct solar radiation

c) Earth's reflected radiation: force

d) Earth's reflected radiation: effects

III. Precise Observations and Geodesy

1. Observation Programs

2. Satellite Geodesy: Methods

3. Satellite Photography

IV. Mathematical Treatment of Observations

1. Analysis of Observations

2. Least Squares and Related Subjects

V. Earth's Gravitational Field

1. Description of the Earth's Gravitational Field]